STUDENT STUDY GUIDE
AND SOLUTIONS MANUAL

BO LOU

COLLEGE
PHYSICS

FOURTH EDITION

WILSON

BUFFA

PRENTICE HALL, Upper Saddle River, NJ 07458

Executive Editor: Alison Reeves
Project Manager: Elizabeth Kell
Special Projects Manager: Barbara A. Murray
Production Editor: Jonathan Boylan
Supplement Cover Manager: Paul Gourhan
Supplement Cover Designer: Liz Nemeth
Manufacturing Manager: Trudy Pisciotti

ISBN 0-13-084365-2

Prentice-Hall International (UK) Limited, London
Prentice-Hall of Australia Pty. Limited, Sydney
Prentice-Hall Canada, Inc., Toronto
Prentice-Hall Hispanoamericana, S.A., Mexico
Prentice-Hall of India Private Limited, New Delhi
Prentice-Hall (Singapore) Pte. Ltd.
Prentice-Hall of Japan, Inc., Tokyo
Editora Prentice-Hall do Brazil, Ltda., Rio de Janeiro

TABLE OF CONTENTS

Preface

This **Study and Guide and Student Solutions Manual** for **College Physics**, fourth edition, was prepared to help students gain a greater understanding of the principles of their introductory physics courses. Most of us learn by summary and examples, and this manual has been organized along these lines. For each chapter you will find:

- **Chapter Objectives**

    states the learning goals for the chapter. The objectives tell you what you should know upon completion of chapter study. Your instructor may omit some topics.

- **Key Terms**

    lists the key terms for the chapter. The definitions and/or explanations of the most important key terms can be found in the next section.

- **Chapter Summary and Discussion**

    outlines the important concepts and provide a brief overview of the major chapter contents. Extra worked out examples are included to further strengthen the concepts and principles. This review allows you to check the thoroughness of your study and serves as a last minute quick review (about 30 minutes) before quizzes or tests. Common students' mistakes and misconceptions are noted.

- **Mathematical Summary**

    lists the important mathematical equations in the chapter. The purpose is for self-review. You should be identify each symbol in an equation and explain what relationship the equation describes. The equation number in the text is included for reference. A last glance of this section is helpful before taking a test or quiz.

- **Solutions to Selected Exercises and Paired/Trio Exercises**

    provides the worked out solutions of the even-numbered annotated (red dot) end-of-chapter text exercises. The paired/trio exercises are similar in nature. You should try to work out the even-numbered paired/trio exercise independently and then check your work with solutions in this manual. After working the odd-numbered exercise of a pair/trio, you can check your answer in the Answers to Odd-Numbered Exercises at the back of the text. Solutions for some additional end-of-the-chapter exercises are also included.

- **Practice Quiz**

  consists of multiple choice questions and problems of the most fundamental concepts and problem solving skills in the chapter. This allows you to self-check your understanding and knowledge of the chapter. The answers to the quizzes are also given.

As you can see, this manual provides a through review for each chapter. The conscientious student can make good use of the various sections to assist in understanding and mastering the course contents and preparing for exams. I certainly hope you find this manual helpful in your learning.

ACKNOWLEDGMENT

I would like to thank the following people and organizations for their enormous help and support.

First to the authors, Jerry D. Wilson and Anthony J. Buffa, for their meticulous checking of my work and numerous constructive and helpful comments and discussions.

To John Kinard and Jerry Wilson for their solutions manual of the second edition Wilson book. Some of the solutions are still from their work.

To the editors of Prentice Hall, Paul corey, Alison Reeves, Karen Karlin, Wendy Rivers, and Liz Kell, for their guidance, and Gillian Kieff for assistance.

To Carolyn Gauntt for proof-reading this manual.

Last but not the least, to my family, Lingfei and Alina, for their essential and generous support and love. I dedicate this manual to them.

Bo Lou, Ph.D., Professor of Physics (http://instruction.ferris.edu/loub)
Department of Physical Sciences
Ferris State University
Big Rapids, MI 49307

CHAPTER 1

Units and Problem Solving

I.   Chapter Objectives

Upon completion of this chapter, you should be able to:

1.   distinguish standard units and system of units.

2.   describe the SI and specify the references for the three main base quantities of this system.

3.   use common metric prefixes and nonstandard metric units.

4.   explain the advantage of and apply dimensional analysis and unit analysis.

5.   explain conversion factor relationships and apply them in converting units within a system or from one system of units to another.

6.   determine the number of significant figures in a numerical value and report the proper number of significant figures after performing simple calculations.

7.   establish a problem-solving procedure and apply it to typical problems.

II.   Key Terms

Upon completion of this chapter, you should be able to define and/or explain the following key terms:

standard unit	fps system
system of units	liter (L)
International System of Units (SI)	dimensional analysis
SI base units	unit analysis
SI derived units	density (ρ)
meter (m)	conversion factor
kilogram (kg)	exact number
second (s)	measured number
mks system	significant figures (sf).
cgs system	

The definitions and/or explanations of the most important key terms can be found in the following section:
III. Chapter Summary and Discussion.

III. Chapter Summary and Discussion

1. International System of Units (SI) (Sections 1.1 – 1.3)

Objects and phenomena are measured and described using **standard units**, a group of which makes up a **system of units**.

(1) The International System of Units (SI), or the metric system, has only seven base quantities (see Table 1.1). The base units for the base quantities length, mass, and time are the **meter** (m), the **kilogram** (kg), and the **second** (s), respectively. A derived quantity (unit) is a combination of the base quantity (units). For example, the units of the derived quantity speed, **meters per second**, are a combination of **meter** and **second**. There are many derived units.

(2) The metric system is a base–10 (decimal) system, which is very convenient in changing measurements from one unit to another. Metric multiples are designated by prefixes, the most common of which are **kilo–**(1000), **centi–**(1/100), and **milli–**(1/1000). For example, a centimeter is 1/100 of a meter, etc. A complete list of the metric prefixes is given in Table 1.2. A unit of volume or capacity is the **liter** (L), and 1 L = 1000 mL = 1000 cm^3 (cubic centimeters).

2. Dimensional Analysis (Section 1.4)

The fundamental or base quantities, such as length, mass, and time are called **dimensions**. These are commonly expressed by bracketed symbols [L], [M], and [T], respectively. **Dimensional analysis** is a procedure by which the dimensional correctness of an equation may be checked. Both sides of an equation must not only be equal in numerical value, but also in dimension; and dimensions can be treated like algebraic quantities. Units, instead of dimensional symbols, may be used in **unit analysis**.

Dimensional analysis can be used to

(1) check whether an equation is **dimensionally correct**, i.e., if an equation has the same dimension (unit) on both sides.

(2) find out dimension or units of derived quantities.

Example 1.1: Check whether the equation $x = at^2$ is dimensionally correct, where x is length, a is acceleration, and t is time interval.

Solution:

The dimensions and units of x, a, and t, are [L], m; $\dfrac{[L]}{[T]^2}$, m/s^2; and [T], s; respectively.

Dimensional analysis: Dimension of left side of the equation is [L].

Dimension of right side of the equation is $\dfrac{[L]}{[T]^2} \times [T]^2 = [L]$.

So the dimension of the left side is equal to the dimension of the right side and the equation is dimensionally correct. Note: *dimensionally correct does not necessarily mean the equation is correct*. For example, the "equation," 2 tables = 3 tables, is dimensionally correct, but not numerically correct.

Unit analysis: Units of the left side are m.

Units of the right side are $(m/s^2)(s)^2 = m$.

So the units of the left side are equal to the units of the right side and the equation is dimensionally correct.

Example 1.2:   Einstein's famous energy–mass equivalence states that the energy a mass has is equal to its mass times the speed of light squared. Determine the dimension and units of energy.

Solution:

Since energy is equal to mass times speed squared, the dimension (units) of energy must be equal to the dimension (units) of mass times the dimension (units) of speed squared.

So the dimension of energy is $[M] \times \left(\dfrac{[L]}{[T]} \right)^2 = \dfrac{[M] \cdot [L]^2}{[T]^2}$,

and the units of energy are $kg \times (m/s)^2 = kg \cdot m^2/s^2$, which is called joule (J).

---

3.   Conversions of Units (Section 1.5)

A quantity may be expressed in other units through the use of **conversion factors** such as (1 mi/1609 m) or (1609 m/1 mi). Note that any conversion factor is equal to 1 (because 1 mi = 1609 m, for example) and so they can be multiplied or divided to any quantity without altering the quantity. The appropriate form of a conversion factor is easily determined by dimensional (unit) analysis.

Example 1.3:   A jogger walks 3200 meters everyday. What is this distance in miles?

Solution:

Here we need to convert meters to miles. We can accomplish this by using the conversion factor (1 mi/1609 m). The result units will be m × (mi/m) = mi. We do not multiply (1609 m/1 mi) because the result units would be m × (m/mi) = m²/mi.

$$(3200 \text{ m}) \times 1 = (3200 \text{ m}) \times \frac{1 \text{ mi}}{1609 \text{ m}} = 1.99 \text{ mi} \approx 2.0 \text{ mi} . \quad \text{(here} \quad \frac{1 \text{ mi}}{1609 \text{ m}} = 1\text{)}$$

Note the cancellation of the units m.

We can also use the conversion factor (1609 m/1 mi). Then we have to divide 3200 m by (1609 m/1 mi) in order to get mi. $\dfrac{3200 \text{ m}}{(1609 \text{ m}) / (1 \text{ mi})} \approx 2.0 \text{ mi}$. Note that unit cancellation tells you if you get the desired result unit.

Example 1.4:   A car travels with a speed of 25 m/s. What is this speed in mi/h (miles per hour)?

Solution:

Here we need to convert meters to miles *and* seconds to hours. We can use the conversion factor (1 mi/1609 m) to convert meters to miles and (3600 s/1 h) to convert 1/s to 1/h.

[Why can't we multiply by (1 h/3600 s)?]

$$(25 \text{ m/s}) \times 1 \times 1 = \left(25 \text{ m/s}\right) \times \frac{1 \text{ mi}}{1609 \text{ m}} \times \frac{3600 \text{ s}}{1 \text{ h}} = 56 \text{ mi/h} .$$

We can also use the direct conversion (1 mi/h = 0.447 m/s).

$$(25 \text{ m} / \text{s}) \times \frac{1 \text{ mi} / \text{h}}{0.447 \text{ m} / \text{s}} = 56 \text{ mi/h}.$$

---

4.   Significant Figures (Section 1.6)

The number of **significant figures** (sf) in a quantity is the number of reliably known digits it contains. For example, the quantity 15.2 m has 3 sf, 0.052 kg has 2 sf, and 3.0 m/s has 2 sf. In general,

- *the final result of a multiplication and/or division should have the same number of significant figures as the quantity with the least number of significant figures used in the calculation*, and
- *the final result of an addition and/or subtraction should have the same number of decimal places as the quantity with the least number of decimal places used in the calculation.*

The proper number of figures or digits is obtained by rounding off a result.

Example 1.5:   Perform the following operations:

(a) $0.586 \times 3.4 =$

(b) $13.90 \div 0.580 =$

(c) $(13.59 \times 4.86) \div 2.1 =$

(d) $4.8 \times 10^5 \div 4.0 \times 10^{-3} =$

(e) $(3.2 \times 10^8)(4.0 \times 10^4) =$

Solution:

The final result of the multiplication and/or division should have the same number of significant figures as the quantity with the least number of significant figures.

(a) $0.586 \times 3.4 = 2.0$.

(b) $13.90 \div 0.580 = 24.0$.

(c) $13.59 \times 4.86 \div 2.1 = 31$.

(d) $4.8 \times 10^5 \div 4.0 \times 10^{-3} = 1.2 \times 10^8$.

(e) $(3.2 \times 10^8)(4.0 \times 10^4) = 1.3 \times 10^{13}$.

Example 1.6:   Perform the following operations:

(a) $23.1 + 45 + 0.68 + 100 =$

(b) $157 - 5.689 + 2 =$

(c) $23.5 + 0.567 + 0.85 =$

(d) $4.69 \times 10^{-6} - 2.5 \times 10^{-5} =$

(e) $8.9 \times 10^4 + 2.5 \times 10^5 =$

Solution:

The final result of the addition and/or subtraction should have the same number of decimal places as the quantity with the least number of decimal places.

(a) $23.1 + 45 + 0.68 + 100 = 169$.

(b) $157 - 5.689 + 2 = 153$.

(c) $23.5 + 0.567 + 0.85 = 24.9$.

(d) $4.69 \times 10^{-6} - 2.5 \times 10^{-5} = 0.469 \times 10^{-5} - 2.5 \times 10^{-5} = -2.0 \times 10^{-5}$.

(e) $8.9 \times 10^4 + 2.5 \times 10^5 = 8.9 \times 10^4 + 25 \times 10^4 = 34 \times 10^4 = 3.4 \times 10^5$.

5. Problem Solving (Section 1.7)

Problem solving is a skill that has to be learned and accumulated gradually over a period of time. You can not learn this skill in a lecture or overnight. It takes practice, lots of practice, and the exact procedure you adopt will probably be unique to you. The point is to develop one that works for you. However, there are some suggested problem solving procedures that can be followed.

(1) *Say it in words.*

Read the problem carefully and analyze it. Write down the given data and what you are to find.

(2) *Say it in pictures.*

Draw a diagram, if appropriate, as an aid in visualizing and analyzing the physical situation of the problem.

(3) *Say it in equations.*

Determine which equation(s) are applicable to this situation and how they can be used to get from the information given to what is to be found.

(4) *Simplify the equations.*

Simplifying mathematical expressions as much as possible through algebraic manipulations before inserting actual numbers.

(5) *Check the units.*

check units before doing calculations.

(6) Insert numbers and calculate; check significant figures.

substitute given quantities into equation(s) and perform calculations. Report the result with proper units and the proper number of significant figures.

(7) *Check the answer: is it reasonable?*

Consider whether the result is reasonable.

The details of these procedures can be found on page 21 and 22 in the text.

Example 1.7: Starting from city A, an airplane flies 250 miles east to city B, then 300 miles north to city C, and finally 700 miles west to city D. What is the distance from city A to city D?

Solution: Given: the distances and directions of each trip.

Find: the distance from city A to D.

Following the problem statement, we draw a diagram. It is easy to see that the distance from A to D is the hypotenuse of the shaded right-angle triangle. The sides perpendicular to each other are 300 mi and (700 mi − 250 mi) = 450 mi.

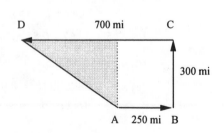

To find the hypotenuse of a right-angle triangle, we use Pythagorean theorem:

$c^2 = a^2 + b^2$, where a and b are the sides perpendicular to each other and c is the hypotenuse.

$$c = \sqrt{a^2 + b^2} = \sqrt{(300 \text{ mi})^2 + (450 \text{ mi})^2} = 541 \text{ mi}.$$

Obviously, the units, miles, are right; the answer, 541, is reasonable; and the number of significant figures, 3, in the final result, is the same as the least number of significant figures between 300 mi and 450 mi. These checking make us feel confident that we did the right calculations.

Example 1.8:   The density of the metal aluminum is 2700 kg/m^3. Find the mass of a solid aluminum cylinder of radius 10 cm and height 1.0 ft.

Solution: Given: the dimensions of the cylinder and the density of aluminum.

Find: the mass of the aluminum.

First we find the volume of the cylinder. Be aware of the different units in the problem; so we first convert all cylinder dimensions to meters.

$$r = 10 \text{ cm } (1 \text{ m}/100 \text{ cm}) = 0.10 \text{ m}, \quad h = 1.0 \text{ ft } (0.3048 \text{ m/ft}) = 0.3048 \text{ m}.$$

The volume of a cylinder is base area times height $= \pi r^2 h = \pi (0.10 \text{ m})^2 (0.3048 \text{ m}) = 9.58 \times 10^{-3} \text{ m}^3$.
Next we find the mass of the cylinder.

From $\quad \rho = \dfrac{m}{V}, \quad$ multiplying V on both sides yields $\quad m = \rho V$.

So $\quad m = (2700 \text{ kg/m}^3)(9.58 \times 10^{-3} \text{ m}^3) = 26 \text{ kg}.$

IV.   Mathematical Summary

Density			Defines density in terms of mass and volume.
	$\rho = \dfrac{m}{V} \left(\dfrac{\text{mass}}{\text{volume}} \right)$	(1.1)	

V.   Solutions of Selected Exercises and Paired/Trio Exercises

6.   (a) Two *Different* ounces are used. One is for volume measurement and another for weight measurement.

(b) Again, two different pound units are used. Avoirdupois lb = 16 oz and troy lb = 12 oz.

12.   The dimension of the left side of the equation is   [L].

The dimension of the right side of the equation is $[L] + \dfrac{[L]}{[T]} \times [T] = [L] + [L]$.

So the dimension of the left side is equal to the dimension of the right side and the equation is dimensionally correct.

18.   From $x = \dfrac{gt^2}{2}$,   we have $g = \dfrac{2x}{t^2}$.

So the units of g are the units of x divided by the units of t^2, that is $\boxed{m/s^2}$.

24.   (a) Since $F = ma$, newton $= (kg)(m/s^2) = \boxed{kg \cdot m/s^2}$.

(b) $\boxed{Yes}$.

From $F = m\dfrac{v^2}{r}$, the units of force F are $(kg) \times \dfrac{(m/s)^2}{m} = kg \cdot m/s^2$.

34.   (a) 0.5 gal $= (0.5 \text{ gal}) \times \dfrac{3.785 \text{ L}}{1 \text{ gal}} = 1.89$ L. 2 L $- 1.89$ L $= 0.11$ L.

So $\boxed{2 \text{ L by } 0.11 \text{ L more}}$.

(b) 16 oz $= (1 \text{ pt}) \times \dfrac{946 \text{ mL}}{2 \text{ pt}} = 473$ mL. 500 mL $- 473$ mL $= 27$ mL.

So $\boxed{500 \text{ mL}}$ gives the most for your money and you get $\boxed{27 \text{ mL more}}$.

38.   Using the $\dfrac{1 \text{ mi}}{1609 \text{ m}}$ and $\dfrac{3600 \text{ s}}{1 \text{ h}}$ conversion factors.

15 m/s $= (15 \text{ m/s}) \times 1 \times 1 = (15 \text{ m/s}) \times \dfrac{1 \text{ mi}}{1609 \text{ m}} \times \dfrac{3600 \text{ s}}{1 \text{ h}} = 34$ mi/h.

So it travels $\boxed{34 \text{ mi}}$ in one hour.

40.   (a) represents the greatest speed.

46.   (a) $x = (19 \text{ in.}) \cos 37° = 15.2$ in.

$y = (19 \text{ in.}) \sin 37° = 11.4$ in.

So the area $= xy = (15.2 \text{ in.})(11.4 \text{ in.}) = \boxed{1.7 \times 10^2 \text{ in.}^2}$.

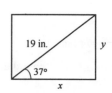

(b) Using the $\dfrac{2.54 \text{ cm}}{1 \text{ in.}}$ conversion factor.

$$1.7 \times 10^2 \text{ in.}^2 = (1.7 \times 10^2 \text{ in.}^2) \times 1 = (1.7 \times 10^2 \text{ in.}^2) \times \left(\frac{2.54 \text{ cm}}{1 \text{ in.}}\right)^2 = \boxed{1.1 \times 10^3 \text{ cm}^2}.$$

The conversion factor is squared because we are converting in.^{2} to cm^2.

48.　(a) Using the $\dfrac{1 \text{ kg}}{1000 \text{ g}}$ and $\dfrac{100 \text{ cm}}{1 \text{ m}}$ conversion factors.

$$13.6 \text{ g/cm}^3 = (13.6 \text{ g/cm}^3) \times 1 \times 1 = (13.6 \text{ g/cm}^3) \times \frac{1 \text{ kg}}{1000 \text{ g}} \times \left(\frac{100 \text{ cm}}{1 \text{ m}}\right)^3 = \boxed{1.36 \times 10^4 \text{ kg/m}^3}.$$

The conversion factor $\dfrac{100 \text{ cm}}{1 \text{ m}}$ is cubed because we are converting cm^3 to m^3.

(b) $\rho = \dfrac{m}{V}$,　☞　$m = \rho V = (13.6 \text{ g/cm}^3)(0.250 \text{ L}) \times \dfrac{1000 \text{ cm}^3}{1 \text{ L}} = 3.40 \times 10^3 \text{ g} = \boxed{3.40 \text{ kg}}.$

54.　The last digit is estimated so the smallest division is the third decimal place, that is $\boxed{0.001 \text{ m or } 1 \text{ mm}}$.

62.　From $\quad V = a^3, \quad$ we have $\quad a = \sqrt[3]{V} = \sqrt[3]{2.5 \times 10^2 \text{ cm}^3} = \boxed{6.3 \text{ cm}}.$

63.　The area is the sum of that of the top, the bottom, and the side. The side of the can is a rectangle with a length equal to the circumference and width equal to the height of the can.

$$A = \frac{\pi d^2}{4} + \frac{\pi d^2}{4} + Ch = \frac{\pi d^2}{4} + \frac{\pi d^2}{4} + (\pi d)h$$

$$= \frac{\pi(12.559 \text{ cm})^2}{4} + \frac{\pi(12.559 \text{ cm})^2}{4} + \pi(12.559 \text{ cm})(5.62 \text{ cm}) = \boxed{470 \text{ cm}^2}.$$

68.　$\rho = \dfrac{m}{V} = \dfrac{6.0 \times 10^{25} \text{ kg}}{1.1 \times 10^{21} \text{ m}^3} = \boxed{5.5 \times 10^3 \text{ kg/m}^3}.$

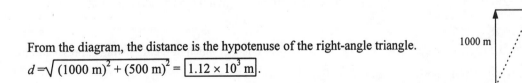

74.　From the diagram, the distance is the hypotenuse of the right-angle triangle.
$d = \sqrt{(1000 \text{ m})^2 + (500 \text{ m})^2} = \boxed{1.12 \times 10^3 \text{ m}}.$

81. By drawing a perpendicular line from the island to the shore, the distance from the island to the shore is

$d = x \tan 30° = (50 \text{ m} - x) \tan 40° = (50 \text{ m}) \tan 40° - x \tan 40°$.

Solving, $\quad x = \dfrac{(50 \text{ m}) \tan 40°}{\tan 30° + \tan 40°} = 29.6 \text{ m}$.

Therefore $\quad d = (29.6 \text{ m}) \tan 30° = \boxed{17 \text{ m}}$.

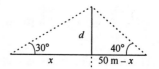

94. By drawing a diagram of the situation, we can see that the distance is the hypotenuse of the shaded right-angle triangle.

$d = \sqrt{(200 \text{ mi})^2 + (300 \text{ mi} - 100 \text{ mi})^2} = \boxed{283 \text{ mi}}$.

$\tan\theta = \dfrac{300 \text{ mi} - 100 \text{ mi}}{200 \text{ mi}} = 1.0$.

So $\quad \theta = \tan^{-1}(1.0) = \boxed{45° \text{ north of east}}$.

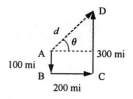

VI. Practice Quiz

1. In the SI, the base units for length, mass, and time are

(a) meters, grams, seconds.   (b) kilometers, kilograms, seconds.

(c) centimeters, kilograms, seconds.   (d) meters, kilograms, seconds.

(e) kilometers, grams, seconds.

2. If v has units of m/s and t has the units of s. What are the units of the quantity t/v?

(a) m   (b) s^2/m (c) s/m   (d) s/m^2 (e) s

3. Which one of the following has the same dimension as time?

(x is length, v is velocity, and a is acceleration)

(a) $\dfrac{x}{a}$ (b) $\sqrt{\dfrac{2x}{a}}$ (c) $\sqrt{\dfrac{v}{x}}$ (d) vx (e) xa

4. Which one of the following is *not* equivalent to 2.50 miles?

(a) 1.32×10^4 ft   (b) 1.58×10^5 in.   (c) 4.02×10^3 km   (d) 4.02×10^5 cm   (e) 4.40×10^3 yd

5. When (3.51×10^4) is multiplied by (4.00×10^2), the product is which of the following expressed to the correct number of significant figures?

(a) 1.40×10^7 (b) 1.4×10^7 (c) 1×10^7 (d) 88   (e) 87.8

6.   The density of water is 1.0×10^3 kg/m^3. Find the mass of water needed to fill a 2.0 L soft drink bottle.

    (a) 0.020 kg   (b) 0.20 kg   (c) 2.0 kg   (d) 20 kg   (e) 200 kg

7.   The area of a room floor is 25 ft^2. How many m^2 are there on the floor?

    (a) 7.6 m^2 (b) 2.3 m^2 (c) 82 m^2 (d) 2.6×10^2 m^2 (e) none of these

8.   An aluminum cube has a mass of 30 kg. What is the length of each side of the cube?

    (the density of aluminum is 2.7×10^3 kg/m^3)

    (a) 0.011 m   (b) 0.11 m   (c) 1.4×10^{-6} m   (d) 0.022 m   (e) 0.22 m

9.   A person stands 35.0 m from a flag pole. With a protractor at eye-level, he finds that the angle the top of

    the flag pole makes with the horizontal is 25.0°. How high is the flag pole? (the distance from his feet to his

    eyes is 1.70 m)

    (a) 14.8 m   (b) 16.3 m   (c) 16.5 m   (d) 18.0 m   (e) 75.1 m

10.   A rectangular garden measures 15 m long and 13.7 m wide. What is the length of a diagonal from one

    corner of the garden to the other?

    (a) 29 m   (b) 1.0 m   (c) 18 m   (d) 4.1×10^2 m   (e) 20 m

Answers to Practice Quiz:

1. d 2. b 3. b 4. c 5. a 6. c 7. b 8. d 9. d 10. e

CHAPTER 2

Kinematics: Description of Motion

I.   Chapter Objectives

Upon completion of this chapter, you should be able to:

1.   define distance and calculate speed, and explain what is meant by a scalar quantity.
2.   define displacement and calculate velocity, and explain the difference between scalar and vector quantities.
3.   explain the relationship between velocity and acceleration and perform graphical analyses of acceleration.
4.   explain the constant acceleration kinematic equations and apply them to physical situations.
5.   use the kinetic equations to analyze free fall.

II.   Key Terms

Upon completion of this chapter, you should be able to define and/or explain the following key terms:

mechanics	vector (quantity)
kinematics	velocity
dynamics	average velocity
motion	instantaneous velocity
distance	acceleration
scalar (quantity)	average acceleration
speed	instantaneous acceleration
average speed	acceleration due to gravity
instantaneous speed	free fall.
displacement	

The definitions and/or explanations of the most important key terms can be found in the following section: **III. Chapter Summary and Discussion**.

III.   Chapter Summary and Discussion

1.   Distance and Speed; Displacement and Velocity (Sections 2.1 – 2.2)

Motion is related to change of position. The length traveled in changing position may be expressed in terms of **distance**, the actual path length between two points. Distance is a scalar quantity, which has only a magnitude with no direction. The direct straight line pointing from the initial point to the final point is called **displacement** (change in position). Displacement only measures the change in position, not the details involved in the change in position. Displacement is a vector quantity, which has both magnitude and direction. In the Figure shown, an object goes from point A to point C by following paths AB and BC. The distance traced is 3.0 m + 4.0 m = 7.0 m, and the displacement is 5.0 m in the direction of the arrow.

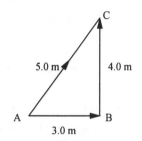

For motion in one dimension along the x axis, the displacement between two points, x_1 and x_2, is simply the vector subtraction between x_1 and x_2. $\Delta x = x_2 - x_1$, where the Greek letter Δ (delta) is used to represent a change or difference in a quantity. For example, if an object moves from a point at $x_1 = 2.0$ m to another point at $x_2 = 4.0$ m, its displacement is $\Delta x = 4.0$ m $- 2.0$ m $= +2.0$ m. The positive sign here indicated the direction of the displacement as in the positive x axis. However, if the motion is reversed, then $\Delta x = 2.0$ m $- 4.0$ m $= -2.0$ m (what is the meaning of the negative sign here?).

Example 2.1:   In a soccer game, a midfielder kicks the ball back 10 yards to a goalkeeper. The goalkeeper then kicks the ball straight up the field 50 yards to a forward. What is the distance traveled by the soccer ball? What is the displacement of the soccer ball?

Solution:

Sketch a diagram of the situation. For clarity, the arrows are laterally displaced.

It is obvious from the diagram that the soccer ball traveled 10 yards first and then 50 yards. So the *distance traveled* is 10 yards + 50 yards = 60 yards.

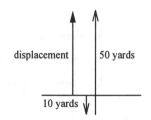

Displacement is the straight line from the initial point to the final point. The ball displaced only 50 yards − 10 yards = 40 yards **straight up the field**.

Try to solve this problem without the diagram. You will find it is very difficult to do. That is why you are encouraged to try to draw a diagram to help solve a problem.

In describing motion, the rate of change of position may be expressed in terms of speed and velocity.

Average speed is defined as the distance traveled divided by the time interval to travel that distance.

$avg.\ sp. = \dfrac{d}{t}$, where $avg.\ sp.$ is average speed, d is distance traveled, and t is time interval (change in time).

Instantaneous speed is the speed at a particular time instant (t is infinitesimally small or close to zero). Since distance is a scalar quantity with no direction, so are average speed and instantaneous speed. Both tell us only how fast objects are moving.

Average velocity is defined as *displacement* divided by the time interval, $\bar{v} = \dfrac{\Delta x}{\Delta t}$, where $\bar{v}$ is average velocity, Δx is displacement (change in position), and Δt is time interval. (Direction of displacement indicated by sign, $+$ or $-$ for one-dimensional motion.) **Instantaneous velocity**, v, is the velocity (magnitude and direction) at a particular instant of time (Δt is close to zero). Since displacement is a vector quantity, so are average velocity and instantaneous velocity. Both tell us not only how fast, but in which directions objects are moving. (Direction of velocity indicated by sign, $+$ or $-$ for one-dimensional motion.) The SI units of speed and velocity are m/s.

Example 2.2:   If the play described in Example 2.1 lasts 5.0 s, what is the average speed of the soccer ball? What is the average velocity of the soccer ball?

Solution:

Average speed: $avg.\ sp. = \dfrac{d}{t} = \dfrac{60\ \text{yd}}{5.0\ \text{s}} = 12\ \text{yd/s}.$

Average velocity: $\bar{v} = \dfrac{\Delta x}{\Delta t} = \dfrac{40\ \text{yd straight up the field}}{5.0\ \text{s}} = 8.0\ \text{yd/s straight up the field}$

---

2.   Acceleration (Section 2.3)

Acceleration is the rate of change of velocity of time. Note: it is *velocity* (a vector), not speed (a scalar). Hence acceleration is also a vector.

Average acceleration is defined as the change in velocity divided by the time interval to make the change,

$\bar{a} = \dfrac{\Delta v}{\Delta t} = \dfrac{v - v_o}{t - t_o}$, where $\bar{a}$ is average acceleration, Δv is change in velocity, and Δt is time interval.

Instantaneous acceleration is the acceleration at a particular instant of time (Δt is close to zero). As noted, velocity is a vector quantity, so are average acceleration and instantaneous acceleration. The SI units of acceleration are m/s/s or m/s^2.

A common misconception about velocity and acceleration has to do with their directions. Since velocity has both magnitude and direction, a change in either magnitude (speed) and/or direction will result in a change in velocity, therefore an acceleration. We can accelerate objects either by speeding them up or down (change magnitude) and/or by changing their directions of travel. We often call the gas pedal of a car an accelerator. Can we call the brake pedal an accelerator? Can we call the steering wheel an accelerator? The answers are yes for both questions. (Why?)

For motion in one-dimension, when the velocity and acceleration of an object are in the same direction (they have the same directional signs), the velocity increases and the object speeds up (acceleration). When the velocity and acceleration are in opposite directions, the velocity decreases and the object slows down (deceleration). The Learn BY Drawing on page 42 in the textbook graphically illustrates this point.

Example 2.3:   An object moving to the right has a decrease in velocity from 5.0 m/s to 1.0 m/s in 2.0 s. What is the average acceleration? What does your result mean?

Solution: Given: $v_0 = +5.0$ m/s, $v = +1.0$ m/s, $t = 2.0$ s.
Find: $\bar{a}$.

According to the definition of average acceleration,

$$\bar{a} = \frac{\Delta v}{\Delta t} = \frac{v - v_0}{t} = \frac{+1.0 \text{ m/s} - (+5.0 \text{ m/s})}{2.0 \text{ s}} = \frac{-4.0 \text{ m/s}}{2.0 \text{ s}} = -2.0 \text{ m/s}^2.$$

The negative sign means the acceleration is opposite to velocity (deceleration). The result means that the object *decreases* its velocity by 2.0 m/s every s or 2.0 m/s^2.

---

3.   Graphical Interpretation (Sections 2.2 - 2.3)

Graphical analysis is often helpful in understanding motion and its related quantities. In algebra, we learned that if $y = mx + b$, m is the slope of the graph y vs. x. If we take $t_0 = 0$, then $\bar{a} = \dfrac{v - v_0}{t - t_0} = \dfrac{v - v_0}{t}$, or

$v = v_0 + \bar{a} \, t$. That is, the slope of a velocity vs. time graph gives the average acceleration.

In general, on a position versus time graph, we can extract the average velocity by finding the slope of a line connecting two points. Instantaneous velocity is equal to the slope of a straight line tangent to the curve at a specific point. For a velocity versus time graph, the average acceleration is the slope of a straight line connecting two points and instantaneous acceleration is the slope of a straight line tangent to the curve at a specific point. The area under the curve in a v vs. t gives displacement and the area under the curve in a a vs. t graph yields velocity.

Example 2.8:   The graph represents the position of a particle as a function of time.

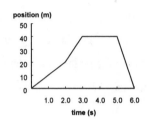

(a) What is the velocity at 1.0 s?

(b) What is the velocity at 2.5 s?

(c) What is the velocity at 4.0 s?

(d) What is the average velocity from 0 to 4.0 s?

(e) What is the average velocity for the 6.0 s interval?

Solution:

(a) Velocity is the slope of the line. $v = \dfrac{\Delta x}{\Delta t} = \dfrac{20 \text{ m} - 0 \text{ m}}{2.0 \text{ s} - 0 \text{ s}} = 10 \text{ m/s}.$

(b) Velocity is the slope of the line. $v = \dfrac{\Delta x}{\Delta t} = \dfrac{40 \text{ m} - 20 \text{ m}}{3.0 \text{ s} - 2.0 \text{ s}} = 20 \text{ m/s}.$

(c) The slope of the line is zero and so $v = 0$.

(d) $\bar{v} = \dfrac{\Delta x}{\Delta t} = \dfrac{40 \text{ m} - 0 \text{ m}}{4.0 \text{ s} - 0 \text{ s}} = 10 \text{ m/s}.$

(e) $\bar{v} = \dfrac{\Delta x}{\Delta t} = \dfrac{0 \text{ m} - 0 \text{ m}}{6.0 \text{ s} - 0 \text{ s}} = 0 \text{ m/s}.$

Example 2.9:   The graph represents the velocity of a particle as a function of time.

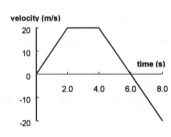

(a) What is the acceleration at 1.0 s?

(b) What is the acceleration at 3.0 s?

(c) What is the average acceleration between 0 and 5.0 s?

(d) What is the average acceleration for the 8.0 s interval?

(e) What is the displacement for the 8.0 s interval?

Solution:

(a) Acceleration is the slope of the line. $a = \dfrac{\Delta v}{\Delta t} = \dfrac{20 \text{ m/s} - 0 \text{ m/s}}{2.0 \text{ s} - 0 \text{ s}} = 10 \text{ m/s}^2.$

(b) The slope of the line is zero and $a = 0$.

(c) $\bar{a} = \dfrac{\Delta v}{\Delta t} = \dfrac{10 \text{ m/s} - 0 \text{ m/s}}{5.0 \text{ s} - 0 \text{ s}} = 2.0 \text{ m/s}^2.$

(d) $a = \dfrac{\Delta v}{\Delta t} = \dfrac{-20 \text{ m/s} - 0 \text{ m/s}}{8.0 \text{ s} - 0 \text{ s}} = -2.5 \text{ m/s}^2$.

(e) The net area equals the displacement.

The area of a rectangle is length × width and the area of a triangle is $\frac{1}{2}$× base × height.

$$\Delta x_{0\text{-}2} = \tfrac{1}{2}(2.0 \text{ s} - 0 \text{ s})(20 \text{ m/s}) = 20 \text{ m}; \qquad \Delta x_{2\text{-}4} = (4.0 \text{ s} - 2.0 \text{ s})(20 \text{ m/s}) = 40 \text{ m};$$

$$\Delta x_{4\text{-}6} = \tfrac{1}{2}(6.0 \text{ s} - 4.0 \text{ s})(20 \text{ m/s}) = 20 \text{ m}; \qquad \Delta x_{6\text{-}8} = \tfrac{1}{2}(8.0 \text{ s} - 6.0 \text{ s})(-20 \text{ m/s}) = -20 \text{ m}.$$

So $\Delta x = 20 \text{ m} + 40 \text{ m} + 20 \text{ m} + (-20 \text{ m}) = 60 \text{ m}.$

---

4.   Kinematic Equations (Section 2.4)

Our discussion is restricted to **motions with constant accelerations**. In a motion with constant acceleration, the acceleration is not changing with time or it is a constant. However, the constant could be zero, a negative or a positive non-zero constant. Zero acceleration simply means the velocity is a constant (no acceleration). For positive velocity, a negative acceleration means deceleration (speed decrease) and a positive acceleration means acceleration (speed increase). For negative velocity, a negative acceleration means acceleration (speed increase) and a positive acceleration means deceleration (speed decrease).

The symbols used in the kinematic are: v_0, initial velocity; v, final velocity; a, acceleration; x, displacement; t, time interval. Be aware that the terms initial and final are relative. The end of one event is always the beginning of another. There are three general equations and two algebraic combinations of these equations that provide calculation convenience.

$x = \bar{v} t$ displacement = average velocity × times interval,

$\bar{v} = \dfrac{v + v_0}{2}$,   average velocity = (final velocity + initial velocity)/2,

$v = v_0 + at$,   final velocity = initial velocity + acceleration × time interval,

$x = v_0 t + \frac{1}{2} at^2$,   displacement = initial velocity × time interval + $\frac{1}{2}$ × acceleration × time interval squared,

$v^2 = v_0^2 + 2ax$,   final velocity squared = initial velocity squared + 2 × acceleration × displacement.

Among the five equations listed, the last three can be used to solve the majority of kinematic problems. Which equation should you select in solving a particular problem? The equation you select must have the unknown quantity in it and everything else must be given, because we can only solve for one unknown in one equation.

Example 2.4:   An object starts from rest and accelerates with a constant acceleration of 5.0 m/s^2. Find its velocity and displacement at a time of 3.4 s.

Solution:　　Given: $v_0 = 0$ (starts from rest),　$a = 5.0$ m/s^2,　$t = 3.4$ s.

　　　　　　　　Find:　v and x.

From $v = v_0 + at$, we have $v = 0 + (5.0 \text{ m/s}^2)(3.4 \text{ s}) = 17$ m/s.

Also $x = v_0 t + \frac{1}{2}at^2 = (0)(3.4 \text{ s}) + \frac{1}{2}(5.0 \text{ m/s}^2)(3.4 \text{ s})^2 = 29$ m.

Both velocity and displacement are in the direction of the motion.

Example 2.5:　　An automobile accelerates uniformly from rest to 25 m/s while traveling 100 m. What is the acceleration of the automobile?

Solution:　　Given: $v_0 = 0$ (rest),　$v = 25$ m/s,　$x = 100$ m.

　　　　　　　　Find:　a.

Since $v^2 = v_0^2 + 2ax$,　$a = \dfrac{v^2 - v_0^2}{2x} = \dfrac{(25 \text{ m/s})^2 - (0)^2}{2(100 \text{ m})} = 3.1$ m/s^2

Since a is positive, it is in the direction of the velocity or motion.

5.　Free Fall (Section 2.5)

Objects in motion solely under the influence of gravity are said to be in **free fall**. A free fall does not necessarily mean a falling object. A vertically rising object is also said to be in free fall. The magnitude of **acceleration due to gravity** is often expressed with a symbol g. Near the surface of the Earth, the acceleration due to gravity is $g = 9.80$ m/s^2 (downward) and near the surface of the Moon, it is $g = 1.7$ m/s^2.

Note that g itself is a positive quantity, 9.80 m/s^2. If you use the upward direction as your positive reference direction, then we say the acceleration due to gravity is $-g = -9./80$ m/s^2 (downward 9.80 m/s^2). However, if you use the downward direction as your positive reference direction, then the acceleration due to gravity is $+g = +9./80$ m/s^2 (still downward 9.80 m/s^2).

Since free fall is in the vertical direction and we often choose the upward direction as the $+y$ axis, we replace the x's by y's and a's by $-g$'s in the kinematic equation. The results are

$$y = \bar{v}t, \quad \bar{v} = \frac{v + v_0}{2}, \quad v = v_0 - gt, \quad y = v_0 t - \frac{1}{2}gt^2, \quad v^2 = v_0^2 - 2gy, \text{ where } g = 9.80 \text{ m/s}^2.$$

Example 2.6: A ball is thrown upward with an initial velocity near the surface of the Earth. When it reaches the highest point

(a) its velocity is zero and its acceleration is non-zero,

(b) its velocity is zero and its acceleration is zero,

(c) its velocity is non-zero and its acceleration is zero,

(d) its velocity is non-zero and its acceleration is non-zero.

Solution:

The answer is (a) , not (b) as you might think. The velocity has to change its direction at the highest point (goes from positive to negative) and so it is zero. However, the acceleration is not zero there. The acceleration is a constant 9.80 m/s^2 downward, independent of velocity. Stop and think, what if both the velocity and acceleration are zero? Will the ball fall down after it reaches the highest point?

Example 2.7: A ball is thrown upward with an initial velocity of 10.0 m/s from the top of a 50.0 m tall building.

(a) With what velocity will the ball strike the ground?

(b) How long does it take the ball to strike the ground?

Solution: Given: $y = -50.0$ m (displacement), $v_0 = +10.0$ m/s.

Find: (a) t (b) v.

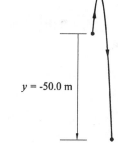

$y = $ -50.0 m

The y in the kinematic equations stands for displacement from the launch point, not distance. When the ball strikes the ground, it will displace -50.0 m, or 50 m below the launch point.

(a) $v^2 = v_0^2 - 2gy = (+10.0 \text{ m/s})^2 - 2(9.80 \text{ m/s}^2)(-50.0 \text{ m}) = 1.08 \times 10^3 \text{ m}^2/\text{s}^2$.

So $v = \sqrt{1.08 \times 10^3 \text{ m}^2/\text{s}^2} = \pm32.9$ m / s.

The positive answer is discarded since the ball is falling when it lands (moving downward).

Therefore $v = -32.9$ m/s.

(b) From $v = v_0 - gt$, we have

$$t = \frac{v_0 - v}{g} = \frac{(+10.0 \text{ m/s} - (-32.9 \text{ m/s})}{9.80 \text{ m/s}^2} = \frac{42.9 \text{ m/s}}{9.80 \text{ m/s}^2} = 4.38 \text{ s}.$$

Try to solve this problem without using the overall displacement concept. You could break it into two phases. First, you would have to find out how high the ball goes, then secondly determine the velocity when it strikes the ground, and the total time it is in the air.

IV.   Mathematical Summary

Average speed	$avg.\ sp. = \dfrac{d}{t}$ (2.2)	Defines average speed.
Average velocity	$\bar{v} = \dfrac{\Delta x}{\Delta t}$ (2.3)	Defines average velocity.
Kinematic equation 1	$x = \bar{v}t$ (2.4)	Relates displacement with average velocity and time.
Kinematic equation 2	$\bar{v} = \dfrac{v + v_0}{2}$ (2.8)	Defines average velocity for motion with constant acceleration.
Kinematic equation 3	$v = v_0 + at$ (2.7)	Relates final velocity with initial velocity, acceleration, and time (constant acceleration only).
Kinematic equation 4	$x = v_0 t + \frac{1}{2}at^2$ (2.9)	Relates displacement with initial velocity, acceleration, and time (constant acceleration only).
Kinematic equation 5	$v^2 = v_0^2 + 2ax$ (2.10)	Relates final velocity with initial velocity, acceleration, and displacement (constant acceleration only).
Equation 1 (free fall)	$y = \bar{v}t$ (2.4')	Relates displacement with average velocity and time.
Equation 2 (free fall)	$\bar{v} = \dfrac{v + v_0}{2}$ (2.8')	Defines average velocity.
Equation 3 (free fall)	$v = v_0 - gt$ (2.7')	Relates final velocity with initial velocity, acceleration, and time.
Equation 4 (free fall)	$y = v_0 t - \frac{1}{2}gt^2$ (2.9')	Relates displacement with initial velocity, acceleration, and time.
Equation 5 (free fall)	$v^2 = v_0^2 - 2gy$ (2.10')	Relates final velocity with initial velocity, acceleration, and displacement.

V.   Solutions of Selected Exercises and Paired/Trio Exercises

8.   Displacement is the change in position. After half a lap, the car is at the opposite end of a diameter. So the magnitude of the displacement is $\boxed{300 \text{ m}}$.

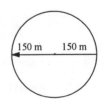

12.	(a) First trip: $\quad avg.\ sp. = \dfrac{d}{t} = \dfrac{150 \text{ km}}{2.5 \text{ h}} = \boxed{60 \text{ km/h}}$.

Return trip: $\quad avg.\ sp. = \dfrac{150 \text{ km}}{2.0 \text{ h}} = \boxed{75 \text{ km/h}}$.

(b) Total trip: $\quad avg.\ sp. = \dfrac{150 \text{ km} + 150 \text{ km}}{2.5 \text{ h} + 2.0 \text{ h}} = \boxed{67 \text{ km/h}}$.

15.	Draw a diagram of the situation.

$d = \sqrt{(40 \text{ m})^2 + (50\text{m} - 30 \text{ m})^2} = \boxed{45 \text{ m}}$.

$\theta = \tan^{-1}\left(\dfrac{50 \text{ m} - 30 \text{ m}}{40 \text{ m}}\right) = \boxed{27° \text{ west of north}}$.

16.	(a) $avg.\ sp. = \dfrac{d}{t} = \dfrac{2(10 \text{ m})}{2.4 \text{ s}} = \boxed{8.3 \text{ m/s}}$.

(b) Since the ball is caught at the initial height, the displacement is zero.

So the average velocity is $\boxed{\text{zero}}$.

20.	(a) From the definition of average velocity, $\bar{v} = \dfrac{\Delta x}{\Delta t}$, we have

$\bar{v}_{AB} = \dfrac{1.0 \text{ m} - 1.0 \text{ m}}{1.0 \text{ s} - 0} = \boxed{0}$; $\qquad \bar{v}_{BC} = \dfrac{7.0 \text{ m} - 1.0 \text{ m}}{3.0 \text{ s} - 1.0 \text{ s}} = \boxed{3.0 \text{ m/s}}$;

$\bar{v}_{CD} = \dfrac{9.0 \text{ m} - 7.0 \text{ m}}{4.5 \text{ s} - 3.0 \text{ s}} = \boxed{1.3 \text{ m/s}}$; $\qquad \bar{v}_{DE} = \dfrac{7.0 \text{ m} - 9.0 \text{ m}}{6.0 \text{ s} - 4.5 \text{ s}} = \boxed{-1.3 \text{ m/s}}$;

$\bar{v}_{EF} = \dfrac{2.0 \text{ m} - 7.0 \text{ m}}{9.0 \text{ s} - 6.0 \text{ s}} = \boxed{-1.7 \text{ m/s}}$; $\ \bar{v}_{FG} = \dfrac{2.0 \text{ m} - 2.0 \text{ m}}{11.0 \text{ s} - 9.0 \text{ s}} = \boxed{0}$;

$\bar{v}_{BG} = \dfrac{2.0 \text{ m} - 1.0 \text{ m}}{11.0 \text{ s} - 1.0 \text{ s}} = \boxed{0.10 \text{ m/s}}$.

(b) $\boxed{\text{The motion of BC, CD, and DE are not uniform}}$ since they are not straight lines.

(c) The object changes its direction of motion at point D. So it has to stop momentarily $v = \boxed{0}$.

26.	To the runner on right, the runner on left is running at a velocity of $+4.50 \text{ m/s} - (-3.50 \text{ m/s}) = +8.00 \text{ m/s}$.

So it takes $\Delta t = \dfrac{\Delta x}{\bar{v}} = \dfrac{100 \text{ m}}{8.00 \text{ m/s}} = \boxed{12.5 \text{ s}}$.

They meet at $(4.50 \text{ m/s})(12.5 \text{ s}) = \boxed{56.3 \text{ m from the initial position of the runner on left}}$.

28.	(d). Any change in either magnitude or direction results in a change in velocity. The brakes and gear shift change the magnitude and the steering wheel and changes the direction.

34. $$60 \text{ mi/h} = (60 \text{ mi/h}) \times \frac{1609 \text{ m}}{1 \text{ mi}} \times \frac{1 \text{ h}}{3600 \text{ s}} = 26.8 \text{ m/s}.$$

$$\bar{a} = \frac{\Delta v}{\Delta t} = \frac{26.8 \text{ m/s} - 0}{3.9 \text{ s}} = \boxed{6.9 \text{ m/s}^2}$$

40.   (a) $\bar{a}_{0\text{-}1.0 \text{ s}} = \dfrac{\Delta v}{\Delta t} = \dfrac{0 - 0}{1.0 \text{ s} - 0} = \boxed{0}$;

$$\bar{a}_{1.0 \text{ s-}3.0 \text{ s}} = \frac{8.0 \text{ m/s} - 0}{3.0 \text{ s} - 1.0 \text{ s}} = \boxed{4.0 \text{ m/s}^2};$$

$$\bar{a}_{3.0 \text{ s-}8.0 \text{ s}} = \frac{-12 \text{ m/s} - 8.0 \text{ m/s}}{8.0 \text{ s} - 3.0 \text{ s}} = \boxed{-4.0 \text{ m/s}^2};$$

$$\bar{a}_{8.0 \text{ s-}9.0 \text{ s}} = \frac{-4 \text{ m/s} - (-12.0 \text{ m/s})}{9.0 \text{ s} - 8.0 \text{ s}} = \boxed{8.0 \text{ m/s}^2};$$

$$\bar{a}_{9.0 \text{ s-}13.0 \text{ s}} = \frac{-4.0 \text{ m/s} - 4.0 \text{ m/s}}{13.0 \text{ s} - 9.0 \text{ s}} = \boxed{0}.$$

(b) $\boxed{\text{Constant velocity of } -4.0 \text{ m/s}}$.

46.   Given: $v_\text{o} = 0$, $a = 2.0 \text{ m/s}^2$, $t = 5.00 \text{ s}$.   Find: v and x.

(a) $v = v_\text{o} + at = 0 + (2.0 \text{ m/s}^2)(5.0 \text{ s}) = \boxed{10 \text{ m/s}}$.

(b) $x = v_\text{o} t + \frac{1}{2} at^2 = 0 + \frac{1}{2}(2.0 \text{ m/s}^2)(5.0 \text{ s})^2 = \boxed{25 \text{ m}}$.

48. $\boxed{-3.1 \text{ m/s}^2}$.

50.   Given: $v_\text{o} = 0$, $v = 560 \text{ km/h} = 155.6 \text{ m/s}$, $x = 400 \text{ m}$.   Find: t.

$$x = \bar{v}t = \frac{v_\text{o} + v}{2}\, t, \quad \mathcal{F} \quad t = \frac{2x}{v_\text{o} + v} = \frac{2(400 \text{ m})}{0 + 155.6 \text{ m/s}} = \boxed{5.14 \text{ s}}.$$

54.   Given: $v_\text{o} = 330 \text{ m/s}$, $v = 0$, $x = 30 \text{ cm} = 0.30 \text{ m}$.   Find: a.

$$v^2 = v_\text{o}^2 + 2ax, \quad \mathcal{F} \quad a = \frac{v^2 - v_\text{o}^2}{2x} = \frac{(0)^2 - (330 \text{ m/s})^2}{2(0.30 \text{ m})} = -\boxed{1.8 \times 10^5 \text{ m/s}^2}.$$

The negative sign here indicates that the acceleration vector is in opposite direction of velocity.

56. $\boxed{2.0 \times 10^2 \text{ m/s}^2}$.

58. $40 \text{ km/h} = (40 \text{ km/h}) \times \dfrac{1000 \text{ m}}{1 \text{ km}} \times \dfrac{1 \text{ h}}{3600 \text{ s}} = 11.11 \text{ m/s}.$

During reaction, the car travels $d = (11.11 \text{ m/s})(0.25 \text{ s}) = 2.78 \text{ m}.$ So the car really has only 13 m − 2.78 m = 10.2 m to come to rest. Let's calculate the stopping distance of the car.

Given: $v_0 = 11.1 \text{ m/s},$ $v = 0,$ $a = -8.0 \text{ m/s}^2.$ Find: $x.$

$v^2 = v_0^2 + 2ax,$ ☞ $x = \dfrac{v^2 - v_0^2}{2a} = \dfrac{0 - (11.1 \text{ m/s})^2}{2(-8.0 \text{ m/s}^2)} = 7.70 \text{ m}.$

Therefore it takes the car only 2.78 m + 7.70 m = $\boxed{10.5 \text{ m} < 13 \text{ m}}$ to stop.

$\boxed{\text{Yes}}$, the car will stop before hitting the child.

66. (c). It accelerates at 9.80 m/s² so it increases its speed by 9.80 m/s in each second.

70. (a) A straight line, slope = −g. (b) A parabola.

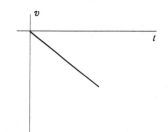

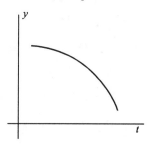

74. Given: $v_0 = 15 \text{ m/s},$ $v = 0$ (maximum height). Find: $y.$

$v^2 = v_0^2 - 2gy,$ ☞ $y = \dfrac{v_0^2 - v^2}{2g} = \dfrac{(15 \text{ m/s})^2 - (0)^2}{2(9.80 \text{ m/s}^2)} = \boxed{11 \text{ m}}.$

76. (a) Given: $v_0 = 21 \text{ m/s},$ $t = 3.0 \text{ s}.$ Find: $y.$

$y = v_0 t - \tfrac{1}{2} g t^2 = (21 \text{ m/s})(3.0 \text{ s}) - \tfrac{1}{2}(9.80 \text{ m/s}^2)(3.0 \text{ s})^2 = \boxed{19 \text{ m}}.$

(b) $12 \text{ m} = (21 \text{ m/s})t - \tfrac{1}{2}(9.80 \text{ m/s}^2)t^2,$ or $4.90t^2 - 21t + 12 = 0.$

Solving the quadratic equation, $t = \boxed{0.68 \text{ s (on the way up) or 3.6 s (on the way down)}}.$

86. (a) Given: $v_0 = 12.50 \text{ m/s}$ (ascending), $y = -60.0 \text{ m}.$ Find: $t.$

$y = v_0 t - \tfrac{1}{2} g t^2,$ ☞ $-60.0 \text{ m} = (12.50 \text{ m/s})t - (4.90 \text{ m/s}^2)t^2.$

Reduce to a quadratic equation: $4.90t^2 - 12.50t - 60.0 = 0.$

Solving, $t = \boxed{5.00 \text{ s}}$ or −2.45 s which is physically meaningless.

(b) $v = v_0 - gt = 12.50 \text{ m/s} - (9.80 \text{ m/s}^2)(5.00 \text{ s}) = -36.5 \text{ m/s} = \boxed{36.5 \text{ m/s downward}}.$

89.　(a) Since 25.0 m is a distance, we need to find the maximum height first.

Given:　$v_o = 7.25$ m/s,　$v = 0$.　Find:　y.

$$v^2 = v_o^2 - 2gy, \quad ☞ \quad y = \frac{v_o^2 - v^2}{2g} = \frac{(7.25 \text{ m/s})^2 - 0}{2(9.80 \text{ m/s}^2)} = 2.68 \text{ m}.$$

So if it has traveled a distance of 25.0 m, it has traveled

25.0 m − 2.68 m = 22.3 m downward after reaching maximum height.

So the displacement is　$y = -(22.3 \text{ m} - 2.68 \text{ m}) = -19.6$ m.

Now　$v^2 = v_o^2 - 2gy = (7.25 \text{ m/s})^2 - 2(9.80 \text{ m/s}^2)(-19.6 \text{ m})$

$$= 4.37 \times 10^2 \text{ m}^2/\text{s}^2. \quad \text{So} \quad v = -\sqrt{v^2} = \boxed{-20.9 \text{ m/s}}.$$

(b) $v = v_o - gt$,　☞　$t = \dfrac{v_o - v}{g} = \dfrac{7.25 \text{ m/s} - (-20.9 \text{ m/s})}{9.80 \text{ m/s}^2} = \boxed{2.87 \text{ s}}$.

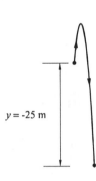

95.　From the definition of displacement,

$$d = \sqrt{(50 \text{ m})^2 + (50 \text{ m})^2} = \boxed{71 \text{ m}}.$$

101.　(a) Given:　$v_o = 15$ m/s,　$y = -25$ m.　Find:　v.

$$v^2 = v_o^2 - 2gy = (15 \text{ m/s})^2 - 2(9.80 \text{ m/s}^2)(-25 \text{ m}) = 715 \text{ m}^2/\text{s}^2.$$

So　$v = -\sqrt{715 \text{ m}^2/\text{s}^2} = \boxed{-27 \text{ m/s}}$

(b) $v = v_o - gt$,　☞　$t = \dfrac{v_o - v}{g} = \dfrac{15 \text{ m/s} - (-27 \text{ m/s})}{9.80 \text{ m/s}^2} = \boxed{4.3 \text{ s}}$.

VI.　Practice Quiz

1.　If you run a full lap around a circular track of radius 25 m in 100 s, the magnitude of your average velocity is

(a) zero.　(b) 0.20 m/s.　(c) 0.50 m/s.　(d) 1.0 m/s.　(e) 3.14 m/s.

2.　An object moving in the $+x$ axis experiences an acceleration of $+5.0$ m/s². This means the object is

(a) traveling 5.0 m in every second.

(b) traveling at 5.0 m/s in every second.

(c) changing its velocity by 5.0 m/s.

(d) increasing its velocity by 5.0 m/s in every second.

3. A car starts from rest and travels 100 m in 5.0 s. What is the magnitude of the constant acceleration.

(a) zero (b) 5.0 m/s^2 (c) 8.0 m/s^2 (d) 10 m/s^2 (e) 40 m/s^2

4. An object is thrown straight up. When it is at the highest point

(a) both its velocity and acceleration are zero.

(b) neither its velocity nor its acceleration is zero.

(c) its velocity is zero and its acceleration is not zero.

(d) its velocity is not zero and its acceleration is zero.

5. Human reaction time is usually greater than 0.10 s. If your lab partner holds a ruler between your fingers and releases it without warning, how far can you expect the ruler to fall before you catch it?

(a) at least 3.0 cm (b) at least 4.9 cm (c) at least 6.8 cm (d) at least 9.8 cm (e) at least 11.0 cm

6. Which one of the following quantities is an example of a vector?

(a) distance (b) acceleration (c) speed (d) mass

7. A ball is thrown vertically upward with a speed v. An identical second ball is thrown upward with a speed $2v$ (twice as fast). What is the ratio of the height of the second ball to that of the first ball? (How many times higher does the second ball go than the first ball?)

(a) 4:1 (b) 2:1 (c) 1.7:1 (d) 1.4:1 (e) 1:1

8. A car starts from rest and accelerates for 4.0 m/s^2 for 5.0 s, then maintains that velocity for 10 s and then decelerates at the rate of 2.0 m/s^2 for 4.0 s. What is the final speed of the car?

(a) 20 m/s (b) 16 m/s (c) 12 m/s (d) 10 m/s (e) 8.0 m/s

9. An object moves 5.0 m north and then 3.0 m east. Find both the distance traveled and the magnitude of the displacement.

(a) 8.0 m; 5.8 m (b) 5.8 m; 8.0 m (c) 8.0 m; 4.0 m (d) 4.0 m; 8.0 m (e) 5.8 m, 34 m

10. A car with a speed of 25.0 m/s brakes to a stop. If the maximum deceleration of the car is 10.0 m/s^2, what is the minimum stopping distance?

(a) 0.032 m (b) 0.80 m (c) 1.3 m (d) 31 m (e) 6.3 × 10^2 m

Answers to Practice Quiz:

1. a 2. d 3. c 4. c 5. b 6. b 7. a 8. c 9. a 10. d

CHAPTER 3

I.   Chapter Objectives

Upon completion of this chapter, you should be able to:

1.   analyze motion in terms of its components and apply the kinematic equations to components of motion.

2.   add and subtract vectors graphically and analytically.

3.   determine relative velocities through vector addition and subtraction.

4.   analyze projectile motion to find position, time of flight, and range.

II.   Key Terms

Upon completion of this chapter, you should be able to define and/or explain the following key terms:

components of motion	unit vector
vector addition (subtraction)	component form
triangle method	analytical component method
parallelogram method	relative velocity
polygon method	projectile motion
component method	parabola
magnitude-angle form	range

The definitions and/or explanations of the most important key terms can be found in the following section:
III. Chapter Summary and Discussion.

III. Chapter Summary and Discussion

1. Components of Motion (Section 3.1)

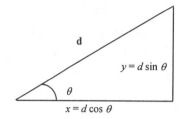

Motion in two dimensions, or curvilinear motion, is motion in which an object moves in a plane which can be described by a rectangular coordinate system. To analyze such motion, quantities are usually resolved into rectangular **components**. The diagram on the right shows the displacement vector being resolved into $x = d \cos\theta$ and $y = d \sin\theta$, the rectangular coordinates. Similarly, velocity and acceleration vectors, **v** and **a**, are resolved into $v_x = v \cos\theta$, $v_y = v \sin\theta$, and $a_x = a \cos\theta$, $a_y = a \sin\theta$, their horizontal and vertical components, respectively.

Note: the angle θ used in the above calculations is the angle relative to the x axis.

Once the displacement, velocity, and acceleration vectors are resolved into their respective components, we can apply the kinematic equations from Chapter 2 to the motion in the x and y directions. For example:

$$v_x = v_{xo} + a_x t, \quad v_y = v_{yo} + a_y t, \quad x = v_{xo} t + \tfrac{1}{2} a_x t^2, \quad y = v_{yo} t + \tfrac{1}{2} a_y t^2, \quad \text{etc.}$$

The key to success in solving two dimensional motion is to resolve the motion in components. Remember to treat the components as independent, i.e., a_x has nothing to do with a_y, etc. However, the time in all the equations is the same, providing a common link. Always think about resolving vectors into components when working problems in two dimensions.

Example 3.1: An airplane is moving at 250 mi/h in a direction of 35° N of E. Find the components of the plane's velocity in the eastward and northward directions.

Solution: Given: **v** = 250 mi/h in a direction of 35° N of E,

or, v = 250 mi/h and θ = 35°.

Find: v_x and v_y.

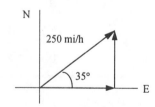

$v_x = v \cos\theta = (250 \text{ mi/h}) \cos 35° = 205 \text{ mi/h}$,

$v_y = v \sin\theta = (250 \text{ mi/h}) \sin 35° = 143 \text{ mi/h}$.

Example 3.2: A boat travels with a speed of 5.0 m/s in a straight path on a still lake. Suddenly, a steady wind pushes the boat perpendicularly to its straight line path with a speed of 3.0 m/s for 5.0 s. Relative to its position just when the wind started to blow, where is the boat at the end of this time?

Solution: Given: $v_{xo} = 5.0$ m/s, $a_x = 0$, $v_{yo} = 3.0$ m/s, $a_y = 0$, $t = 5.0$ s.

Find: x and y.

Both motions are motion with constant velocity. Choose the straight path of the boat as x axis and the direction of wind as the y axis.

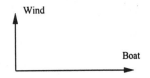

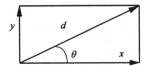

$x = v_{xo} t + \frac{1}{2} a_x t^2 = (5.0 \text{ m/s})(5.0 \text{ s}) + 0 = 25 \text{ m}$,

$y = v_{yo} t + \frac{1}{2} a_y t^2 = (3.0 \text{ m/s})(5.0 \text{ s}) + 0 = 15 \text{ m}$.

Or $d = \sqrt{x^2 + y^2} = \sqrt{(25 \text{ m})^2 + (15 \text{ m})^2} = 29 \text{ m}$,

And $\theta = \tan^{-1} \dfrac{y}{x} = \dfrac{15 \text{ m}}{25 \text{ m}} = 31°$.

2. Vector Addition and Subtraction (Section 3.2)

Vector addition can be done graphically with the triangle method or the parallelogram method for two vectors and the polygon method for more than two vectors. **Vector subtraction** is a special case of vector addition because $\mathbf{A} - \mathbf{B} = \mathbf{A} + (-\mathbf{B})$, and a negative vector is defined as a vector having the same magnitude but opposite in direction to the positive vector. For example, the negative vector of a velocity vector at 45 m/s north is simply 45 m/s south. When adding or subtracting vectors graphically, a convenient scale must be used.

Example 3.3: Two vectors **A** and **B** are given. Show

(a) **A** + **B** with the triangle method.

(b) **A** + **B** with the parallelogram method.

(c) **A** − **B** with the triangle method.

(d) **A** − **B** with the parallelogram method.

Solution: (a)   (b)

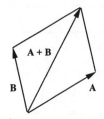

(c)

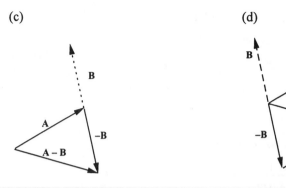

(d)

Vector addition is conveniently done by the **analytical component method**. The recommended procedure is as follows:

(1)   Resolve the vectors to be added into their x and y components. Include directional signs (plus or minus) in the components.

(2)   Add, algebraically, all the x components together and all the y components together to get the x and y components of the resultant vector.

(3)   Express the resultant vector using   (a) the component form, e.g., $\mathbf{A} = A_x\,\hat{\mathbf{x}} + A_y\,\hat{\mathbf{y}}$, or

$$\text{(b) in magnitude-angle form, e.g., } A = \sqrt{A_x^2 + A_y^2},$$

$$\theta = \tan^{-1}\frac{A_y}{A_x} \text{ (relative to } x \text{ axis)}.$$

A detailed treatment of this procedure can be found on page 75 to 76 in the textbook.

Example 3.4:   Find the resultant velocity of the sum of

(i) $\mathbf{v}_1 = 35$ m/s 30° N of E;   (ii) $\mathbf{v}_2 = 55$ m/s 45° N of W.

Solution:

(1)   Resolve the vectors to be added into their x and y components.

$v_{1x} = v_1 \cos 30° = (35 \text{ m/s}) \cos 30° = 30.3$ m/s,

$v_{1y} = (35 \text{ m/s}) \sin 30° = 17.5$ m/s.

$v_{2x} = -v_2 \cos 45° = (55 \text{ m/s}) \cos 45° = -38.9$ m/s, ($-x$ direction)

$v_{2y} = (55 \text{ m/s}) \sin 45° = 38.9$ m/s.

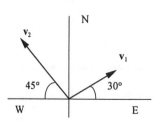

(2)   Add components.

$v_x = v_{1x} + v_{2x} = 30.3$ m/s $+ (-38.9$ m/s$) = -8.6$ m/s,

$v_y = v_{1y} + v_{2y} = 17.5$ m/s $+ 38.9$ m/s $= 56.4$ m/s.

(3)   Express the resultant vector.

First we draw the resultant velocity vector based on the

components obtained in the previous procedure. We know that the

x-component is -8.6 m/s and the y-component is 56.4 m/s.

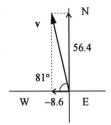

In component form: $\mathbf{v} = \mathbf{v}_1 + \mathbf{v}_2 = -8.6$ m/s $\hat{\mathbf{x}} + 56$ m/s $\hat{\mathbf{y}}$;

In magnitude-angle form:

$$v = \sqrt{v_x^2 + v_y^2} = \sqrt{(-8.6 \text{ m/s})^2 + (56.4 \text{ m/s})^2} = 57 \text{ m/s},$$

and $\theta = \tan^{-1} \dfrac{v_y}{v_x} = \tan^{-1} \dfrac{56.4 \text{ m/s}}{-8.6 \text{ m/s}} = 81° \text{ N of W.}$

Note: Once you have the components for the resultant [at the end of step (2)], you need to draw a diagram like the one above to determine the angle and in which quadrant the vector is. It is very difficult for you to determine the location of the vector without the diagram.

---

3.   Relative Velocity (Section 3.3)

Physical phenomena can be observed from different **frames of reference**. The velocity of a ball tossed by a passenger in a moving car will be measured differently by a passenger on the car than by an observer on the Earth. As a matter of fact, any velocity we measure is **relative**. The velocity of a moving car is measured relative to the ground, and the revolving motion of our Earth around the Sun is relative to the Sun, etc. Relative velocity can be determined with vector addition or subtraction. The symbols used in relative velocity such as $\mathbf{v}_{cg}$ (where c stands for car and g stands for ground) means the velocity of *a car relative to the ground*.

Example 3.5:   A river has a current with a velocity of 1.0 m/s south. A boat, whose speed in still water is 5.0 m/s, is directed east across the 100 m wide river.

(a) How long does it take the boat to reach the opposite shore?

(b) How far downstream will the boat land?

(c) What is the velocity of the boat relative to the shore?

Solution: Given: $v_{rs} = 1.0$ m/s, $v_{br} = 5.0$ m/s, $y = 100$ m.

Find: (a) t (b) x (c) $\mathbf{v}_{bs}$.

We use the following subscripts:

r = river, b = boat, s = shore.

We chose the coordinate system as shown.

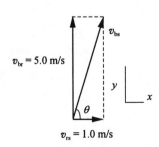

(a) From the concept of components of motion, the time it takes the boat to reach the opposite shore is

simply $t = \dfrac{y}{v_{br}} = \dfrac{100 \text{ m}}{5.0 \text{ m/s}} = 20$ s.

(b) $x = v_{rs} t = (1.0 \text{ m/s})(20 \text{ s}) = 20$ m.

(c) The velocity of the boat *relative to shore* is the vector sum of the velocity of the boat *relative to the river* and the velocity of the river *relative to the shore* (current). $\mathbf{v}_{bs} = \mathbf{v}_{br} + \mathbf{v}_{rs}$.

Note: The pattern of the subscripts are helpful in problem solving. On the right side of the equation, the two inner subscripts are the same (r). The outer subscripts (b and s) are sequentially the same as those for the relative velocity on the left side of the equation. This pattern is a good check to see if you write the relative velocity equation correctly!

$v_{bs} = \sqrt{v_{br}^2 + v_{rs}^2} = \sqrt{(5.0 \text{ m/s})^2 + (1.0 \text{ m/s})^2} = 5.1$ m/s,

$\theta = \tan^{-1}\left(\dfrac{5.0 \text{ m/s}}{1.0 \text{ m/s}}\right) = 79°$ measured from shoreline.

The velocity of the *boat relative to the shore* (or an observer standing on the shore) is 5.1 m/s in a direction of 79° measured from the shoreline.

Example 3.6: If the person on the boat in the previous example (Example 3.5) wants to travel directly across the river,

    (a) what angle upstream must the boat be directed?

    (b) with what speed will the boat cross the river?

    (c) how long will it take the boat to reach the opposite shore?

Solution:

To travel directly across the river, the velocity of the boat *relative to the shore* must be directly across the river. The vector form of the relative velocity equation in Example 3.5,

$\mathbf{v}_{bs} = \mathbf{v}_{br} + \mathbf{v}_{rs}$ is still valid.

(a) $\theta = \sin^{-1}\dfrac{v_{rs}}{v_{br}} = \sin^{-1}\dfrac{1.0 \text{ m/s}}{5.0 \text{ m/s}} = 12°$ upstream from a line straight

across the river.

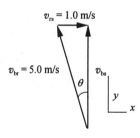

(b) From the triangle in the diagram, we have $v_{bs}^2 + v_{rs}^2 = v_{br}^2$,

so $v_{bs} = \sqrt{v_{br}^2 - v_{rs}^2} = \sqrt{(5.0 \text{ m/s})^2 - (1.0 \text{ m/s})^2} = 4.9$ m/s.

(c) The time is then $t = \dfrac{y}{v_{bs}} = \dfrac{100 \text{ m}}{4.9 \text{ m/s}} = 20$ s.

---

4. Projectile Motion (Section 3.4)

Projectile motion is motion in two dimensions, horizontal and vertical, with the vertical motion under the action of gravity only (downward), Since the action of gravity is in the vertical direction, the horizontal motion has zero acceleration, if air resistance is ignored. The vertical motion is a free fall and so the acceleration is the acceleration due to gravity, $a_y = -g = -9.80$ m/s^2, if the upward direction is chosen positive. This two-dimensional motion is analyzed using components, that is, the horizontal quantities are independent of the vertical quantities and vice versa. However, the time of flight, the time the projectile spends in the air, is the common quantity for both the horizontal and vertical motions.

Applying the general kinematic equations in component form to projectile motion ($a_x = 0$ and $a_y = -g$ with the upward direction chosen positive), we have $v_x = v_{xo}$, $v_y = v_{yo} - gt$, $x = v_{xo}\,t$, $y = v_{yo}\,t - \frac{1}{2}gt^2$,   etc. Again, the key to success in solving projectile motion is to resolve the motion into components, treat the components as independent, and use the same time of flight for both motions. Always think about resolving vectors into components. Usually, the time of flight is something you have to find first, since it is a common quantity for both motions and acts as a linkage between the two motions.

Example 3.7: A package is dropped from an airplane traveling with a constant horizontal speed of 120 m/s at an altitude of 500 m. What is the horizontal distance the package travels before hitting the ground (range)?

Solution: Given: horizontal motion   vertical motion

(taken in the x direction)   (up as positive)

$v_{xo} = 120$ m/s, $v_{yo} = 0$,

$y = -500$ m.

Find: x (range).

Since the range is given by $x = v_{xo}\,t$, we have to find the time of flight t first.

From the vertical motion, we use $y = v_{yo}\,t - \frac{1}{2}gt^2$.

So -500 m $= 0 - \frac{1}{2}(9.80$ m/s$^2)t^2$,   solving, $t = 10.1$ s.

Therefore $x = (120$ m/s$)(10.1$ s$) = 1.21 \times 10^3$ m $= 1.21$ km.

Note: The quantities such as initial velocities and displacements have to be treated independently.   For example, the initial horizontal velocity is 120 m/s and the initial vertical velocity is zero. The 12.0 m/s can *only* be used in the horizontal motion and the 0 m/s can *only* be used in the vertical motion. A common mistake is to mix up these quantities or not treat them as independent.

Example 3.8: A golfer hits a golf ball with a velocity of 35 m/s at an angle of 25 degrees above the horizontal. If the point where the ball is hit and the point where the ball lands are at the same level,

(a) how long is the ball in the air?

(b) what is the range of the ball?

Solution: Given: horizontal motion   vertical motion

$$v_{xo} = v_o \cos\theta \qquad\qquad v_{yo} = v_o \sin\theta$$

$$= (35 \text{ m/s}) \cos 25° \qquad = (35 \text{ m/s}) \sin 25°$$

$$= 31.7 \text{ m/s}, \qquad\qquad = 14.8 \text{ m/s}.$$

Find: (a) t (b) x.

(a) On landing, $y = 0$; and from $y = v_{yo}t - \frac{1}{2}gt^2$, we have $0 = (14.8 \text{ m/s})t - \frac{1}{2}(9.80 \text{ m/s}^2)t^2$.

Solving, $t = 0$ or 3.02 s. The $t = 0$ root corresponds to the position at the start ($x = 0$ and $y = 0$) and the $t = 3.02$ s corresponds to the landing position ($x = $ range and $y = 0$). So the time of flight is 3.0 s, (to two significant figures).

(b) $x = v_{xo}t = (31.7 \text{ m/s})(3.02 \text{ s}) = 96$ m.

Or, using the range equation, we have $x = R = \dfrac{v_o^2 \sin 2\theta}{g} = \dfrac{(35 \text{ m/s})^2 \sin 2(25°)}{9.80 \text{ m/s}^2} = 96$ m

IV.   Mathematical Summary

Components of initial velocity	$v_{xo} = v\cos\theta$ (3.1a)	Relates the x and y components to the magnitude and the
	$v_{yo} = v\sin\theta$ (3.1b)	angle of the initial velocity. (θ is from x axis)
Components of displacement	$x = v_{xo}t + \frac{1}{2}a_x t^2$ (3.3a)	Relates the displacement components to initial velocity
	$y = v_{yo}t + \frac{1}{2}a_y t^2$ (3.3b)	components, acceleration components and time (constant acceleration only).
Components of velocity	$v_x = v_{xo} + a_x t$ (3.3c)	Relates the velocity components to initial velocity
	$v_y = v_{yo} + a_y t$ (3.3d)	components and acceleration components (constant acceleration only).
Vector Representation	$C = \sqrt{C_x^2 + C_y^2}$, (3.4a)	Magnitude-angle form.
	$\theta = \tan^{-1}\left|\dfrac{C_y}{C_x}\right|$ (3.4b)	
Vector Representation	$\mathbf{C} = C_x\,\hat{\mathbf{x}} + C_y\,\hat{\mathbf{y}}$ (3.7)	Component form.

V.   Solutions of Selected Exercises and Paired/Trio Exercises

4.   Horizontal component: $v_x = v \cos\theta = (35 \text{ m/s}) \cos 37° = \boxed{28 \text{ m/s}}$.

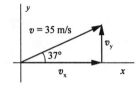

Vertical component: $v_y = v \sin\theta = (35 \text{ m/s}) \sin 37° = \boxed{21 \text{ m/s}}$.

6. $\boxed{\pm 6.3 \text{ m/s}}$.

8.   The displacement that will bring the student back to the starting point is pointing from the finishing point to the starting point.

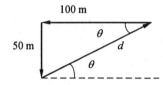

$d = \sqrt{(50 \text{ m})^2 + (100 \text{ m})^2} = \boxed{1.1 \times 10^2 \text{ m}}$, $\theta = \tan^{-1}\left(\dfrac{50 \text{ m}}{100 \text{ m}}\right) = \boxed{27° \text{ north of east}}$.

12.   From the kinetic equations for components of motion:

$x = v_x t = (0.60 \text{ m/s})(2.5 \text{ s}) = 1.5 \text{ m}$, $y = v_y t = (0.80 \text{ m/s})(2.5 \text{ s}) = 2.0 \text{ m}$.

$d = \sqrt{x^2 + y^2} = \sqrt{(1.5 \text{ m})^2 + (2.0 \text{ m})^2} = \boxed{2.5 \text{ m}}$. $\theta = \tan^{-1}\left(\dfrac{2.0 \text{ m}}{1.5 \text{ m}}\right) = \boxed{53° \text{ above } +x \text{ axis}}$.

18.   (c), because the magnitude of the resultant is 1 if the two vectors are opposite and 4 if they are in the same direction.

23. $\boxed{\text{Yes}}$, vector addition is associative.

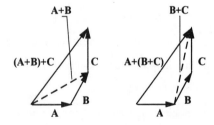

28.   (a) See diagram.

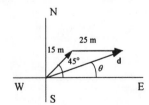

(b) For the 15 m vector: $d_{1x} = (15 \text{ m}) \cos 45° = 10.6 \text{ m}$,

$$d_{1y} = (15 \text{ m}) \sin 45° = 10.6 \text{ m}.$$

For the 25 m vector: $d_{2x} = 25 \text{ m}$, $\qquad d_{2y} = 0$.

So $d_x = d_{1x} + d_{2x} = 10.6 \text{ m} + 25 \text{ m} = 35.6 \text{ m}$, $\quad d_y = d_{1y} + d_{2y} = 10.6 \text{ m} + 0 = 10.6 \text{ m}$.

Therefore $\qquad d = \sqrt{(35.6 \text{ m})^2 + (10.6 \text{ m})^2} = \boxed{37 \text{ m}}$

and $\qquad \theta = \tan^{-1}\left(\dfrac{10.6 \text{ m}}{35.6 \text{ m}}\right) = \boxed{17° \text{ north of east}}$.

30. $\boxed{145 \text{ N } 50.1° \text{ north of east}}$.

33. (a) $\mathbf{F}_1 = [(12.0 \text{ N}) \cos 37°]\,\hat{x} + [(12.0 \text{ N}) \sin 37°]\,\hat{y} = (9.58 \text{ N})\,\hat{x} + (7.22 \text{ N})\,\hat{y}$.

$\mathbf{F}_2 = [-(12.0 \text{ N}) \cos 37°]\,\hat{x} + [(12.0 \text{ N}) \sin 37°]\,\hat{y} = (-9.58 \text{ N})\,\hat{x} + (7.22 \text{ N})\,\hat{y}$.

So $\quad \mathbf{F}_1 + \mathbf{F}_2 = \boxed{(14.4 \text{ N})\,\hat{y}}$.

(b) $\mathbf{F}_1 = [(12.0 \text{ N}) \cos 27°]\,\hat{x} + [(12.0 \text{ N}) \sin 27°]\,\hat{y} = (10.7 \text{ N})\,\hat{x} + (5.45 \text{ N})\,\hat{y}$.

So $\quad \mathbf{F}_1 + \mathbf{F}_2 = (1.1 \text{ N})\,\hat{x} + (12.7 \text{ N})\,\hat{y}$. $\quad F_1 + F_2 = \sqrt{(1.1 \text{ N})^2 + (12.7 \text{ N})^2} = \boxed{12.7 \text{ N}}$.

$\theta = \tan^{-1}\left(\dfrac{12.7 \text{ N}}{1.1 \text{ N}}\right) = \boxed{85.0° \text{ above } +x \text{ axis}}$

38. From $\mathbf{F}_1 + \mathbf{F}_2 + \mathbf{F}_3 = 0$,

$\mathbf{F}_3 = -\mathbf{F}_1 - \mathbf{F}_2 = -(3.0 \text{ N})\,\hat{x} - (3.0 \text{ N})\,\hat{y} - [(-6.0 \text{ N})\,\hat{x} + (4.5 \text{ N})\,\hat{y}] = (3.0 \text{ N})\,\hat{x} - (1.5 \text{ N})\,\hat{y}$.

So $\qquad F_3 = \sqrt{(3.0 \text{ N})^2 + (-1.5 \text{ N})^2} = \boxed{3.4 \text{ N}}$,

and $\qquad \theta = \tan^{-1}\left(\dfrac{-1.5 \text{ N}}{3.0 \text{ N}}\right) = \boxed{27° \text{ below the } +x \text{ axis}}$.

44. $\mathbf{d}_1 = (60 \text{ mi})[(\cos 45°)\,\hat{x} + (\sin 45°)\,\hat{y}] = (42.4 \text{ mi})\,\hat{x} + (42.4 \text{ mi})\,\hat{y}$. $\quad \mathbf{d}_2 = (75 \text{ mi})\,\hat{y}$.

$\mathbf{d} = \mathbf{d}_2 - \mathbf{d}_1 = (75 \text{ mi})\,\hat{y} - [(42.4 \text{ mi})\,\hat{x} + (42.4 \text{ mi})\,\hat{y}] = (-42.4 \text{ mi})\,\hat{x} + (32.6 \text{ mi})\,\hat{y}$.

So $\quad d = \sqrt{(-42.4 \text{ mi})^2 + (32.6 \text{ mi})^2} = 53.48 \text{ mi}$.

Therefore $\quad v = \dfrac{53.48 \text{ mi}}{2.0 \text{ h}} = \boxed{26.7 \text{ mi/h}}$.

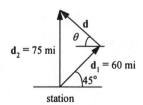

$\theta = \tan^{-1}\left(\dfrac{32.6 \text{ mi}}{-42.4 \text{ mi}}\right) = \boxed{37.6° \text{ north of west}}$.

46. $\boxed{23 \text{ km } 27° \text{ north of west}}$.

52. Use the following subscripts: t = truck, b = ball, and o = observer.

So v_{tg} = 70 km/h, v_{bt} = −15 km/h.

(a) $v_{bo} = v_{bt} + v_{to}$ = −15 km/h + 70 km/h = $\boxed{+55 \text{ km/h}}$.

(b) $v_{bt} = v_{bo} - v_{to}$ = 55 km/h − 90 km/h = $\boxed{-35 \text{ km/h}}$.

58. Use the following subscripts: s = swimmer, c = current, and b = bank.

$\mathbf{v}_{sb} = \mathbf{v}_{sc} + \mathbf{v}_{cb}$.

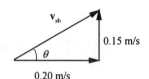

So $v_{sb} = \sqrt{(0.20 \text{ m/s})^2 + (0.15 \text{ m/s})^2} = \boxed{0.25 \text{ m/s}}$.

$\theta = \tan^{-1}\left(\dfrac{0.15 \text{ m/s}}{0.20 \text{ m/s}}\right) = \boxed{37° \text{ north of east}}$.

60. (a) $\boxed{0.075 \text{ m/s}}$. (b) $\boxed{30°}$

66. The horizontal motion does not affect the vertical motion. The vertical motion of the ball projected horizontally is identical to that of the ball dropped.

70. Given: v_{xo} = 15 m/s, v_{yo} = 0, y = −6.0 m. Find: x.

First we need to find the time of flight from the vertical motion.

From $y = v_{yo}t - \frac{1}{2}gt^2 = 0 - \frac{1}{2}gt^2$, we have $t = \sqrt{-\dfrac{2y}{g}} = \sqrt{-\dfrac{2(-6.0 \text{ m})}{9.80 \text{ m/s}^2}} = 1.11$ s.

$x = v_{xo}t = (15 \text{ m/s})(1.11 \text{ s}) = \boxed{17 \text{ m}}$.

72. $\boxed{7.6 \text{ m/s}}$.

78. $v_{xo} = v_o \cos\theta = (20.0 \text{ m/s}) \cos 15.0° = 19.32$ m/s,

$v_{yo} = v_o \sin\theta = (20.0 \text{ m/s}) \sin 15.0° = 5.176$ m/s.

(a) At maximum height, $v_y = 0$.

From $v_y^2 = v_{yo}^2 - 2gy$, we have $y = \dfrac{(5.176 \text{ m/s})^2}{2(9.80 \text{ m/s}^2)} = \boxed{1.37 \text{ m}}$.

(b) At impact, $y = 0$. From $y = v_{yo}t - \frac{1}{2}gt^2$, we have $t = \dfrac{5.176 \text{ m/s}}{\frac{1}{2}(9.80 \text{ m/s}^2)} = 1.056$ s.

So $R = x = v_{xo}t = (19.32 \text{ m/s})(1.056 \text{ s}) = \boxed{20.4 \text{ m}}$.

(c) Since the range depends on the initial speed and the angle so the player can either kick the ball harder to increase v_o and/or increase the angle to as close to 45° as possible.

84. The range $R = 15$ m.   So $R = \dfrac{v_o^2 \sin 2\theta}{g} = 15$ m.

or $R = \dfrac{(55 \text{ m/s})^2 \sin 2\theta}{9.80 \text{ m/s}^2} = 15$ m.   Solving, $\sin 2\theta = 0.0486$.

Therefore $2\theta = \sin^{-1}(0.0486) = 2.79°$,   hence $\theta = \boxed{1.4°}$.

89. (a) $\alpha = \tan^{-1}\left(\dfrac{12.0 \text{ m}}{150 \text{ m}}\right) = 45.7°$, so the launch angle $\theta = 4.57° + 10.0° = 14.6°$.

$x = v_{xo}\, t = (v_o \cos\theta)t,$ ☞ $t = \dfrac{x}{v_{xo}} = \dfrac{x}{v_o \cos\theta},$

$y = v_{yo}\, t - \tfrac{1}{2}gt^2 = v_o \sin\theta \times \dfrac{x}{v_o \cos\theta} - \tfrac{1}{2}g\left(\dfrac{x}{v_o \cos\theta}\right)^2$

$= x \tan\theta - \dfrac{gx^2}{2v_o^2 \cos^2\theta}.$

$v_o = \dfrac{x}{\cos\theta}\sqrt{\dfrac{g}{2(x \tan\theta - y)}} = \dfrac{150 \text{ m}}{\cos 14.57°}\sqrt{\dfrac{9.80 \text{ m/s}^2}{2[(150 \text{ m}) \tan 14.57° - 12.0 \text{ m}]}}$

$= \boxed{66.0 \text{ m/s}}$.

(b) $\theta = 4.57° + 10.5° = 15.07°$.

First we need to find the time of flight from the vertical motion.

$12.0 \text{ m} = (66.0 \text{ m/s})(\sin 15.07°)t - (4.90 \text{ m/s}^2)t^2,$

or $4.90\, t^2 - 17.16\, t + 12.0 = 0.$ Solving, $t = 0.97$ s or 2.54 s.

The 0.97 s answer is the time it takes to reach 12 m on the way up.

Therefore $x = (66.0 \text{ m/s})(\cos 15.07°)(2.54 \text{ s}) = 162$ m > 150 m,

$y = (66.0 \text{ m/s})(\sin 15.07°)(2.54 \text{ s}) - (4.90 \text{ m/s}^2)(2.54 \text{ s})^2 = 12.0$ m.

So the shot is $\boxed{\text{too long for the hole}}$.

94. Use the following subscripts: b = boat, w = water, and g = ground.
For the boat to make the trip straight across, v_{bw} must be the hypotenuse of the right-angle triangle. So it must be greater in magnitude than v_{wg}. So if the reverse is true, that is, if $v_{wg} > v_{bw}$, the boat cannot make the trip directly across the river.

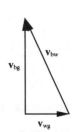

VI. Practice Quiz

1. The resultant of two vectors is greatest when the angle between them is

   (a) 0°   (b) 45°   (c) 60°   (d) 90°   (e) 180°

2. If a ball is thrown with a velocity of 25 m/s at an angle of 37° above the horizontal, what is the vertical component of the velocity?

   (a) 25 m/s   (b) 20 m/s   (c) 18 m/s   (d) 15 m/s   (e) 10 m/s

3. An object is moving in the x–y plane. The acceleration in the x and y directions are 1.0 m/s^2 and 3.0 m/s^2, respectively. If the object starts from rest, what are its coordinates at $t = 4.0$ s?

   (a) (8.0 m, 24 m) (b) (24 m, 8.0 m) (c) (4.0 m, 12 m) (d) (12 m, 4.0 m) (e) (8.0 m, 12 m)

4. A boat, whose speed in still water is 8.0 m/s, is directed across a river with a current of 6.0 m/s along the shore. What is the speed of the boat relative to the shore as it crosses the river?

   (a) 2.7 m/s   (b) 5.3 m/s   (c) 6.0 m/s   (d) 8.0 m/s   (e) 10 m/s

5. Find the resultant of the following vectors:

   $\mathbf{v}_1 = 2.0$ m/s $\mathbf{x} + 3.0$ m/s $\mathbf{y}$, $\mathbf{v}_2 = -4.0$ m/s $\mathbf{x} + 7.0$ m/s $\mathbf{y}$, $\mathbf{v}_3 = 10$ m/s, 37° above the $-x$ axis.

   (a) 19 m/s, 58° above $+x$ axis   (b) 19 m/s, 58° above $-x$ axis   (c) 19 m/s, 22° above $+x$ axis

   (d) 19 m/s, 22° above $-x$ axis   (e) none of the above

6. A stone is thrown horizontally with an initial speed of 8.0 m/s from the edge of a cliff. A stop watch measures the stone's trajectory time from top of cliff to bottom to be 3.4 s. What is the height of the cliff?

   (a) 17 m   (b) 27 m   (c) 34 m   (d) 57 m   (e) 1.1×10^2 m

7. An Olympic long-jumper goes into the jump with a speed of 10 m/s at an angle of 30° above the horizontal. How far is the jump?

   (a) 1.0 m   (b) 5.0 m   (c) 8.8 m   (d) 9.8 m   (e) 10 m

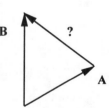

8. The resultant shown in the vector diagram on the right is for

   (a) $\mathbf{A} + \mathbf{B}$.   (b) $\mathbf{A} - \mathbf{B}$.   (c) $\mathbf{B} + \mathbf{A}$.   (d) $\mathbf{B} - \mathbf{A}$.   (e) $\mathbf{A} \times \mathbf{B}$.

9. You are traveling at +55 mi/h relative to a straight, level road and pass a car traveling at +45 mi/h. The velocity of your car relative to the other car is

   (a) 10 mi/h.   (b) −10 mi/h.   (c) 65 mi/h.   (d) 35 mi/h.   (e) 100 mi/h.

10.   The football field for your college team is 10 km at 30° north of east from your residence. You first drive

3.0 km north to go to a store. What should be your next displacement so you can go to the football game?

    (a) 13 km at 8.9° north of east   (b) 8.9 km at 13° north of east   (c) 7.0 km at 30° north of east

    (d) 7.0 km at 60° north of east   (e) 12 km at 43° north of east

CHAPTER 4

Force and Motion

I.   Chapter Objectives

Upon completion of this chapter, you should be able to:

1.   relate force and motion and explain what is meant by a net or unbalanced force.

2.   state and explain Newton's first law of motion and describe inertia and its relationship to mass.

3.   state and explain Newton's second law of motion, apply it to physical situations, and distinguish weight and mass.

4.   state and explain Newton's third law of motion and identify action–reaction force pairs.

5.   apply Newton's second law in analyzing various situations, using free-body diagrams, and understand the concept of translational equilibrium.

6.   explain the causes of friction and how it is described by using coefficients of friction.

II.   Key Terms

Upon completion of this chapter, you should be able to define and/or explain the following key terms:

force	translational equilibrium
net (unbalanced) force	condition for translational equilibrium
inertia	force of friction
Newton's first law of motion	static friction
(law of inertia)	kinetic (sliding) friction
newton (unit)	rolling friction
Newton's second law of motion	coefficient of static friction
weight	coefficient of kinetic friction
Newton's third law of motion	air resistance
normal force	terminal velocity
free-body diagram	

The definitions and/or explanations of the most important key terms can be found in the following section:
III. Chapter Summary and Discussions.

III. Chapter Summary and Discussion

1. Force and Net Force (Section 4.1)

Force F is the cause of *acceleration* or change in velocity, and it is the technical term for what we commonly call a push, pull, kick, or shove. Force is a vector quantity (it has both magnitude and direction). There must be a **net force** (unbalanced force) acting on an object for the object to change its velocity (either magnitude and/or direction), or to accelerate.

Net Force ΣF or F_{net} is the vector sum, the resultant, or the unbalanced force acting on an object. Here the symbol Σ means "the sum of." ΣF means the vector sum of forces. The unit of force in SI is a combination of the fundamental units of mass, length, and time and it is called **newton** (N). 1 N is the net force required to accelerate a mass of 1 kg mass with an acceleration of 1 m/s^2 or, $1\ N = 1\ kg \cdot m/s^2$. For example, as illustrated in the diagram above, two forces, 10 N and 30 N, are acting on an object in opposite directions. There are two forces on the object. However, there is only one net force (the resultant of the vector sum of the two forces) of 20 N (to the right) acting on the object.

Note: The net force is *not* a separate force. It is simply the vector sum of the individual forces.

To analyze the forces acting on an object, you should draw a **free-body diagram** (more on free-body diagram later in Section 4.5). This is done by

(1) Isolating the object of interest (this should be the only object in the diagram).

(2) Drawing all the forces acting on the object as vectors (including directions).

Example 4.1: Draw a free-body diagram of a book sitting on a horizontal desk.

Solution:

First we isolate the book and analyze the forces acting on the book. There are two forces acting on the book, the gravitational force (or weight) and the supporting force on the book by the desk. First of all, the gravitational force (an action-at-a-distance force with no physical contact) is always present if we are dealing with objects on the Earth. Secondly, whenever an object makes a physical contact with another object, a force results. Here the book makes a contact with the desk and so there is a supporting force. In most cases, these contact forces are perpendicular to the contact surface and therefore are called **normal forces**. As a follow up to this example, try to draw a free-body diagram of an object sliding freely on a frictionless inclined ramp.

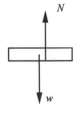

---

2. Newton's Laws of Motion (Sections 4.2 – 4.4)

(1) **Newton's first law** describes the state of motion of an object when there is *no net force* (net force equals zero) acting on the object.

An object at rest will remain at rest and an object in motion will keep moving with constant velocity if the net force on the object is zero.

An object at rest has no change in velocity and so its acceleration is zero. An object moving with constant velocity (same speed and same direction) has no change in velocity and its acceleration is also zero. Therefore, Newton's first law can be summarized in a very simple relation:

If $\Sigma\mathbf{F} = 0$, then $\mathbf{v}$ = constant (zero is just a special case of velocity being constant), or $\mathbf{a} = 0$.

Note: Newton's first law is often called the law of inertia. **Inertia** is the natural tendency of an object to resist acceleration or change in motion, and it is measured quantitatively by its mass.

A common misconception about Newton's first law is that a force is required to keep an object in motion. This is not so. Experiments done on air tracks (where there is negligible friction) show that no force is required to keep an object moving with constant velocity. We get this misconception because friction is always present in our everyday lives. To maintain constant velocity of a moving car, you have to push the accelerator. Does this contradict Newton's first law? No! It is seen clearly in the diagram on the right that the forward force on the car is equal in magnitude and opposite to the backward friction force (air friction plus ground friction) and so they cancel each other, resulting in a zero net force Therefore, there is still no net force and the car moves away with constant velocity.

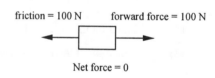

friction = 100 N   forward force = 100 N

Net force = 0

What happens if the *net force is not zero*? Newton's second law describes the relation among force, mass, and acceleration when there is a non-zero net force on an object.

(2) **Newton's second law** states that the acceleration depends on the net force $\Sigma\mathbf{F}$ and on the mass m of the object (note it is net force here, not just force). Mathematically, it is equivalent to

$$\mathbf{a} = \frac{\Sigma\mathbf{F}}{m} \quad \text{or} \quad \Sigma\mathbf{F} = m\,\mathbf{a}.$$

If you think of inertia as the qualitative term for the tendency of a body that resists acceleration, then **mass** (a scalar quantity) is the quantitative measure of inertia. If the mass is large, the acceleration produced by a given net force will be small.

Newton's second law is a vector equation. Note acceleration **a** is in the direction of the net force $\Sigma \mathbf{F}$, not necessarily the direction of velocity **v**. You must include *all* the forces acting on an object to determine the net force, and then the acceleration. However, many times you will find that a force on an object is balanced by an equal and opposite force, such as weight balanced by a normal force. Since these pair of forces cancel each other, you do not need to include them in your calculation.

Example 4.2:   A 20-kilogram box sitting on a horizontal surface is pulled by a horizontal force of 5.0 N. A friction force of 3.0 N retards the motion. What is the acceleration of the object?

Solution: Given: $m = 20$ kg, F (pulling) $= 5.0$ N,

f (friction) $= 3.0$ N.

Find: a.

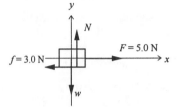

First we draw a free-body diagram of the box. The two vertical forces w, and N are equal and opposite (if they were not, the box would accelerate up or down). So they cancel out or $\Sigma F_y = ma_y = 0$ and we do not need to include them in our calculation. Thus we have only two forces, 5.0 N to the right and 3.0 N to the left, to deal with. The net force in this case is (with F direction taken as positive)

$\Sigma \mathbf{F} = F - f = 5.0$ N $- 3.0$ N $= 2.0$ N (to the right or $+x$).

Hence the acceleration of the object is $\mathbf{a} = \dfrac{\Sigma \mathbf{F}}{m} = \dfrac{2.0 \text{ N}}{20 \text{ kg}} = 0.10$ m/s^2 (to the right or $+x$).

If you apply Newton's second law to gravity near the Earth's surface, you get the relation among weight w (gravitational force), mass m, and gravitational acceleration g: $w = mg$.

Example 4.3:   Find the weight of a 3.50-kg object.

Solution: Given: $m = 3.50$ kg.   Find: w.

We use the relation: $w = mg = (3.50$ kg$)(9.80$ m/s^2 downward$) = 34.3$ N downward.

Note: Weight and mass are two very different physical quantities. Mass is a measure of the inertia or resistance to change in motion of an object. Mass is a constant for a given object and so it is independent of where the mass is located. For example, a 5.0-kilogram mass on the Earth is still 5.0 kg on the Moon. Weight is the gravitational force acting on an object, and so depends on the acceleration due to gravity and mass. A 5.0-kilogram mass has a weight of $(5.0$ kg$)(9.80$ m/s$^2) = 49$ N on the Earth, $(5.0$ kg$)(1.67$ m/s$^2) = 8.4$ N on the Moon, and *zero* newton in deep space (why?).

(3)   **Newton's Third Law** states that if object 1 exerts a force on object 2, then object 2 exerts an equal and opposite force on object 1. In mathematical terms, it can be written as $\mathbf{F}_{12} = -\mathbf{F}_{21}$.

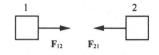

The negative sign in the above relation simply means that $\mathbf{F}_{12}$ is opposite $\mathbf{F}_{21}$. The notation $\mathbf{F}_{12}$ stands for the force on object 1 by object 2.

The force by object 1 is *on* object 2, and the force by object 2 is *on* object 1, and so the two forces are always acting on *two different* objects. Therefore, even though these two forces are equal and opposite, they *do not* cancel out each other because of this fact. If you analyze only object 1 in the diagram, there is only *one* force acting on it and object 1 will accelerate.

Newton's third law is often called action and reaction. The first force, for example, the force by object 1, is called the action force and the second force is then called the reaction force.

There is another misconception concerning the third law. The third law states that the two forces are *equal* no matter what. For example, an egg and a stone collide with each other, the egg breaks and the stone is intact. Since the egg breaks, we often conclude that the force by the stone on the egg is greater than the force by the egg on the stone. This is not so. The forces are always equal. The egg breaks because it is simply easier to break. It takes a smaller force to break the egg than to break the stone.

Example 4.4:   A large truck collides head-on with a small car and causes a lot of damage to the small car. Since there is more damage on the small car than on the large truck,

(a) the force on the truck is greater in magnitude than the force on the car,

(b) the force on the truck is equal in magnitude to the force on the car,

(c) the force on the truck is smaller in magnitude than the force on the car,

(d) the force on the truck is in the same direction as the force on the car.

(e) the truck did not slow down during the collision.

Solution:

According to Newton's third law, the answer is (b). Why isn't answer (c) correct since the truck causes more damage on the car? It takes a smaller force to damage a small car than to damage a large truck. When the truck and the car collide, they exert equal in magnitude but opposite in direction forces on each other, say 15 000 N. It may take only 10 000 N to damage the bumper on the car and 20 000 N to damage the bumper on the truck. So the car gets damaged and the truck remains basically undamaged, even though the forces by the car on the truck is the same magnitude as the force on the car by the truck.

(e) is also wrong as the truck *did* slow down from the force by the car is opposite its motion.

---

3. Application of Newton's Laws (Section 4.5)

When it comes to applying Newton's laws to various mechanical systems, there is no short cut to take. You *must* follow a certain set of procedures to analyze and solve the unknown physical quantities. Here is a simplified version of the procedures outlined in the text:

(1) Draw a free-body diagram for each object involved in the analysis.

(2) Select a rectangular coordinate system. The solutions will be much easier if you select the +x axis in the direction of acceleration and the +y axis perpendicular to the x axis.

(3) Resolve all forces not pointing in the x or y directions to their x and y components, respectively.

(4) Add, algebraically, all the x components and y components of the forces, respectively.

(5) Set $\Sigma F_x = ma$ and $\Sigma F_y = 0$ and solve for the unknown quantities.

Since you have chosen the +x axis in the direction of acceleration, the object will not accelerate in the y direction. So its acceleration in the y direction is zero and $\Sigma F_y = ma_y = 0$.

Do these steps read familiar? In essence, these are the same procedures we followed when we added vectors using the component method. Our goal here is to find the net force and then the acceleration. Since force is a vector, it really boils down to add vectors.

Example 4.5: A student pulls a box of books on a smooth horizontal floor with a force of 100 N in a direction of 37° above the horizontal surface. If the mass of the box and the books is 40.0 kg, what is the acceleration of the box and the normal force on the box by the floor?

Solution: Given: $F = 100$ N, $\theta = 37°$, $m = 40.0$ kg.

Find: a and N.

(1) Free-body diagram:

Since the box is on the Earth, it has a weight of $w = mg$ pointing toward the center of the earth (or perpendicular to the horizontal direction). The box is also making physical contacts with the floor and the student. So there will be two contact forces. The contact force with the floor is a normal force and is directed straight upward perpendicular to the contact surface. The force by the student is directed at 37° above the horizontal direction (smooth floor implies negligible friction).

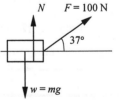

(2) Coordinate system:

Even though the student is pulling at an angle above the horizontal direction, the box will still move or accelerate along the horizontal direction. So we choose the horizontal direction (to the right) as the $+x$ axis and the vertical direction (upward) as the $+y$ axis.

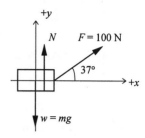

(3) x and y components:

Among the three forces, only the student's pulling force is not completely along either the x or the y axis. So we need to find its x and y components. From trigonometry, we can see that the x component is adjacent to the 37° angle and the y component is opposite to the 37° angle. Therefore we use cos 37° to calculate the x component and sin 37° to calculate the y component.

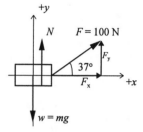

$$F_x = F \cos 37° = (100 \text{ N}) \cos 37° = 80.0 \text{ N},$$
$$F_y = F \sin 37° = (100 \text{ N}) \sin 37° = 60.0 \text{ N}.$$

(4) Adding x and y components, respectively:

In the x direction, there is only one force, the x component of the student's pulling force, 80 N. In the y direction, there are three forces: the upward normal force N, the upward y component of the student's pulling force (60 N), and the downward weight of the box and books of

$$w = mg = (40.0 \text{ kg})(9.80 \text{ m/s}^2) = 392 \text{ N}.$$

So $\Sigma F_x = 80.0 \text{ N}$ and $\Sigma F_y = N + 60.0 \text{ N} - w = N + 60.0 \text{ N} - 392 \text{ N} = N - 332 \text{ N}.$

(5) Setting $\Sigma F_x = ma$ and $\Sigma F_y = 0$ and solving for the unknown quantities:

$$\Sigma F_x = 60.0 \text{ N} = ma = (40.0 \text{ kg})a. \qquad \text{Eq. (1)}$$

and

$$\Sigma F_y = N - 332 \text{ N} = 0. \qquad \text{Eq. (2)}$$

From Eq. (1), we have $a = \dfrac{80.0 \text{ N}}{40.0 \text{ kg}} = 2.00 \text{ m/s}^2.$

From Eq. (2), we have $N = 332 \text{ N}.$

Here the normal force (332 N) is not equal to the weight of the box and the books (392 N). Why?

Example 4.6:   A 5.0-kilogram box, starting from rest, slides down a smooth 37° inclined plane.

(a) Find the acceleration of the mass and the normal force by the inclined plane on the mass.

(b) If the plane is 10 m long, what will be its speed at the bottom of the plane?

Solution: Given: $m = 5.0$ kg, $\theta = 37°$, $v_0 = 0$, $x = 10$ m.

Find: (a) a and N (b) v.

(a) (1) Free-body diagram:

Since the box is on the Earth, it has weight $w = mg$ pointing toward the center of the Earth (or perpendicular to the horizontal direction). The box is also making physical contact with the inclined plane. So there will be a contact force (normal force) perpendicular to the inclined plane.

(2) Coordinate system:

Since the mass slides down the inclined plane, we choose the $+x$ axis down the inclined plane and the $+y$ axis perpendicular to the inclined plane.

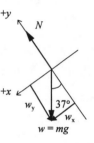

(3) x and y components:

The weight of the mass $w = mg$ is not along either the x axis or the y axis. So we need to find its x and y components. From trigonometry, we can see that the x component is opposite to the 37° angle and the y component is adjacent to the 37° angle.

So $w_x = w \sin 37° = mg \sin 37°$ and $w_y = w \cos 37° = mg \cos 37°$.

(4) Adding x and y components, respectively:

In x direction, there is only one force, w_x.

So $\Sigma F_x = mg \sin 37° = mg \sin 37°$.

In y direction, there are two forces, N and w_y. So $\Sigma F_y = N - mg \cos 37°$.

(5) Setting $\Sigma F_x = ma$ and $\Sigma F_y = 0$ and solving for the unknown quantities:

$$\Sigma F_x = mg \sin 37° = ma. \qquad\qquad \text{Eq. (1)}$$

and $\Sigma F_y = N - mg \cos 37°$.   Eq. (2)

From Eq. (1), we have $a = g \sin 37° = (9.80 \text{ m/s}^2) \sin 37° = 5.9 \text{ m/s}^2$.

From Eq. (2), we have $N = mg \cos 37° = (5.0 \text{ kg})(9.80 \text{ m/s}^2) \cos 37° = 39$ N.

(b) Since velocity is a kinematic quantity, we need to use one of the kinematic equations.

From $v^2 = v_0^2 + 2ax = (0)^2 + 2(5.9 \text{ m/s}^2)(10 \text{ m}) = 118 \text{ m}^2/\text{s}^2$,   so $v = 11$ m/s.

If there are *no forces* acting on an object, or if there are *equal and opposite forces* acting on it, the net force on the object *is zero* and the object *will not* accelerate. This state of motion is called **translational equilibrium**. If the system is in equilibrium, we can set $\Sigma \mathbf{F} = 0$, that is, $\Sigma F_x = 0$ and $\Sigma F_y = 0$ in component form, and then solve for unknown quantities.

Translational equilibrium is a special application of Newton's laws ($a = 0$). The problem solving procedure is the same as described in Examples 4.5 and 4.6. The only minor difference is in step (5). Instead of setting $\Sigma F_x = ma$ and $\Sigma F_y = 0$, we set both $\Sigma F_x = 0$ and $\Sigma F_y = 0$, and solving for the unknown quantities.

4.   Force of Friction (Section 4.6)

Consider one object in contact with another, such as a book on a desk. Suppose that the desk is stationary and that the book is either *sliding* or *on the verge of sliding* on the desk. There is interaction between the book and the desk based on intermolecular attractions and repulsions. It is convenient, however, to consider the interaction as if it were two independent ones: one perpendicular to the contact surface (the *normal* force mentioned earlier) and another one parallel to the contact surface, which is called *force of friction* or *friction force*.

There are two kinds of friction forces: static friction and kinetic friction. Static friction force ($f_s \leq \mu_s N$) is parallel to the contact surface when there is *no relative motion* between the objects in contact and kinetic friction force ($f_k = \mu_k N$) is also parallel to the contact surface when there is *relative motion* between objects in contact. Here μ_s is called the coefficient of static friction and μ_k the coefficient of kinetic friction. They are basically measures of the strengths of the molecular interactions. N is the normal force.

Kinetic friction force doesn't cause much conceptual difficulty, but static friction force is often misunderstood and therefore incorrectly interpreted. If a book is at rest on a desk and no forces with horizontal components are applied to the book, there is *no* static friction force. However, suppose a small horizontal force is applied to the book but the book still remains at rest, then we must conclude that an equal and opposite force acts on the book to prevent it from moving. This is the static friction force. If the horizontally applied force increases, the static friction force will increase by the same amount until its maximum value is reached ($f_s = \mu_s N$). So the static friction force is *not* a fixed value, but is always equal to the applied force, with an upper limit of $f_s = \mu_s N$.

Another misconception about static friction force is about its direction. Somehow we have the idea that friction force is always opposite to the direction of motion. This is always true only if the friction force is kinetic. Static friction force can be in directions other than opposite to motion. Think about a machine part moving on a conveying belt in a factory and a car turning on a flat surface. What is moving the part and turning the car? It is static friction force. The static friction force is in the direction of motion on the machine part and perpendicular to the direction of motion on the car. Think carefully when you walk the next time. What enables you to walk? In which direction is that force? (Hint: why is it difficult to walk on ice? Is it possible to walk on perfectly smooth surface?)

Example 4.7: The coefficients of static and kinetic frictions between a 3.0-kilogram box and a desk are 0.40, and 0.30, respectively. What is the net force on the box when each of the following horizontal forces is applied to the box? (a) 5.0 N, (b) 10 N, (c) 15 N.

Solution: Given: $\mu_s = 0.40$, $\mu_k = 0.30$, $m = 3.0$ kg. $F = 5.0$ N (a), 10 N (b), 15 N (c).

Find: ΣF (net force) in (a), (b), and (c).

From the discussion about friction forces, we know that if the applied force is greater than the maximum static friction force, the box will accelerate. Since both static and kinetic forces depend on the normal force, we must find it first.

Applying Newton's second law in the vertical direction, $\Sigma F_y = N - w = ma_y = 0$ so the normal force between the box and the desk is $N = w = mg = (3.0 \text{ kg})(9.80 \text{ m/s}^2) = 29.4$ N.

The maximum static friction force is then $f_{s\,max} = \mu_s N = 0.40(29.4 \text{ N}) = 11.8$ N.

(a) Since the applied force of 5.0 N is smaller than 11.8 N, the maximum static friction force, the object remains at rest and therefore the net force is zero.

(b) The applied force is still smaller than the maximum static friction force and the net force is still zero.

(c) Now the applied force of 15 N is greater than the maximum static friction force, the object will accelerate and there is a relative motion along the contact surface. Therefore we need to replace the static friction force with the kinetic friction force.

$$f_k = \mu_k N = 0.30(29.4 \text{ N}) = 8.82 \text{ N}.$$

Hence the net force is $\Sigma F = 15 \text{ N} - 8.82 \text{ N} = 6.18$ N and the box will accelerate at a rate of

$$a = \frac{\Sigma F}{m} = \frac{6.18 \text{ N}}{3.0 \text{ kg}} = 2.1 \text{ m/s}^2.$$

The direction of the acceleration is in the same direction as the horizontal force.

IV. Mathematical Summary

Newton's second law	$\Sigma \mathbf{F} = m\mathbf{a}$ (4.1)	Relates acceleration with mass and net force when the
	(Note: it is a vector equation)	net force on it is not zero.
Component form of Newton's second law	$\Sigma F_x = ma_x$ (4.3b) $\Sigma F_y = ma_y$ (4.3b)	Relates acceleration with mass and force in a particular direction.
Newton's third law	$\mathbf{F}_{12} = -\mathbf{F}_{21}$	Action and reaction (on different objects).
Weight and mass	$\mathbf{w} = m\mathbf{g}$ (4.2)	Relates weight (gravitational force) with mass and gravitational acceleration.

Translational equilibrium condition	$\Sigma\mathbf{F} = 0$ (4.4) $\Sigma F_x = 0$ and $\Sigma F_y = 0$ (4.5)	The conditions an object must satisfy if it has zero acceleration.
Force of static friction	$f_s \leq \mu_s N$ (4.6) $f_{smax} = \mu_s N$ (4.7) (Note: f_s is not a fixed value)	Calculates the static friction force when there is no relative motion along the contact surface.
Force of kinetic friction	$f_k = \mu_k N$ (4.8) (Note: f_k is a fixed value)	Calculates the kinetic friction force when there is a relative motion along the contact surface.

V.   Solutions of Selected Exercises and Paired/Trio Exercises

6.   The bubble would move $\boxed{\text{forward}}$ in the direction of velocity or acceleration, because the inertia of the liquid will resist the forward acceleration. So the bubble of negligible mass or inertia moves forward relative to the liquid. Then it moves $\boxed{\text{backward}}$ opposite the velocity (or in the direction of acceleration) for the same reason.

(b) The principle is based on the $\boxed{\text{inertia of the liquid}}$.

12.   (a) $\mathbf{F}_1 = (3.6 \text{ N})[(\cos 74°)\,\hat{\mathbf{x}} - (\sin 74°)\,\hat{\mathbf{y}}] = (0.99 \text{ N})\,\hat{\mathbf{x}} - (3.46 \text{ N})\,\hat{\mathbf{y}}$.

$\mathbf{F}_2 = (3.6 \text{ N})[(-\cos 34°)\,\hat{\mathbf{x}} + (\sin 34°)\,\hat{\mathbf{y}}] = -(2.98 \text{ N})\,\hat{\mathbf{x}} + (2.01 \text{ N})\,\hat{\mathbf{y}}$.

If $a = 0$, then $\Sigma\mathbf{F} = 0$ from Newton's first law.

$\Sigma\mathbf{F} = \mathbf{F}_1 + \mathbf{F}_2 = -(1.99 \text{ N})\,\hat{\mathbf{x}} - (1.45 \text{ N})\,\mathbf{y} \neq 0$, so there must be a third force to make $\Sigma\mathbf{F} = 0$.

So $\Sigma\mathbf{F} = \mathbf{F}_1 + \mathbf{F}_2 + \mathbf{F}_3 = 0$,   and $\mathbf{F}_3 = -(\mathbf{F}_1 + \mathbf{F}_2) = (1.99 \text{ N})\,\hat{\mathbf{x}} + (1.45 \text{ N})\,\mathbf{y}$.

$F_3 = \sqrt{(1.99 \text{ N})^2 + (1.45 \text{ N})^2} = \boxed{2.5 \text{ N}}$. $\theta = \tan^{-1}\left(\dfrac{1.45 \text{ N}}{1.99 \text{ N}}\right) = \boxed{36° \text{ above the } +x \text{ axis}}$.

(b) $\boxed{\text{No}}$, all we can say is that the acceleration is zero. The object could be at rest or moving with constant velocity.

18.   From Newton's second law $\Sigma F = ma$,   we have $a = \dfrac{\Sigma F}{m} = \dfrac{3.0 \text{ N}}{1.5 \text{ kg}} = \boxed{2.0 \text{ m/s}^2}$.

21.   From Newton's second law $\Sigma F = ma = (7.0 \times 10^7 \text{ kg})(0.10 \text{ m/s}^2) = \boxed{7.0 \times 10^6 \text{ N}}$.

26. $150 \text{ lb} = (150 \text{ lb}) \times \dfrac{4.45 \text{ N}}{1 \text{ lb}} = \boxed{668 \text{ N}}.$ $m = \dfrac{w}{g} = \dfrac{668 \text{ N}}{9.80 \text{ m/s}^2} = \boxed{68.2 \text{ kg}}.$

29. (a) $m = \dfrac{w}{g} = \dfrac{98 \text{ N}}{9.80 \text{ m/s}^2} = 10 \text{ kg}.$ So $a = \dfrac{\Sigma F}{m} = \dfrac{12 \text{ N}}{10 \text{ kg}} = \boxed{1.2 \text{ m/s}^2}.$

   (b) For the same object, the mass is still 10 kg. So the acceleration is also the $\boxed{\text{same}}$.

30. The resistive force is opposite the forward force.

   From Newton's second law $\Sigma F = ma,$ we have $a = \dfrac{\Sigma F}{m} = \dfrac{15 \text{ N} - 8.0 \text{ N}}{1.0 \text{ kg}} = \boxed{7.0 \text{ m/s}^2}.$

34. First we need to find acceleration from kinematics.

   $v_0 = 90 \text{ km/h} = 25 \text{ m/s},$ $v = 0,$ $t = 5.5 \text{ s}.$

   So $a = \dfrac{v - v_0}{t} = \dfrac{0 - 25 \text{ m/s}}{5.5 \text{ s}} = -4.56 \text{ m/s}^2.$

   Now from Newton's second law $\Sigma F = ma = (60 \text{ kg})(-4.55 \text{ m/s}^2) = -\boxed{2.7 \times 10^2 \text{ N}}.$

   The negative sign indicates that the force is opposite the motion or the velocity.

36. $\boxed{69 \text{ N}}.$

38. The answer is (d). (a) is incorrect because the forces are always equal in magnitude. (b) is incorrect because the forces cannot be cancelled as they are on two different bodies. (c) is incorrect because the forces always act on two different bodies.

43. From Newton's second law, the force on the female by the male is

   $F = ma = (45 \text{ kg})(2.0 \text{ m/s}^2) = 90 \text{ N}.$

   The force on the male by the female is also 90 N according to Newton's third law.

   So $a_{\text{male}} = \dfrac{90 \text{ N}}{60 \text{ kg}} = \boxed{1.5 \text{ m/s}^2}$ opposite to hers.

46. (a) The scale reading is equal to the normal force on the person. The net force in the vertical direction is $\Sigma F = w - N = ma = 0,$

   so $N = w = mg = (75.0 \text{ kg})(9.80 \text{ m/s}^2) = \boxed{735 \text{ N}}.$

   (b) a is still zero. So $N = \boxed{735 \text{ N}}.$

   (c) From $\Sigma F = N - w = ma,$

   we have $N = w + ma = mg + ma = m(g + a) = (75.0 \text{ kg})(9.80 \text{ m/s}^2 + 2.00 \text{ m/s}^2) = \boxed{885 \text{ N}}.$

50. (a) Although the boy pulls in a non-horizontal direction, the box still accelerates in the horizontal direction, which is chosen as the $+x$ axis.

$\Sigma F_x = F \cos 30° = (25 \text{ N}) \cos 30° = 21.65 \text{ N} = ma_x$,

so $a_x = \dfrac{21.65 \text{ N}}{30 \text{ kg}} = \boxed{0.72 \text{ m/s}^2}$.

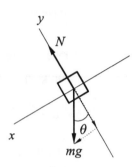

(b) $\Sigma F_y = N + F \sin 30° - w = ma_y = 0$,

so $N = w - F \sin 30° = (30 \text{ kg})(9.80 \text{ m/s}^2) - (25 \text{ N}) \sin 30° = \boxed{2.8 \times 10^2 \text{ N}}$.

52. (a) The x-component of the weight is the side opposite to the angle θ shown, so sine is used. $\Sigma F_x = mg \sin\theta = ma_x$,

So $a_x = g \sin\theta = (9.80 \text{ m/s}^2) \sin 37° = \boxed{5.9 \text{ m/s}^2}$.

(b) From $v^2 = v_0^2 + 2ax$, $v = \sqrt{(5.0 \text{ m/s})^2 + 2(5.9 \text{ m/s}^2)(35 \text{ m})} = \boxed{21 \text{ m/s}}$.

59. The free-body diagrams of the three objects.

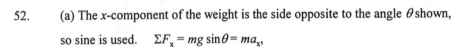

For m_1: $T_1 - m_1 g = m_1 a$,   (1)

For m_3: $T_2 - T_1 = m_3 a$,   (2)

For m_2: $m_2 g - T_2 = m_2 a$,   (3)

$(1) + (2) + (3)$ gives $(m_2 - m_1)g = (m_1 + m_2 + m_3)a$.

Solving, $a = \dfrac{(m_2 - m_1)g}{m_1 + m_2 + m_3}$.

(a) $a = \dfrac{(0.50 \text{ kg} - 0.25 \text{ kg})(9.80 \text{ m/s}^2)}{0.25 \text{ kg} + 0.50 \text{ kg} + 0.25 \text{ kg}} = \boxed{2.5 \text{ m/s}^2 \text{ to right}}$.

(b) $a = \dfrac{(0.15 \text{ kg} - 0.35 \text{ kg})(9.80 \text{ m/s}^2)}{0.35 \text{ kg} + 0.15 \text{ kg} + 0.50 \text{ kg}} = -2.0 \text{ m/s}^2$.

So it is $\boxed{2.0 \text{ m/s}^2 \text{ to left}}$.

60. From Example 4.6 in textbook (The Atwood machine),

$a = \dfrac{(m_2 - m_1)g}{m_1 + m_2} = \dfrac{(0.25 \text{ kg} - 0.20 \text{ kg})(9.80 \text{ m/s}^2)}{0.25 \text{ kg} + 0.20 \text{ kg}} = \boxed{1.2 \text{ m/s}^2, \text{ up}}$.

64. Both at rest and moving with constant velocity corresponds to zero acceleration, so $a_x = a_y = 0$.

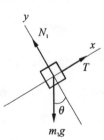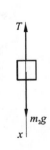

For m_1: $\Sigma F_x = T - m_1 g \sin\theta = m_1 a_x = 0$, that is $T = m_1 g \sin\theta$.

For m_2: $\Sigma F_y = m_2 g - T = m_2 a_y = 0$, or $m_2 g = T = m_1 g \sin\theta$.

Combining the two equations, we have $m_2 = m_1 \sin\theta = (2.0 \text{ kg}) \sin 37° = \boxed{1.2 \text{ kg}}$.

If both are moving at constant velocity, the answer is the $\boxed{\text{same}}$ 1.2 kg because the acceleration is still zero and the forces must still balance out.

68. (c).

74. (a) From the exercise, we know that 275 N is the maximum static friction force and 195 N is the kinetic friction force.

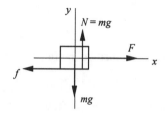

$\Sigma F_y = N - mg = ma_y = 0$, that is $N = mg$.

$\Sigma F_x = F - f_s = ma_x = 0$ (on the verge of moving),

so $f_{s\,max} = \mu_s N = F$,

or $\mu_s = \dfrac{F}{mg} = \dfrac{275 \text{ N}}{(35.0 \text{ kg})(9.80 \text{ m/s}^2)} = \boxed{0.802}$.

(b) Similarly, $\mu_k = \dfrac{195 \text{ N}}{(35.0 \text{ kg})(9.80 \text{ m/s}^2)} = \boxed{0.569}$.

78. (a) First we need to find acceleration from dynamics and use μ_k from Table 4.1.

$\Sigma F_x = -f_k = -\mu_k mg = ma$, so $\mu_k = -\dfrac{a}{g}$.

Therefore $a = -\mu_k g = -0.85(9.80 \text{ m/s}^2) = -8.33 \text{ m/s}^2$,

and $v_0 = 90 \text{ km/h} = 25 \text{ m/s}$, $v = 0$.

From the kinematic equation $v^2 = v_0^2 + 2ax$,

we have $x = \dfrac{v^2 - v_0^2}{2a} = \dfrac{0 - (25 \text{ m/s})^2}{2(-8.33 \text{ m/s}^2)} = \boxed{38 \text{ m}}$.

(b) $a = -0.60(9.80 \text{ m/s}^2) = -5.88 \text{ m/s}^2$. So $x = \dfrac{0 - (25 \text{ m/s})^2}{2(-5.88 \text{ m/s}^2)} = \boxed{53 \text{ m}}$.

82. $\Sigma F_y = N - mg \cos\theta = ma_y = 0$, that is $N = mg \cos\theta$.

For constant velocity, $a_x = 0$. So $\Sigma F_x = mg \sin\theta - f_k = 0$.

Or $mg \sin\theta = \mu_k N = \mu_k (mg \cos\theta)$.

Therefore $\mu_k = \dfrac{\sin\theta}{\cos\theta} = \boxed{\tan\theta}$.

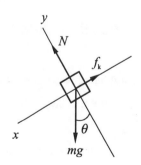

84. $\theta = \boxed{30°}$ and $\mu_s = \boxed{0.58}$.

89. For m_1: $\Sigma F = T_1 - m_1 g = 0,$ (1)

For m_3: $\Sigma F = T_2 - T_1 - f_{smax} = 0,$ (2)

For m_2: $\Sigma F = m_2 g - T_2 = 0,$ (3)

(1) + (2) + (3) gives $(m_2 - m_1)g - f_s = 0,$

or $(m_2 - m_1)g - \mu_s N_3 = (m_2 - m_1)g - \mu_s m_3 g = 0.$

Solving, $\mu_s = \dfrac{m_2 - m_1}{m_3} = \dfrac{0.50 \text{ kg} - 0.25 \text{ kg}}{0.75 \text{ kg}} = \boxed{0.33}.$

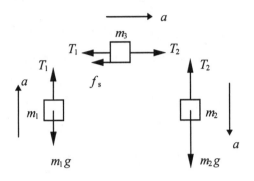

Alternate method (more conceptual).

Since $a = 0,$ $T_1 = m_1 g$ and $T_2 = m_2 g.$

So for m_3, $m_2 g - m_1 g - f_s = 0,$ or $m_2 g - m_1 g - \mu_s m_3 g = 0.$

Therefore $\mu_s = \dfrac{m_2 - m_1}{m_3}.$

97. $\Sigma F_y = N - mg - F \sin\theta = 0,$ so $N = mg + F \sin\theta.$

$\Sigma F_x = F \cos\theta - f_{smax} = 0,$

or $F \cos\theta - \mu_s N = F \cos\theta - \mu_s (mg + F \sin\theta) = 0.$

Therefore $F = \dfrac{\mu_s mg}{\cos\theta - \mu_s \sin\theta}.$

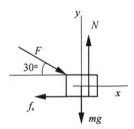

(a) $\mu_s = 0.75$ (Table 4.1),   so $F = \dfrac{0.75(5.0 \text{ kg})(9.80 \text{ m/s}^2)}{\cos 30° - (0.75) \sin 30°} = \boxed{75 \text{ N}}.$

(b) $\mu_s = 0.12,$ so $F = \dfrac{0.12(5.0 \text{ kg})(9.80 \text{ m/s}^2)}{\cos 30° - (0.12) \sin 30°} = \boxed{7.3 \text{ N}}.$

VI.   Practice Quiz

1.   Which of Newton's laws of motion *best* explains why motorists should buckle-up?

    (a) the first law   (b) the second law   (c) the third law   (d) the law of gravitation

2.   A net force F accelerates a mass m with an acceleration a. If the same net force is applied to mass $m/2$, then the acceleration will be

    (a) $4a.$ (b) $2a.$ (c) $a.$ (d) $a/2.$ (e) $a/4.$

3.   A brick hits a window and breaks the glass. Since the brick breaks the glass,

(a) the force on the brick is greater in magnitude than the force on the glass,

(b) the force on the brick is equal in magnitude to the force on the glass,

(c) the force on the brick is smaller in magnitude than the force on the glass,

(d) the force on the brick is in the same direction as the force on the glass.

(e) the brick did not slow down.

4.   An object weighs 100 N on the surface of Earth. What is the mass of this object on the surface of the Moon where the acceleration due to gravity is only 1/6 of that on the Earth?

(a) 1.70 kg   (b) 10.2 kg   (c) 16.7 kg   (d) 58.8 kg   (e) 100 kg

5.   Two horizontal forces act on a 5.0-kilogram object. One force has a magnitude of 8.0 N and is directed due north. The second force toward the east has a magnitude of 6.0 N. What is the acceleration of the object?

(a) 1.6 m/s^2 due north   (b) 1.2 m/s^2 due east   (c) 2.0 m/s^2 at 53° N of E   (d) 2.0 m/s^2 at 53° E of N

6.   A box is placed on a smooth inclined plane with an angle of 20° to the horizontal. If the inclined plane is 5.0-meter long, how long does it take for the mass to reach the bottom of the inclined plane after it is released from rest?

(a) 1.0 s   (b) 1.3 s   (c) 1.5 s   (d) 1.7 s   (e) 1.9 s

7.   During a hockey game, a hockey puck is given an initial speed of 10 m/s. It slides 50 m on the ice before it stops. What is the coefficient of kinetic friction between the puck and ice?

(a) 0.090   (b) 0.10   (c) 0.11   (d) 0.12   (e) 0.13

8.   A person on a scale rides in an elevator. If the mass of the person is 60 kg and the elevator accelerate upward with an acceleration of 4.9 m/s^2, what is the reading on the scale in newtons?

(a) 147 N   (b) 294 N   (c) 588 N   (d) 882 N   (e) 1176 N

9.   A traffic light of weight 100 N is supported by two ropes as shown in the diagram. What are the tensions in the ropes?

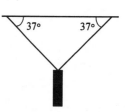

(a) 50 N   (b) 63 N

(c) 66 N   (d) 83 N

(e) 100 N

10. Find the magnitudes of the acceleration of the system and the tension in the connecting string. (Neglect friction and mass of the pulley)

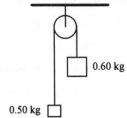

(a) 5.3 m/s^2, 7.5 N   (b) 4.5 m/s^2, 7.1 N

(c) 0.89 m/s^2, 5.3 N   (d) 0.090 m/s^2, 4.9 N

(e) 1.2 m/s^2, 4.3 N.

Answers to Practice Quiz:

1.a 2.b 3.c 4.b 5.c 6.d 7.b 8.d 9.d 10.c

CHAPTER 5

I. Chapter Objectives

Upon completion of this chapter, you should be able to:

1. define mechanical work and compute the work done in various situations.
2. differentiate work done by constant and variable force and compute the work done by a spring force.
3. explain the work–energy theorem and apply it in solving problems.
4. explain how potential energy depends on position and compute values of gravitational potential energy.
5. distinguish between conservative and nonconservative forces and explain their effects on the conservation of energy.
6. define power and describe mechanical efficiency.

II. Key Terms

Upon completion of this chapter, you should be able to define and/or explain the following key terms:

work	nonconservative force
joule (unit)	total mechanical energy
spring (force) constant	conservative system
kinetic energy	law of conservation of mechanical energy
work-energy theorem	power
potential energy	watt (unit)
gravitational potential energy	horsepower (unit)
law of conservation of total energy	efficiency
conservative force	

The definitions and/or explanations of the most important key terms can be found in the following section: **III. Chapter Summary and Discussion.**

III. Chapter Summary and Discussion

1. Work Done by a Constant Force (Section 5.1)

The **work** done by a constant force is $W = (F \cos\theta)\, d$, where F and d are the magnitudes of the force and displacement vectors, respectively, and θ is the angle between these two vectors. Although force and displacement are vectors, work is a scalar quantity. The SI unit of work is N·m, which is called a joule (J).

Note: The angle in the work definition is the *angle between the force vector and the displacement vector*, which is not necessarily the angle from the horizontal.

Graphically, the work done by a force is equal to the area under the curve in a force versus position graph.

Example 5.1 A 500-kilogram elevator is pulled upward by a constant force of 5500 N for a distance of 50.0 m.

        (a) Find the work done by the upward force.

        (b) Find the work done by the gravitational force.

        (c) Find the work done by the net force (the net work done on the elevator).

Solution: Given: $F_{up} = 5500$ N, $w = mg = (500 \text{ kg})(9.80 \text{ m/s}^2) = 4900$ N, $d = 50.0$ m.

        Find: (a) W_{up}, (b) W_{grav}, and (c) W_{net}.

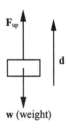

(a) The displacement is upward and the upward force is (of course) upward. So the angle between them is zero.

Therefore $W_{up} = (F \cos\theta)\, d = (F_{up} \cos 0°)\, d = (5500 \text{ N})(1)(50.0 \text{ m}) = 2.75 \times 10^5$ J.

(b) The angle between displacement and the gravitational force (weight vector) is 180°.

So $W_{grav} = (w \cos 180°)\, d = (4900 \text{ N})(-1)(50.0 \text{ m}) = -2.45 \times 10^5$ J.

(c) The work done by the net force is equal to the net work done on the elevator.

$W_{net} = W_{up} + W_{grav} = 2.75 \times 10^5 \text{ J} + (-2.45 \times 10^5 \text{ J}) = 3.0 \times 10^4$ J.

Note: W_{net} is also equal to $W_{net} = (F_{net} \cos\theta)\, d$, where $F_{net} = F - w$.

Example 5.2 A force moves an object in the direction of the force. The graph on the right shows the force versus the object's position. Find the net work done when the object moves

        (a) from 0 to 2.0 m.

        (b) from 2.0 to 4.0 m.

        (c) from 4.0 to 6.0 m.

        (d) from 0 to 6.0 m.

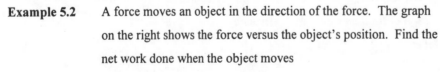
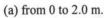

Solution:

Work done is equal to the area under the curve.

(a) The area under the curve from 0 to 2.0 m is the left triangle. The area of a triangle is $\frac{1}{2}$(base × height).

So $W_{0\text{-}2} = \frac{1}{2}(2.0 \text{ m} - 0)(20 \text{ N}) = 20 \text{ J}$.

(b) The area under the curve from 2.0 m to 4.0 m is the rectangle.

So $W_{2\text{-}4} = (4.0 \text{ m} - 2.0 \text{ m})(20 \text{ N}) = 40 \text{ J}$.

(c) The area under the curve from 4.0 m to 6.0 m is the right triangle.

So $W_{4\text{-}6} = \frac{1}{2}(6.0 \text{ m} - 4.0 \text{ m})(20 \text{ N}) = 20 \text{ J}$.

(d) The area under the curve from 0 to 6.0 m is the sum of the areas of the two triangles and the rectangle.

So $W_{0\text{-}6} = W_{0\text{-}2} + W_{2\text{-}4} + W_{4\text{-}6} = 20 \text{ J} + 40 \text{ J} + 20 \text{ J} = 80 \text{ J}$.

2. Work Done by a Variable Force (Section 5.2)

Our discussion of variable force in this section is restricted to spring force described by Hooke's law, $F_s = -kx$ (with $x_0 = 0$), where k is a constant called spring constant or force constant which measures the stiffness of a spring, x is the displacement from the spring's unstretched position, and the negative sign indicates that the spring force and the spring displacement are always opposite to each other. Hooke's law is a linear equation, that is, doubling x will also double F_s.

The work done by an external force in stretching or compressing a spring (overcoming the spring force) is $W = \frac{1}{2}kx^2$, where x is the stretch or compression distance. Note that this expression is a quadratic expression, that is, doubling x will also quadruple W.

Example 5.3   A spring of spring constant 20 N/m is to be compressed by 0.10 m.

   (a) What is the maximum force required?
   (b) What is the work required?

Solution: Given: $k = 20$ N/m   and $x = -0.10$ m.

   Find: (a) $F_{s(max)}$ (b) W.

(a) From Hooke's law, the maximum force corresponds to the maximum compression.

$F_{s(max)} = -kx = -(20 \text{ N/m})(-0.10 \text{ m}) = 2.0 \text{ N}$.

(b) $W = \frac{1}{2}kx^2 = \frac{1}{2}(20 \text{ N/m})(-0.10 \text{ m})^2 = 0.10 \text{ J}$.

---

3. Work-Energy Theorem: Kinetic Energy (Section 5.3)

Kinetic energy is the energy of motion and it is defined as $K = \frac{1}{2}mv^2$, where m is the mass and v is the velocity of the object. Although velocity is a vector quantity, kinetic energy is a scalar quantity because it depends on the square of velocity. The SI unit of kinetic energy is the joule (J). By combining the definition of work, a kinematic equation, and Newton's second law, we can derive the **work–energy theorem**, which states that the net work done on an object is equal to its change in kinetic energy, i.e., $W_{net} = K - K_0 = \Delta K$. In general, work is a measure of energy transfer, and energy is the capacity of doing work.

Example 5.4   An object hits a wall and bounces back with half of its original speed. What is the ratio of the final kinetic energy to the initial kinetic energy?

Solution: Given: $v = \dfrac{v_0}{2}$.   Find: $\dfrac{K}{K_0}$.

$K_0 = \frac{1}{2}mv_0^2$ and $K = \frac{1}{2}mv^2 = \frac{1}{2}m\left(\dfrac{v_0}{2}\right)^2 = \frac{1}{4}\left(\frac{1}{2}mv_0^2\right)$.   So $\dfrac{K}{K_0} = \dfrac{1}{4}$.

Why isn't the result $\frac{1}{2}$?

Since the work–energy theorem is a combination of kinematics and dynamics, it offers a convenient method to solve mechanical problems. Rather than working from kinematics then to dynamics or vice versa, this new theorem can solve these typical "two-step" problems in a single step, as the following example shows.

Example 5.5   The kinetic friction force between a 60.0-kilogram object and a horizontal surface is 50.0 N. If the initial speed of the object is 25.0 m/s, what distance will it slide before coming to a stop?

Solution: Given: $m = 60.0$ kg, $v_0 = 25.0$ m/s, $v = 0$, $f_k = 50.0$ N.
Find: d.

The kinetic friction force f_k is the only unbalanced force and the angle between the friction force and the displacement is 180°. With $W_{net} = K - K_0$, we have

$W_{net} = (F\cos\theta)\, d = (f_k \cos 180°)\, d = (50.0\ \text{N})(-1)d = K - K_0 = \frac{1}{2}mv^2 - \frac{1}{2}mv_0^2$

$= \frac{1}{2}(60.0\ \text{kg})(0)^2 - \frac{1}{2}(60.0\ \text{kg})(25.0\ \text{m/s})^2$.   Solving, $d = 3.75 \times 10^2$ m.

Note: This problem can also be solved with dynamics and kinematics. We can first find the acceleration of the object from Newton's second law and then use kinematic equations to find the distance.

---

4. Potential Energy (Section 5.4)

Potential energy is the energy of position. Two forms of potential energy are considered here, **gravitational potential energy** and **elastic (spring) potential energy**. A zero reference point/level is always required in measuring position and potential energy. For example, a question like "how high is that table?" is really meaningless unless we specify a zero reference level, such as the floor, the ground, etc., to measure that height This choice of zero reference point/level is arbitrary, i.e., you can choose anywhere as your zero reference point/level in your calculation. With the zero potential location chosen as the zero reference point/level, the two forms of potential energy can be written as

Gravitational potential energy: $U = mgh$ (with $h_0 = 0$)

Elastic potential energy: $U = \frac{1}{2}kx^2$ (with $x_0 = 0$)

Example 5.6   A 10.0-kilogram object is moved from the second floor of a house 3.00 m above the ground to the first floor 0.30 m above the ground. What is the change in gravitational potential energy?

Solution: We choose the ground level as our reference level ($h_0 = 0$).

Given: $m = 10.0$ kg, $h_1 = 3.00$ m, $h_2 = 0.30$ m.   Find: ΔU.

$\Delta U = U_2 - U_1 = mgh_2 - mgh_1 = mg(h_2 - h_1) = (10.0 \text{ kg})(9.80 \text{ m/s}^2)(0.30 \text{ m} - 3.00 \text{ m}) = -2.6 \times 10^2$ J.

As expected, ΔU is negative, since the object moves from a higher potential level to a lower potential level.

If we choose the reference level at the first floor level (height is zero at that level), then

$h_1 = 3.00$ m $- 0.30$ m $= 2.70$, and $h_2 = 0$.

So $\Delta U = mg(h_2 - h_1) = (10.0 \text{ kg})(9.80 \text{ m/s}^2)(0 - 2.70 \text{ m}) = -2.6 \times 10^2$ J, independent of reference choice.

Now try to calculate the change in potential energy by choosing the second floor as the reference level.

Example 5.7   A spring with a spring constant of 15 N/m is initially compressed by 3.0 cm. How much work is required to compress the spring an additional 4.0 cm?

Solution: We choose the uncompressed position as $x_0 = 0$.

Given: $k = 15$ N/m, $x_1 = 0.030$ m, $x_2 = x_1 + \Delta x = 0.030$ m $+ 0.040$ m $= 0.070$ m.

Find: W.

The work required goes into the change in elastic potential energy. $W = \Delta K + \Delta U = \Delta U$ ($\Delta K = 0$).

$$W = \Delta U = U_2 - U_1 = \tfrac{1}{2}kx_2^2 - \tfrac{1}{2}kx_1^2 = \tfrac{1}{2}k(x_2^2 - x_1^2) = \tfrac{1}{2}(15\ \text{N/m})[(0.070\ \text{m})^2 - (0.030\ \text{m})^2] = 0.030\ \text{J}.$$

Note: Why $W \neq \tfrac{1}{2}k(x_2 - x_1)^2 = \tfrac{1}{2}(14\ \text{m/s})(0.070\ \text{m} - 0.030\ \text{m})^2 = 0.012\ \text{J}$?

5.   The Conservation of Energy (Section 5.5)

A force is **conservative** if the work done by or against the force is independent of the path, but dependent on only the initial and final locations. Gravitational force is an example of a conservative force. A force is **nonconservative** if the work done by or against it depends on the path. Force of friction is an example of a nonconservative force.

The **total mechanical energy**, E, of a system is defined as the sum of the kinetic energy and potential energy, i.e, $E = K + U$. If the working forces (the forces which are doing work) in a system are conservative, the total mechanical energy of the system is conserved. This is **the conservation of total mechanical energy**.

$$K_0 + U_0 = K + U \quad \text{or} \quad \tfrac{1}{2}mv^2 + U = \tfrac{1}{2}mv_0^2 + U_0.$$

Note: When applying the conservation of mechanical energy, you should choose a zero reference point/level for position to determine U values. The choice is arbitrary, that is, you can choose the zero reference point/level to be anywhere it is convenient. Also clearly identify the initial and final velocities, and the initial and final positions. These are the only four physical quantities involved in mechanical energy conservation so you need to know three of them before you can solve the problem.

Example 5.8   A 70-kilogram skier starts from rest on the top of a 25-meter high slope. What is the speed of the skier on reaching the bottom of the slope? (Neglect friction.)

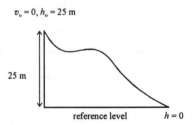

$v_0 = 0, h_0 = 25\ \text{m}$

25 m

reference level $h = 0$

Solution: Given: $v_0 = 0, \quad h_0 = 50\ \text{m}, \quad h = 0$.

Find: v.

We choose the bottom of the slope as the zero reference level ($h = 0$).
From the conservation of mechanical energy

$$K_0 + U_0 = K + U \quad \text{or} \quad \tfrac{1}{2}mv^2 + U = \tfrac{1}{2}mv_0^2 + U_0.$$

we have $\tfrac{1}{2}mv^2 + mgh = \tfrac{1}{2}mv_0^2 + mgh_0$.

So $\tfrac{1}{2}mv^2 + mg(0) = \tfrac{1}{2}m(0)^2 + mgh_0$,

therefore $v = \sqrt{2gh_0} = \sqrt{2(9.80\ \text{m/s}^2)(25\ \text{m})} = 22\ \text{m/s}$.

Note: The mass of the skier cancels out in the equation and we did not even use the mass of the skier!

Example 5.9   A 1500-kilogram car moving at 25 m/s hits an initially uncompressed horizontal spring with spring constant of 2.0×10^6 N/m. What is the maximum compression of the spring? (Neglect the mass of the spring.)

Solution: Given: $m = 1500$ kg, $k = 2.0 \times 10^6$ N/m, $v_0 = 25$ m/s, $v = 0$, $x_0 = 0$.

(Remember we need three known velocity and/or position quantities.)

Find: x.

Here we choose the uncompressed position of the spring as the reference point ($x_0 = 0$). Before the car hits the spring, the car is moving with an initial speed of 25 m/s and the compression of the spring is zero. When the spring is maximally compressed, the car stops (transfer all its kinetic energy to elastic potential energy in the spring) and the spring is compressed by a distance of x. From the conservation of mechanical energy, $K_0 + U_0 = K + U$ or $\frac{1}{2}mv^2 + U = \frac{1}{2}mv_0^2 + U_0$,

we have $\frac{1}{2}mv^2 + \frac{1}{2}kx^2 = \frac{1}{2}mv_0^2 + \frac{1}{2}kx_0^2$.

So $\frac{1}{2}m(0) + \frac{1}{2}kx^2 = \frac{1}{2}mv_0^2 + \frac{1}{2}k(0)^2$,

therefore $x = \sqrt{\dfrac{mv_0^2}{k}} = \sqrt{\dfrac{m}{k}}\, v_0 = \sqrt{\dfrac{1500 \text{ kg}}{2.0 \times 10^6 \text{ N/m}}}\,(25 \text{ m/s}) = 0.68$ m.

If there is a nonconservative force doing work in a system, the total mechanical energy of the system is *not* conserved. However, the total energy (not mechanical!) of the system is still conserved. Some of the total energy is used to overcome the work done by the nonconservative force. The difference in mechanical energy is equal to the work done by the nonconservative force, that is $W_{nc} = E_0 - E = -\Delta E$.

Example 5.10   In Example 5.8, if the work done by the kinetic friction force is -6.0×10^3 J (the work done by kinetic friction force is negative because the angle between the friction force and the displacement is 180°). What is the speed of the skier at the bottom of the slope?

Solution:

Since the kinetic friction force is a nonconservative force, mechanical energy is *not* conserved. However, the difference in mechanical energy is equal to the work done by the friction force.

$W_{nc} = E - E_0 = (\frac{1}{2}mv^2 + mgh) - (\frac{1}{2}m v_0^2 + mgh_0)$.

So -6.0×10^3 J $= \frac{1}{2}(70 \text{ kg})v^2 + mg(0) - \frac{1}{2}m\,(0)^2 - (70 \text{ kg})(9.80 \text{ m/s}^2)(25 \text{ m})$.

Solving, $v = 18$ m/s.

6. Power (Section 5.6)

Average **power** is the average rate of doing work (work done divided by time interval), $\overline{P} = \dfrac{W}{t}$. The SI unit of power is watt (W). A common British unit of power is horsepower (hp) and 1 hp = 746 W. If the work is done by a constant force in the direction of motion or displacement ($\theta = 0°$) and the object is moved through a distance of d, $\overline{P} = \dfrac{Fd}{t} = F\,\overline{v}$, where $\dfrac{d}{t} = \overline{v}$ is the magnitude of the average velocity.

Example 5.11   A 1500-kilogram car accelerates from 0 to 25 m/s in 7.0 s. What is the average power delivered to the car by the engine? Ignore all frictional and other losses.

Solution: Given: $m = 1500$ kg, $v_0 = 0$, $v = 25$ m/s, $t = 7.0$ s.   Find: $\overline{P}$.

Since power is the rate of doing work, we need to calculate the work done to the car first.

From work-energy theorem, $W = \Delta K = \frac{1}{2}mv^2 - \frac{1}{2}mv_0^2 = \frac{1}{2}(1500\text{ kg})[(25\text{ m/s})^2 - (0)^2] = 4.69 \times 10^5$ J.

So $\overline{P} = \dfrac{W}{t} = \dfrac{4.69 \times 10^5\text{ J}}{7.0\text{ s}} = 6.7 \times 10^4$ W ($= 90$ hp).

Alternate method: The average velocity of the car is $\overline{v} = \dfrac{v + v_0}{2} = \dfrac{25\text{ m/s} + 0}{2} = 12.5$ m/s.

From Newton's second law $F = ma = m\dfrac{v - v_0}{t} = (1500\text{ kg}) \times \dfrac{25\text{ m/s} - 0}{7.0\text{ s}} = 5.36 \times 10^3$ N.

So $\overline{P} = F\,\overline{v} = (5.36 \times 10^3\text{ N})(12.5\text{ m/s}) = 6.7 \times 10^4$ W $= 90$ hp.

IV.   Mathematical Summary

Work	$W = (F\cos\theta)d$	(5.2)	Defines work.
Hooke's Law (spring force)	$F_s = -kx$	(5.3)	Relates spring force with spring constant and change in length.
Work done in a Spring	$W = \frac{1}{2}kx^2$	(5.4)	Relates work done by external force in stretching or compressing a spring.
Kinetic Energy	$K = \frac{1}{2}mv^2$	(5.5)	Defines kinetic energy in terms of mass and velocity.
Work-Energy Theorem	$W_{net} = K - K_0 = \Delta K$	(5.6)	The net work done on an object is equal to the change in kinetic energy.

Elastic (spring) Potential Energy	$U = \frac{1}{2}kx^2$ (with $x_0 = 0$) (5.7)	Defines elastic potential energy.
Gravitational Potential Energy	$U = mgh$ (with $h_0 = 0$) 5.8	Defines gravitational potential energy.
Conservation of Mechanical Energy	$\frac{1}{2}mv^2 + U = \frac{1}{2}mv_0^2 + U_0$ (5.10)	States that the total mechanical energy of a system is conserved if only conservative forces are doing work.
Conservation of Energy (with a nonconservative force)	$W_{nc} = E - E_0 = \Delta E$ (5.13)	Equates the work done by nonconservative force to the change in mechanical energy of a system.
Average Power	$\overline{P} = \dfrac{W}{t} = \dfrac{Fd}{t} = F\overline{v}$ (5.15)	Defines and calculates the average power delivered by a constant force in direction of d and v.
	$\overline{P} = \dfrac{Fd\cos\theta}{t}$ (5.16)	For if the force is making an angle θ with d or v.
Efficiency (percent)	$\varepsilon = \dfrac{W_{out}}{W_{in}}$ ($\times 100\%$) (5.17)	Defines the mechanical efficiency of a system.
	$= \dfrac{P_{out}}{P_{in}}$ ($\times 100\%$) (5.18)	

V. Solutions of Selected Exercises and Paired/Trio Exercises

6. From the definition of work $W = (F\cos\theta)d$,

    we have $F = \dfrac{W}{d\cos\theta} = \dfrac{50 \text{ J}}{(10 \text{ m})\cos 0°} = \boxed{5.0 \text{ N}}$.

7. The kinetic friction force is $f_k = \mu_k N = \mu_k mg$ and the angle between the kinetic friction force and displacement is $0°$.

    $W = (F\cos\theta)d = \mu_k mg \cos\theta\, d = 0.20(5.0 \text{ kg})(9.80 \text{ m/s}^2) \cos 180° (10 \text{ m}) = \boxed{-98 \text{ J}}$.

14. $\Sigma F_y = N + F\sin\theta - mg = 0$,   or $N = mg - F\sin\theta$.

    $\Sigma F_x = F\cos\theta - f_k = 0$,   or $F\cos\theta = \mu_k N = \mu_k(mg - F\sin\theta) = 0$.

    So $F = \dfrac{\mu_k mg}{\cos\theta + \mu_k \sin\theta} = \dfrac{0.20(35 \text{ kg})(9.80 \text{ m/s}^2)}{\cos 30° + 0.20 \sin 30°} = 71.0 \text{ N}$.

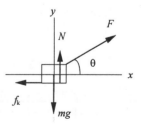

    Therefore $W = (F\cos\theta)d = (71.0 \text{ N})\cos 30° (10 \text{ m}) = \boxed{6.1 \times 10^2 \text{ J}}$.

16.   (a) $\boxed{1.0 \times 10^3 \text{ J}}$.   (b) $\boxed{-6.2 \times 10^2 \text{ J}}$.   (c) $\boxed{3.8 \times 10^2 \text{ J}}$.

21. $\boxed{\text{No}}$, more work is required, because the force increases as the spring stretches, according to Hooke's law:

$F_s = -kx$.

24. $W = \frac{1}{2}kx^2 = \frac{1}{2}(40 \text{ N/m})(0.020 \text{ m})^2 = \boxed{8.0 \times 10^{-3} \text{ J}}$.

28.   (a) $W = \frac{1}{2}kx^2 = \frac{1}{2}(2.5 \times 10^3 \text{ N/m})(0.060 \text{ m})^2 = \boxed{4.5 \text{ J}}$.

       (b) $W = \frac{1}{2}k(x_2^2 - x_1^2) = \frac{1}{2}(2.5 \times 10^3 \text{ N/m})[(0.080 \text{ m})^2 - (0.060 \text{ m})^2] = \boxed{3.5 \text{ J}}$.

33.   Because the heel can lift from the blade, the blade will be in contact with the ice longer. This will make the displacement a bit greater (longer stride) so that the work done is greater. Greater work translates to faster speed according to the work-energy theorem.

34.   (a) $K = \frac{1}{2}mv^2 = \frac{1}{2}(4m)v^2 = 2mv^2$.   (b) $K = \frac{1}{2}(3m)(2v)^2 = 6mv^2$.

       (c) $K = \frac{1}{2}(2m)(3v)^2 = 9mv^2$.   (d) $K = \frac{1}{2}(m)(4v)^2 = 8mv^2$.

       So (a) has the smallest kinetic energy.

38.   90 km/h = 25 m/s.

       (a) $K_o = \frac{1}{2}mv^2 = \frac{1}{2}(1.2 \times 10^3 \text{ kg})(25 \text{ m/s})^2 = \boxed{3.8 \times 10^5 \text{ J}}$.

       (b) From the work-energy theorem,

       $W_{\text{net}} = \frac{1}{2}mv^2 - \frac{1}{2}mv_o^2 = 0 - 3.8 \times 10^5 \text{ J} = \boxed{-3.8 \times 10^5 \text{ J}}$.

40. $-\boxed{1.5 \times 10^3 \text{ N}}$, opposite to direction of velocity.

43.   From the work-energy theorem $W = \frac{1}{2}mv^2 - \frac{1}{2}mv_o^2 = \frac{1}{2}m(v^2 - v_o^2)$,

       $\dfrac{W_2}{W_1} = \dfrac{(30 \text{ km/h})^2 - (20 \text{ km/h})^2}{(20 \text{ km/h})^2 - (10 \text{ km/h})^2} = 1.67$,

       so $W_2 = 1.67W_1 = 1.67(5.0 \times 10^3 \text{ J}) = \boxed{8.3 \times 10^3 \text{ J}}$.

49.   From the definition of gravitational potential energy $U = mgh$,

we have $\Delta U = mg\Delta h = (1.0 \text{ kg})(9.80 \text{ m/s}^2)(1.5 \text{ m} - 0.90 \text{ m}) = \boxed{5.9 \text{ J}}$.

52.   (a) On the board: $U = mgh = (60 \text{ kg})(9.80 \text{ m/s}^2)(5.0 \text{ m}) = \boxed{2.9 \times 10^3 \text{ J}}$.

At the bottom of pool: $U = (60 \text{ kg})(9.80 \text{ m/s}^2)(-3.0 \text{ m}) = \boxed{-1.8 \times 10^3 \text{ J}}$.

(b) To the board: $\Delta U = mg\Delta h = (60 \text{ kg})(9.80 \text{ m/s}^2)(-8.0 \text{ m} - 0) = \boxed{-4.7 \times 10^3 \text{ J}}$.

To the surface: $\Delta U = (60 \text{ kg})(9.80 \text{ m/s}^2)(-3.0 \text{ m} - 5.0 \text{ m}) = \boxed{-4.7 \times 10^3 \text{ J}}$.

To the bottom of pool: $\Delta U = (60 \text{ kg})(9.80 \text{ m/s}^2)(0 - 8.0 \text{ m}) = \boxed{-4.7 \times 10^3 \text{ J}}$.

60.   (a) $E_0 = K_0 + U_0 = 0 + (0.250 \text{ kg})(9.80 \text{ m/s}^2)(115 \text{ m}) = \boxed{282 \text{ J}}$.

(b) $U_1 = (0.250 \text{ kg})(9.80 \text{ m/s}^2)(115 \text{ m} - 75.0 \text{ m}) = \boxed{98.0 \text{ J}}$.

Since $E = E_0$ is conserved, $K_1 = E_0 - U_1 = 282 \text{ J} - 98.0 \text{ J} = \boxed{184 \text{ J}}$.

(c) $E_2 = K_2 + 0 = K_2 = \boxed{282 \text{ J}}$. $K_2 = \frac{1}{2} mv^2$,

so $v = \sqrt{\dfrac{2K_2}{m}} = \dfrac{2(282 \text{ J})}{0.250 \text{ kg}} = \boxed{47.5 \text{ m/s}}$.

(d) For (a) $E_0 = 0 + 0 = \boxed{0}$.

For (b) $U_1 = (0.250 \text{ kg})(9.80 \text{ m/s}^2)(-75.0 \text{ m}) = \boxed{-184 \text{ J}}$. $K_1 = 0 - (-184 \text{ J}) = \boxed{184 \text{ J}}$.

For (c) $E_2 = K_2 + U_2 = \boxed{0}$. $K_2 = 0 - (0.250 \text{ kg})(9.80 \text{ m/s}^2)(-115 \text{ m}) = 282 \text{ J}$.

So $v = \dfrac{2(282 \text{ J})}{0.250 \text{ kg}} = \boxed{47.5 \text{ m/s}}$.

62. $\boxed{5.10 \text{ m}}$.

66.   We choose the bottom of the slope (point B) as the reference for height ($h_0 = 0$).

From the conservation of mechanical energy $\frac{1}{2} mv_B^2 + U_B = \frac{1}{2} mv_A^2 + U_A$,

we have $\frac{1}{2} mv_B^2 + mg(0) = \frac{1}{2} m(5.0 \text{ m/s})^2 + mg(10 \text{ m})$,

so $v_B = \sqrt{(5.0 \text{ m/s})^2 + 2(9.80 \text{ m/s}^2)(10 \text{ m})} = \boxed{15 \text{ m/s}}$.

67. We choose the lowest point on the course (point B) as the reference for height ($h_o = 0$).

(a) From the conservation of mechanical energy $\frac{1}{2}mv_B^2 + U_B = \frac{1}{2}mv_A^2 + U_A$,

we have $\frac{1}{2}mv_B^2 + mg(0) = \frac{1}{2}m(5.0 \text{ m/s})^2 + mg(5.0 \text{ m})$,

so $v_B = \sqrt{(5.0 \text{ m/s})^2 + 2(9.80 \text{ m/s}^2)(5.0 \text{ m})} = \boxed{11 \text{ m/s}}$.

(b) $E_A = E_B = \frac{1}{2}m\,(11 \text{ m/s})^2 = 60.5m$. $E_C = mg(8.0 \text{ m}) = m(9.80 \text{ m/s}^2)(8.0 \text{ m}) = 78.4m > E_A$.

So $\boxed{\text{no}}$, it will not reach point C.

(c) $\frac{1}{2}mv_A^2 + mg(5.0 \text{ m}) = \frac{1}{2}m(0)^2 + mg(8.0 \text{ m})$,   solving, $v_A = \boxed{7.7 \text{ m/s}}$.

71. When a nonconservative force does work, the work done by the nonconservative force is equal to the change in mechanical energy. $W_{nc} = \Delta E = E_o - E = [\frac{1}{2}m(5.0 \text{ m/s})^2 + mg(10 \text{ m})] - [\frac{1}{2}mv_B^2 + mg(0)]$.

So $2500 \text{ J} = [\frac{1}{2}(60 \text{ kg})(5.0 \text{ m/s})^2 + (60 \text{ kg})(9.80 \text{ m/s}^2)(10 \text{ m})] - [\frac{1}{2}(60 \text{ kg})v_B^2 + mg(0)]$.

Solving, $v_B = \boxed{12 \text{ m/s}}$.

78. 90 km/h = 25 m/s.   From work-energy theorem $W = \Delta K = \frac{1}{2}mv^2 - 0$.

$$\overline{P} = \frac{W}{t} = \frac{mv^2}{2t} = \frac{(1500 \text{ kg})(25 \text{ m/s})^2}{2(5.0 \text{ s})} = \boxed{9.4 \times 10^4 \text{ W} = 1.3 \times 10^2 \text{ hp}}.$$

80. (a) $\boxed{4.4 \times 10^2 \text{ W}}$.   (b) $\boxed{0.59 \text{ hp}}$.

85. (a) $\Sigma F_x = F - f - mg \sin\theta = 0$,

so $F = f + mg \sin\theta = 950 \text{ N} + (120 \text{ kg})(9.80 \text{ m/s}^2) \sin 15° = 1254 \text{ N}$.

Also $v = 5.0 \text{ km/h} = 1.389 \text{ m/s}$.

So $P = Fv = (1254 \text{ N})(1.389 \text{ m/s}) = 1742 \text{ W} = \boxed{2.3 \text{ hp}}$.

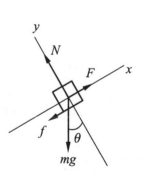

$\boxed{\text{The horse is working hard}}$ (it is working as hard as 2.3 horses. In spurts a horse can be more than 1 hp).

(b) 20 km/h = 5.556 m/s.

$a = \dfrac{5.556 \text{ m/s} - 1.389 \text{ m/s}}{5.0 \text{ s}} = 0.833 \text{ m/s}^2$.

Now $\Sigma F_x = F - f - mg \sin\theta = ma$,

so $F = 1254 \text{ N} + (120 \text{ kg})(0.833 \text{ m/s}^2) = 1354 \text{ N}$.

$P = (1354 \text{ N})(5.556 \text{ m/s}) = 7522 \text{ W} = \boxed{10 \text{ hp}}$.

96. (a) We choose the stopped position as $h = 0$ and assume the ball compresses the spring by x. Then the initial height $h_o = (1.20 \text{ m} + x)$ high.

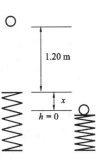

From conservation of mechanical energy: $mgh_o + 0 = 0 + \frac{1}{2}kx^2$,

so $mg(1.20 \text{ m} + x) = \frac{1}{2}kx^2$, simplifying, $175x^2 - 3.53x - 4.23 = 0$.

Solving, $x = \boxed{0.166 \text{ m}}$ or -0.150 m.

The negative answer has no physical meaning and is discarded.

(b) $mg(1.20 \text{ m} + 0.0500 \text{ m}) = \frac{1}{2}k(0.0500 \text{ m})^2 + \frac{1}{2}mv^2$,

or $4.41 \text{ J} = 0.438 \text{ J} + \frac{1}{2}(0.360 \text{ kg})v^2$, so $v = \boxed{4.70 \text{ m/s}}$.

VI. Practice Quiz

1. Which one of the following is the correct unit of work expressed in SI base units?

(a) $kg \cdot m/s$ (b) $kg \cdot m/s^2$ (c) $kg \cdot m^2/s$ (d) $kg \cdot m^2/s^2$ (e) $kg^2 \cdot m/s^2$

2. A 40-newton crate is pulled 5.0 m up along a 37° inclined plane. What is the magnitude of the work done by the weight (gravitational force) of the crate? (Hint: draw a diagram.)

(a) 6.0 J   (b) 12 J   (c) 1.2×10^2 J   (d) 2.0×10^2 J   (e) 1.2×10^3 J.

3. What work is required to stretch a spring of spring constant 40 N/m from $x = 0.20$ m to 0.25 m? (The unstretched position is at $x = 0$.)

(a) 0.45 J   (b) 0.80 J   (c) 1.3 J   (d) 0.050 J   (e) 0.90 J

4. A force of 200 N, directed at 20° above the horizontal, is applied to move a 50-kilogram cart (initially at rest) across a 10 m level surface. What is the speed of the cart at the end of the 10 m distance?

(a) 5.2 m/s   (b) 8.6 m/s   (c) 8.9 m/s   (d) 6.8×10^2 m/s   (e) 2.0×10^3 m/s

5. A roller coaster makes a run down a track from a vertical distance of 25 m. If there is negligible friction and the coaster starts from rest, what is its speed at the bottom of the track?

(a) 5.0 m/s   (b) 16 m/s   (c) 22 m/s   (d) 2.5×10^2 m/s   (e) 4.9×10^2 m/s

6. A 10-newton force is needed to move an object with a constant velocity of 5.0 m/s. What power must be delivered to the object by the force?

(a) 0.50 W   (b) 1.0 W   (c) 2.0 W   (d) 50 W   (e) 100 W

7.   If it takes 50 m to stop a car initially moving at 15 m/s, what distance is required to stop a car moving at 30 m/s? (Assume the same braking force.)

(a) 25 m   (b) 50 m   (c) 100 m   (d) 150 m   (e) 200 m

8.   What is the minimum speed of the ball at the bottom of its swing (point B) in order for it to reach point A, which is 0.20 m above the bottom of the swing?

(a) 0.40 m/s   (b) 1.4 m/s   (c) 2.0 m/s   (d) 3.1 m/s   (e) 3.9 m/s

9.   A force of 10 N is applied horizontally to a 2.0-kilogram mass on a level surface. The coefficient of kinetic friction between the mass and the surface is 0.20. If the mass is moved a distance of 10 m, what is the change in its kinetic energy?

(a) 100 J   (b) 61 J   (c) 46 J   (d) 39 J   (e) 20 J

10.   A 1500-kilogram car is moving with a speed of 25 m/s. How much work is required to stop the car?

(a) 1.5×10^3 J   (b) 1.9×10^4 J   (c) 3.8×10^4 J   (d) 4.7×10^5 J   (e) 9.4×10^5 J

Answers to Practice Quiz:

1. d 2. c 3. a 4. b 5. c 6. d 7. e 8. c 9. b 10. d

CHAPTER 6

Momentum and Collisions

I.   Chapter Objectives

Upon completion of this chapter, you should be able to:

1.   compute linear momentum and the components of momentum.

2.   relate impulse and momentum, and kinetic energy and momentum.

3.   explain the condition for the conservation of linear momentum and apply it to physical situations.

4.   describe the conditions on kinetic energy and momentum in elastic and inelastic collisions.

5.   explain the concept of the center of mass and compute its location for simple systems, and describe how the center of mass and center of gravity are related.

6.   apply the conservation of momentum in the explanation of jet propulsion and the operation of rockets.

II.   Key Terms

Upon completion of this chapter, you should be able to define and/or explain the following key terms:

linear momentum	inelastic collision
total linear momentum	completely inelastic collision
impulse	center of mass (CM)
impulse-momentum theorem	center of gravity (CG)
conservation of linear momentumjet propulsion	
elastic collision	

The definitions and/or explanations of the most important key terms can be found in the following section:
III. Chapter Summary and Discussion.

III. Chapter Summary and Discussion

1. Linear Momentum (Section 6.1)

The **linear momentum** of an object is defined as the product of its mass and velocity, $\mathbf{p} = m\,\mathbf{v}$. Since velocity is a vector, so is momentum. The SI unit of momentum is kg·m/s.

Note: Momentum depends on *both* the mass and velocity, not just either mass or velocity.

The **total linear momentum** of a system is the vector sum of the momenta of the individual particles.

$$\mathbf{P} = \mathbf{p}_1 + \mathbf{p}_2 + \mathbf{p}_3 + \ldots = \sum_i \mathbf{p}_i$$

Example 6.1 Which has more linear momentum:

(a) a 1500-kilogram car moving at 25.0 m/s or

(b) a 40 000-kilogram truck moving at 1.00 m/s?

Solution: Given: mass, m, and velocity, v.

Find: magnitude of momentum, p.

(a) $p = mv = (1500 \text{ kg})(25.0 \text{ m/s}) = 3.75 \times 10^4$ kg·m/s.

(b) $p = (40\,000 \text{ kg})(1.00 \text{ m/s}) = 4.00 \times 10^4$ kg·m/s.

So the much slower truck has more momentum than the faster car because the truck has more mass.

Example 6.2 Two identical 1500-kilogram cars are moving perpendicular to each other. One moves with a speed of 25.0 m/s due north and the other moves at 15.0 m/s due east. What is the total linear momentum of the system?

Solution: Given: $m_1 = m_2 = 1500$ kg, $v_1 = 25.0$ m/s, $v_2 = 15.0$ m/s.

Find: **P**.

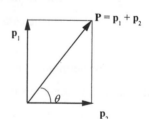

$p_1 = m_1 v_1 = (1500 \text{ kg})(25.0 \text{ m/s}) = 3.75 \times 10^4$ kg·m/s.

$p_2 = m_2 v_2 = (1500 \text{ kg})(15.0 \text{ m/s}) = 2.25 \times 10^4$ kg·m/s.

So $P = \sqrt{p_1^2 + p_2^2} = \sqrt{(3.75 \times 10^4 \text{ kg·m/s})^2 + (2.25 \times 10^4 \text{ kg·m/s})^2}$

$= 4.4 \times 10^4$ kg·m/s.

$\theta = \tan^{-1} \dfrac{p_1}{p_2} = \dfrac{3.75 \times 10^4 \text{ kg·m/s}}{2.25 \times 10^4 \text{ kg·m/s}} = 59.1°$ north of east.

Note: Momentum is a vector quantity. You must use vector addition when adding momenta.

Newton's second law can also be expressed in terms of momentum: $\overline{\mathbf{F}} = \dfrac{\Delta \mathbf{p}}{\Delta t}$. It is equivalent to $\mathbf{F} = m\mathbf{a}$ if mass is a constant. (Actually, Newton used the momentum form of his second law when he first started formatting the law.) For system in which mass is not a constant, such rocket propulsion, $\overline{\mathbf{F}} = \dfrac{\Delta \mathbf{p}}{\Delta t}$ should be used. The impulse-momentum theorem discussed later is a variation of this form of Newton's second law.

2.   Impulse (Section 6.2)

From Newton's second law, $\overline{\mathbf{F}} = \dfrac{\Delta \mathbf{p}}{\Delta t}$, we can derive the **impulse-momentum theorem**. This theorem states that *impulse is equal to the change in momentum*, or $\overline{\mathbf{F}}\,\Delta t = \Delta \mathbf{p} = \mathbf{p} - \mathbf{p}_o$, where $\overline{\mathbf{F}}\,\Delta t$ is called impulse ($\overline{\mathbf{F}}$ is the average force and Δt is the time interval the force is in action).

Impulse-momentum is very useful in explaining some everyday phenomena. For example, why do tennis players follow through? Why do football players wear pads? Why can some not-so-strong martial artists break objects like bricks with their bare fists? Try to answer some of these questions and the answers may surprise you.

Note: Again, momentum is a vector quantity and so must be treated as a vector.

Example 6.3   A 0.10-kilogram ball is dropped onto a table top. The speeds of the ball right before hitting the table top and right after hitting the table top are 5.0 m/s and 4.0 m/s, respectively. If the collision between the ball and the tabletop lasts 0.15 s, what is the average force exerted on the ball by the table top?

Solution: Given: $m = 0.10$ kg, $v_o = -5.0$ m/s (downward), $v = +4.0$ m/s (upward), $\Delta t = 0.15$ s.
Find: $\overline{\mathbf{F}}$.

The velocity of the ball before the collision is downward and the velocity of the ball after the collision is upward. Since these two velocities are opposite, we have to assign signs (+ or −) to them to find the change in velocity, $\Delta \mathbf{v}$, properly. Conventionally, we chose the upward direction as positive so the downward direction is negative.

From impulse-momentum theorem, we have $\overline{\mathbf{F}}\,\Delta t = \Delta \mathbf{p} = \mathbf{p} - \mathbf{p}_o$.

So $\overline{\mathbf{F}} = \dfrac{\mathbf{p} - \mathbf{p}_o}{\Delta t} = \dfrac{m\,v - m\,v_o}{\Delta t} = \dfrac{(0.10 \text{ kg})(4.0 \text{ m/s}) - (0.10 \text{ kg})(-5.0 \text{ m/s})}{0.15 \text{ s}} = +6.0$ N.

The positive sign indicates that the force on the ball by the tabletop is upward, which makes sense.

---

3. The Conservation of Linear Momentum (Section 6.3)

The total linear momentum of a system is conserved if the net external force on the system is zero.

$$\mathbf{P} = \mathbf{P_o}, \quad \text{or} \quad \mathbf{p_1} + \mathbf{p_2} + \ldots = \mathbf{p_{1o}} + \mathbf{p_{2o}} + \ldots.$$

Note: Momentum is a vector quantity. You must use vector addition when applying the conservation of linear momentum.

When applied to a "collision" involving two objects, a more convenient form of the conservation of momentum can be written as $m_1 v_1 + m_2 v_2 = m_1 v_{1o} + m_2 v_{2o}$, where the m's are the masses of the two objects, the v_o's are the initial velocities, and the v's are the final velocities.

Note: The terms initial and final are always relative and here they are relative to the "collision" process. A collision is generally defined in physics as a process in which forces are exerted.

Example 6.4   A 50-kilogram pitching machine (excluding the baseball) is placed on a frozen pond. The machine fires a 0.40-kilogram baseball with a speed of 35 m/s in the horizontal direction. What is the recoil velocity of the pitching machine? (Assume negligible friction.)

Solution: Given: $m_1 = 50$ kg (machine) $m_2 = 0.40$ kg (ball)

$v_{1o} = 0$ $v_{2o} = 0$,

$v_2 = 35$ m/s

Find: v_1.

Here the "collision" process is the firing of the baseball. Before the baseball is fired, neither the machine nor the baseball is moving so both have zero velocity and momentum (the total initial momentum $P_o = 0$). After the ball is fired, the ball moves in one direction and the machine must recoil in the opposite direction so that the total momentum $P = 0$.

From momentum conservation, we have $\mathbf{P} = \mathbf{P_o}$,   or $\mathbf{p_1} + \mathbf{p_2} = \mathbf{p_{1o}} + \mathbf{p_{2o}}$.

So $m_1 v_1 + m_2 v_2 = m_1 v_{1o} + m_2 v_{2o} = 0$,

therefore $v_1 = -\dfrac{m_2 v_2}{m_1} = -\dfrac{(0.40 \text{ kg})(35 \text{ m/s})}{50 \text{ kg}} = -0.28$ m/s.

The negative sign here indicates the machine is moving in exactly the opposite direction to the ball.

Example 6.5   A 10-gram bullet moving at 300 m/s is fired into a 1.0-kilogram block. The bullet emerges (does not stay embedded in the block) with half of its original speed. What is the velocity of the block right after the collision?

Solution: Given: $m_1 = 0.010$ kg (bullet) $m_2 = 1.0$ kg (block)

$v_{1o} = 300$ m/s $v_{2o} = 0$

$v_1 = 150$ m/s

Find: v_2.

Right before the bullet hits the block, the bullet is moving at 300 m/s and the block is at rest. Right after the bullet hits the block, it emerges with (300 m/s)/2 = 150 m/s and the block should be moving in the same direction as the bullet. (Why?)

From momentum conservation, we have $\mathbf{P} = \mathbf{P_o}$,   or $\mathbf{p_1} + \mathbf{p_2} = \mathbf{p_{1o}} + \mathbf{p_{2o}}$,

we have $m_1 v_1 + m_2 v_2 = m_1 v_{1o} + m_2 v_{2o} = m_1 v_{1o} + 0 = m_1 v_{1o}$.

So $v_2 = \dfrac{m_1 v_{1o} - m_1 v_1}{m_2} = \dfrac{(0.010 \text{ kg})(300 \text{ m/s} - 150 \text{ m/s})}{1.0 \text{ kg}} = +1.5$ m/s.

The result is positive because the block is moving in the same direction as the bullet.

4.   Elastic and Inelastic Collision (Section 6.4)

Linear momentum is always conserved in a collision, as long as the net external force is zero on the system, which is approximately true at least during the small time interval Δt of the collision. However, in an **elastic collision**, kinetic energy is also conserved while in an **inelastic collision**, kinetic energy is not conserved. (**Note:** when kinetic energy is not conserved as is the usual case, this does not mean the total energy is not conserved. Some of the kinetic energy is converted to heat, sound, etc. during a collision. Total energy is always conserved as we shall see!). If the objects in a system stick together after collision, the collision is called **completely inelastic**. In an inelastic collision, the kinetic energy of the system before the collision is always greater than the kinetic energy of the system after the collision. Why?

Example 6.6   While standing on skates on a frozen pond, a student of mass 70.0 kg catches a 2.00-kilogram ball travelling horizontally at 15.0 m/s toward him.

(a) What is the speed of the student and the ball immediately after he catches it?

(b) How much kinetic energy is lost in the process?

Solution: Given: $m_1 = 70.0$ kg $m_2 = 2.00$ kg

$v_{1o} = 0$ $v_{2o} = 15.0$ m/s

Find: v_1 v_2.

Since the student catches the ball, they must have the same final velocity, $v_1 = v_2 = v$. Here the "collision" is the catch action.

(a) From momentum conservation, we have $\mathbf{P} = \mathbf{P}_o$, or $\mathbf{p}_1 + \mathbf{p}_2 = \mathbf{p}_{1o} + \mathbf{p}_{2o}$,

we have $m_1 v_1 + m_2 v_2 = m_1 v + m_2 v = (m_1 + m_2)v = m_1 v_{1o} + m_2 v_{2o}$.

So $v = \dfrac{m_1 v_{1o} + m_2 v_{2o}}{m_1 + m_2} = \dfrac{(70.0 \text{ kg})(0) + (2.00 \text{ kg})(15.0 \text{ m/s})}{70.0 \text{ kg} + 2.00 \text{ kg}} = 0.417 \text{ m/s}.$

(b) Initial kinetic energy $K_o = \frac{1}{2} m_1 v_{1o}^2 + \frac{1}{2} m_2 v_{2o}^2 = \frac{1}{2}(70.0 \text{ kg})(0)^2 + \frac{1}{2}(2.00 \text{ kg})(15.0 \text{ m/s})^2 = 225$ J.

Final kinetic energy $K = \frac{1}{2} m_1 v_1^2 + \frac{1}{2} m_2 v_2^2 = \frac{1}{2}[(70.0 \text{ kg}) + (2.00 \text{ kg})](0.417 \text{ m/s})^2 = 207$ J.

So the kinetic energy lost in the collision is $|\Delta K| = K_o - K = 225$ J 207 J $= 18$ J.

The percentage of kinetic energy loss is $\dfrac{K - K_o}{K_o} = \dfrac{-18 \text{ J}}{225 \text{ J}} = -0.080 = -8.0\%.$

Example 6.7   A rubber ball with a speed of 5.0 m/s collides head on elastically with an identical ball at rest. Find the velocity of each object after the collision.

Solution: Given: $m_1 = m$ $m_2 = m$ (identical ball)

$v_{1o} = 5.0$ m/s $v_{2o} = 0$

Find: v_1 v_2.

We use Equations (6.15) and (6.16).

$$v_1 = \left(\frac{m_1 - m_2}{m_1 + m_2}\right) v_{1o} = \frac{m - m}{m + m}(5.0 \text{ m/s}) = 0. \qquad v_2 = \left(\frac{2m_1}{m_1 + m_2}\right) v_{1o} = \frac{2m}{m + m}(5.0 \text{ m/s}) = 5.0 \text{ m/s}.$$

The two balls exchange velocities. This is always the case if $m_1 = m_2$ in such collisions.

---

5.   Center of Mass (Section 6.5)

The **center of mass** of a system is the point at which all the mass of the system may be considered to be concentrated. If the acceleration due to gravity, g, is a constant, then the **center of gravity**, or the point at which all the weight of the system may be considered to be concentrated, is at the center of mass. For a system of particles, the

coordinates of the center of mass are calculated by $X_{CM} = \dfrac{\sum\limits_i m_i x_i}{M}$ and $Y_{CM} = \dfrac{\sum\limits_i m_i y_i}{M}$, where x_i and y_i are

the coordinates of the particle m_i, and $M = \sum\limits_i m_i$ is the total mass of the system.

The motion of the center of mass also obeys Newton's second law: $\mathbf{F}_{net\ external} = M \mathbf{A}_{CM}$, where $\mathbf{F}_{net\ external}$ is the net external force, M is the total mass of the system, and $\mathbf{A}_{CM}$ is the acceleration of the center of mass.

Example 6.8   Find the location of the center of mass of the following three mass system.

mass	location
$m_1 = 1.0$ kg	(0, 0)
$m_2 = 2.0$ kg	(1.0 m, 1.0 m)
$m_3 = 3.0$ kg	(2.0 m, −2.0 m)

Solution:

$$X_{CM} = \frac{\sum_i m_i x_i}{M} = \frac{(1.0 \text{ kg})(0) + (2.0 \text{ kg})(1.0 \text{ m}) + (3.0 \text{ kg})(2.0 \text{ m})}{1.0 \text{ kg} + 2.0 \text{ kg} + 3.0 \text{ kg}} = 1.3 \text{ m.}$$

$$Y_{CM} = \frac{\sum_i m_i y_i}{M} = \frac{(1.0 \text{ kg})(0) + (2.0 \text{ kg})(1.0 \text{ m}) + (3.0 \text{ kg})(-2.0 \text{ m})}{1.0 \text{ kg} + 2.0 \text{ kg} + 3.0 \text{ kg}} = -0.67 \text{ m.}$$

So the location of the center of mass is at $(X_{CM}, Y_{CM}) = (1.3 \text{ m}, -0.67 \text{ m})$.

IV.   Mathematical Summary

Linear Momentum	$\mathbf{p} = m\mathbf{v}$	(6.1)	Defines linear momentum.
Total Linear Momentum	$\mathbf{P} = \mathbf{p}_1 + \mathbf{p}_2 + \mathbf{p}_3 + \ldots$ $= \sum_i \mathbf{p}_i$	(6.2)	Computes total linear momentum of a system.
Newton's Second Law in terms of Momentum	$\overline{\mathbf{F}} = \dfrac{\Delta \mathbf{p}}{\Delta t}$	(6.3)	Rewrites Newton's second law in terms of momentum.
Impulse-momentum theorem	Impulse = $\overline{\mathbf{F}} \Delta t = \Delta \mathbf{p} = m\mathbf{v} - m\mathbf{v}_0$	(6.5)	States that the impulse is equal to the change in momentum.
Conditions for an Elastic Collision	$\mathbf{P}_f = \mathbf{P}_i, \quad K_f = K_i$	(6.8)	Define the conditions for an elastic collision.
Conditions for an Inelastic Collision	$\mathbf{P}_f = \mathbf{P}_i, \quad K_f < K_i$	(6.9)	Define the conditions for an inelastic collision.
Final Velocities in Head-On Elastic Collision ($v_{2_0} = 0$)	$v_1 = \left(\dfrac{m_1 - m_2}{m_1 + m_2} \right) v_{1_0}$ (6.15) $v_2 = \left(\dfrac{2m_1}{m_1 + m_2} \right) v_{1_0}$ (6.16)		Express the final velocities in head-on elastic collisions.

Coordinate of center of Mass (using sign for directions)	$$X_{CM} = \frac{\sum_i m_i x_i}{M} \qquad (6.20)$$	Calculates the coordinate the center of mass of a system (using signs for directions).

V. Solutions of Selected Exercises and Paired/Trio Exercises

6.   (a) $p = mv = (7.1 \text{ kg})(12 \text{ m/s}) = \boxed{85 \text{ kg·m/s}}$.

(b) 90 km/h = 25 m/s. $p = (1200 \text{ kg})(25 \text{ m/s}) = \boxed{3.0 \times 10^4 \text{ kg·m/s}}$.

10.   Since the ball moves in the opposite direction, the initial velocity must have opposite sign as the final velocity. If $v_0 = 4.50$ m/s then $v = -34.7$ m/s.

$\Delta p = mv - mv_0 = (0.150 \text{ kg})(-34.7 \text{ m/s}) - (0.150 \text{ kg})(4.50 \text{ m/s}) = -5.88 \text{ kg·m/s}$

$= \boxed{5.88 \text{ kg·m/s in the direction opposite } v_0}$.

13.   (a) From $v = v_0 - gt$,

$p = mv = m(v_0 - gt) = (0.50 \text{ kg})[0 - (9.80 \text{ m/s}^2)(0.75 \text{ s})] = -1.8 \text{ kg·m/s} = \boxed{3.7 \text{ kg·m/s down}}$.

(b) From $v^2 = v_0^2 - 2gy = 0 - 2(9.80 \text{ m/s}^2)(-10 \text{ m}) = 196 \text{ m}^2/\text{s}^2$, we have $v = -14$ m/s.

So $p = (0.50 \text{ kg})(-14 \text{ m/s}) = -7.0 \text{ kg·m/s} = \boxed{7.0 \text{ kg·m/s down}}$.

18.   From Newton's second law in terms of momentum,

$\bar{F} = \dfrac{\Delta p}{\Delta t} = \dfrac{mv - mv_0}{\Delta t} = \dfrac{(10 \text{ kg})(4.0 \text{ m/s} - 0)}{2.5 \text{ s}} = \boxed{16 \text{ N}}$.

20. $\boxed{68 \text{ N}}$.

26.   We consider only the horizontal motion and use the impulse-momentum theorem.

Since $\bar{F}\Delta t = mv - mv_0 = mv$, $v = \dfrac{\bar{F}\Delta t}{m} = \dfrac{3.0 \text{ N·s}}{0.20 \text{ kg}} = \boxed{15 \text{ m/s}}$.

28. $\boxed{13 \text{ m/s}}$.

33. From the impulse-momentum theorem, $\bar{F}\Delta t = mv - mv_o = -mv_o$,

we have $\bar{F} = -\dfrac{mv_o}{\Delta t}$.   So the magnitude is $\dfrac{mv_o}{\Delta t}$.

$\bar{F}_1 = \dfrac{(0.16 \text{ kg})(25 \text{ m/s})}{3.5 \times 10^{-3} \text{ s}} = \boxed{1.1 \times 10^3 \text{ N}}$;

$\bar{F}_2 = \dfrac{(0.16 \text{ kg})(25 \text{ m/s})}{8.5 \times 10^{-3} \text{ s}} = \boxed{4.7 \times 10^3 \text{ N}}$.

37. 40 km/h = 11.1 m/s,   240 lb = 1068 N.

The force on the infant is opposite to velocity because the infant is decelerating.

From the impulse-momentum theorem, $F\Delta t = mv - mv_o$,

we have $\Delta t = \dfrac{mv - mv_o}{F} = \dfrac{(5.5 \text{ kg})(0 - 11.1 \text{ m/s})}{-1068 \text{ N}} = \boxed{0.057 \text{ s}}$.

42. According to the conservation of momentum, the astronaut moves in the opposite direction.

Given: $m_1 = 0.50$ kg, $m_2 = 60$ kg, $v_{1o} = 0$, $v_{2o} = 0$, $v_1 = 10$ m/s.   find: v_2.

From momentum conservation, $\mathbf{P}_o = \mathbf{P}$,

we have $m_1 v_{1o} + m_2 v_{2o} = m_1 v_1 + m_2 v_2$.

So $v_2 = \dfrac{m_1 v_{1o} + m_2 v_{2o} - m_1 v_1}{m_2} = \dfrac{0 + 0 - (0.50 \text{ kg})(10 \text{ m/s})}{60 \text{ kg}} = \boxed{0.083 \text{ m/s}}$.

46. Apply momentum conservation $\mathbf{P}_o = \mathbf{P}$

in x axis: (2.0 kg)(0) = (0.50 kg)(−2.8 m/s) + (1.3 kg)(0) + (1.2 kg)v_x,   ☞ $v_x = 1.17$ m/s;

in y axis: (3.0 kg)(0) = (0.50 kg)(0) + (1.3 kg)(−1.5 m/s) + (1.2 kg)v_y,   ☞ $v_y = 1.63$ m/s.

Therefore $v = \sqrt{(1.17 \text{ m/s})^2 + (1.63 \text{ m/s})^2} = \boxed{2.0 \text{ m/s}}$,

and $\theta = \tan^{-1}\left(\dfrac{1.63}{1.17}\right) = \boxed{54° \text{ above } +x \text{ axis}}$.

53. We first use energy conservation to find the velocity of the bullet and the bob right after collision from the swing motion. The velocity right after the collision is the same as the velocity at the start of the swing.

$\frac{1}{2}(m + M)v^2 + (m + M)g(0) = \frac{1}{2}(m + M)(0)^2 + (m + M)g(h)$,   Solving, $v = \sqrt{2gh}$.

Now apply momentum conservation $\mathbf{P}_o = \mathbf{P}$,

we have $mv_o + M(0) = (m + M)v = (m + M)\sqrt{2gh}$, so $v_o = \dfrac{m + M}{m}\sqrt{2gh}$.

58. From Equations (6.15) and (6.16),

$$v_1 = \frac{m_1 - m_2}{m_1 + m_2}\, v_{1o} = \frac{4.0 \text{ kg} - 2.0 \text{ kg}}{4.0 \text{ kg} + 2.0 \text{ kg}} \,(4.0 \text{ m/s}) = \boxed{+1.3 \text{ m/s}}.$$

$$v_2 = \frac{2m_1}{m_1 + m_2}\, v_{1o} = \frac{2(4.0 \text{ kg})}{4.0 \text{ kg} + 2.0 \text{ kg}} \,(4.0 \text{ m/s}) = \boxed{+5.3 \text{ m/s}}.$$

62. We first find the velocity of the 6.0-kilogram ball right after collision from momentum conservation.

Given: $m_1 = 2.0$ kg, $\quad m_2 = 6.0$ kg, $\quad v_{1o} = 12$ m/s, $\quad v_{2o} = -4.0$ m/s ("toward each other"),

$\qquad\qquad v_1 = -8.0$ m/s ("recoil").

$(2.0 \text{ kg})(12 \text{ m/s}) + (6.0 \text{ kg})(-4.0 \text{ m/s}) = (2.0 \text{ kg})(-8.0 \text{ m/s}) + (6.0 \text{ kg})v_2$, $\quad$ so $\quad v_2 = 2.67$ m/s.

The initial kinetic energy $K_o = \frac{1}{2}(2.0 \text{ kg})(12 \text{ m/s})^2 + \frac{1}{2}(6.0 \text{ kg})(4.0 \text{ m/s})^2 = 192$ J.

The final kinetic energy $K = \frac{1}{2}(2.0 \text{ kg})(8.0 \text{ m/s})^2 + \frac{1}{2}(6.0 \text{ kg})(2.67 \text{ m/s})^2 = 84.5$ J.

So the kinetic energy lost is $K_o - K = \boxed{1.1 \times 10^2 \text{ J}}$.

64. (a) Apply momentum conservation $\mathbf{P}_o = \mathbf{P}$

in x: $\quad mv + M(0) = (m + M)v'_x$, $\quad$ so $\quad v'_x = \frac{mv}{m + M} = \frac{(2.0 \text{ kg})(3.0 \text{ m/s})}{(2.0 \text{ kg}) + 4.0 \text{ kg}} = \boxed{1.0 \text{ m/s}}$;

in y: $\quad m(0) + M(V) = (m + M)v'_y$, $\quad$ so $\quad v'_y = \frac{MV}{m + M} = \frac{(4.0 \text{ kg})(5.0 \text{ m/s})}{(2.0 \text{ kg}) + 4.0 \text{ kg}} = \boxed{3.3 \text{ m/s}}$.

(b) $\theta = \tan^{-1}\left(\frac{3.3}{1.0}\right) = \boxed{73°}$.

66. $\boxed{15°}$; $\boxed{0.57 \text{ m/s}}$.

71. (a) From momentum conservation, $\mathbf{P}_o = \mathbf{P}$,

we have $mv_0 + M(0) = (m + M)v$, $\quad$ so $\quad v = \frac{m\,v_0}{m + M} = \frac{0.010 \text{ kg}}{0.010 \text{ kg} + 0.890 \text{ kg}}\,v_0 = \boxed{\frac{v_0}{90}}$.

(b) From energy conservation: $\quad v = \sqrt{2gh} = \sqrt{2(9.80 \text{ m/s}^2)(0.40 \text{ m})} = 2.8$ m/s.

So $\quad v_0 = 90v = \boxed{2.5 \times 10^2 \text{ m/s}}$.

(c) The fraction of kinetic energy lost is

$$\frac{|\Delta K|}{K_o} = \frac{K_o - K}{K_o} = 1 - \frac{K}{K_o} = 1 - \frac{\frac{1}{2}\dfrac{(m\,v_0)^2}{m + M}}{\frac{1}{2}m\,v_0^2} = 1 - \frac{m}{m + M} = \frac{M}{m + M}.$$

$$= \frac{0.890 \text{ kg}}{0.010 \text{ kg} + 0.890 \text{ kg}} = \boxed{99\%}.$$

75.   This pole will ⌈lower the center of mass⌉ of the walker-pole system. The pole will also increase the moment of inertia of the system, and with a torque that rotates around the rope, the angular acceleration is smaller, giving the walker more time to recover.

80.   We choose the less massive mass as the origin ($x = 0$).

$$X_{CM} = \frac{\Sigma_i \, m_i \, x_i}{M} = \frac{(4.0 \text{ kg})(0) + (7.5 \text{ kg})(1.5 \text{ m})}{4.0 \text{ kg} + 7.5 \text{ kg}} = \boxed{0.98 \text{ m}} \text{ from the less massive sphere.}$$

88.   Due to the lack of external force, the CM is stationary and is at where they meet.

$$X_{CM} = \frac{\Sigma_i \, m_i \, x_i}{M} = \frac{(3000 \text{ kg})(0) + (100 \text{ kg})(5.0 \text{ m})}{3000 \text{ kg} + 100 \text{ kg}}$$

$$= \boxed{0.16 \text{ m from capsule's original position}}.$$

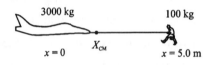

95.   (a) $p_x = p_y = (0.50 \text{ kg})(3.3 \text{ m/s}) = 1.65 \text{ kg·m/s}$.   So

$$p = \sqrt{(1.65 \text{ kg·m/s})^2 + (1.65 \text{ kg·m/s})^2} = \boxed{2.3 \text{ kg·m/s}},$$

$$\theta = \tan^{-1}\left(\frac{1.65}{1.65}\right) = \boxed{45° \text{ below the } -x \text{ axis}}.$$

(b) ⌈Not necessarily⌉, a collision does not have to occur; momentum would be the same.

99.   (a) The stunt man has zero horizontal velocity before he jumps onto the sled.
Apply momentum conservation $P_o = P$ in the horizontal direction.

$(75 \text{ kg})(0) + (50 \text{ kg})(10 \text{ m/s}) = (75 \text{ kg} + 50 \text{ kg})v = (125 \text{ kg})v$,   so $v = \boxed{4.0 \text{ m/s}}$.

(b) The stunt man's momentum is still conserved, so he continues to move with a speed of $\boxed{4.0 \text{ m/s}}$.

105.   If the triangle is suspended at one corner, the CM will be on a line perpendicular to the base and through the corner. If the triangle is suspended at a second corner, the CM will be where the two lines cross (see diagram). From symmetry,

$Y_{CM} = 15 \text{ cm}$, and $X_{CM} = (15 \text{ cm})\tan 30° = 8.7 \text{ cm}$.

Thus the CM is at $\boxed{(8.7 \text{ cm}, 15 \text{ cm})}$.

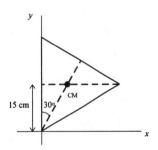

VI. Practice Quiz

1. The SI unit of impulse is which one of the following?

    (a) N·m   (b) N/s   (c) N·s   (d) N/m   (e) kg·m/s^2

2. A force of 10 N acts on a 5.0-kilogram object, initially at rest, for 2.5 s. What is the final speed of the object?

    (a) 1.0 m/s   (b) 2.0 m/s   (c) 3.0 m/s   (d) 4.0 m/s   (e) 5.0 m/s

3. A 70-kilogram astronaut is space walking outside the space capsule and is stationary when the tether line breaks. As a means of returning to the capsule he throws his 2.0-kilogram space wrench at a speed of 14 m/s away from the capsule. At what speed does the astronaut move toward the capsule?

    (a) 0.40 m/s   (b) 1.5 m/s   (c) 3.5 m/s   (d) 5.0 m/s   (e) 7.0 m/s

4. A 1500-kilogram, is allowed to coast along a level track at a speed of 8.0 m/s. It collides and couples with a 2000-kilogram truck, initially at rest and with brakes released. What is the speed of the two vehicles after they collide?

    (a) 0.75 m/s   (b) 3.4 m/s   (c) 4.6 m/s   (d) 6.0 m/s   (e) 11 m/s

5. A 0.060-kilogram tennis ball, initially moving at a speed of 12 m/s, is struck by a racket causing it to rebound in the opposite direction at a speed of 18 m/s. What is the change in momentum of the ball?

    (a) 0.18 kg·m/s   (b) 0.36 kg·m/s   (c) 0.72 kg·m/s   (d) 1.1 kg·m/s   (e) 1.8 kg·m/s.

6. The center of mass of a two-particle system is at the origin. One particle is located at (3.0 m, 0 m) and has a mass of 2.0 kg. What is the location of the second mass of 4.0 kg?

    (a) (−3.0 m, 0 m)   (b) (−2.0 m, 0 m)   (c) (−1.5 m, 0 m)   (d) (−0.75 m, 0 m)   (e) (−0.50 m, 0 m)

7. In an elastic collision, if the momentum of the system is conserved, then which of the following statements is true about kinetic energy?

    (a) kinetic energy is also conserved.   (b) kinetic energy is gained.

    (c) kinetic energy is lost.   (d) kinetic energy is halved.

    (e) none of the above.

8. A 0.10-kilogram object with a velocity of 0.20 m/s in the +x direction makes a head-on elastic collision with a 0.15-kilogram object initially at rest. What is the velocity of the 0.10-kilogram object after collision?

    (a) −0.16 m/s   (b) +0.16 m/s   (c) 0   (d) −0.040 m/s   (e) +0.040 m/s

---

9.   If the momentum of an object is halved, by what factor will its kinetic energy change?

(a) 2   (b) 4   (c) 0   (d) 1/2   (e) 1/4 ·

10.   A model rocket sits on the launch pad and its fuel is ignited, blasting the rocket upward. What happens to the center of mass of the rocket-fuel system?

(a) it goes up   (b) it is stationary   (c) it follows the path of the rocket

(d) it follows the path of the fuel   (e) not enough information given

Answers to Practice Quiz:

1. c 2. e 3. a 4. b 5. e 6. c 7. a 8. d 9. e 10. b

CHAPTER 7

Circular Motion and Gravitation

I.   Chapter Objectives

Upon completion of this chapter, you should be able to:

1.   define units of angular measure and show how angular measure is related to circular arc length.

2.   describe and compute angular speed and velocity and explain their relationship to tangential speed.

3.   explain why there is a centripetal acceleration in constant or uniform circular motion and compute centripetal acceleration.

4.   define angular acceleration and analyze rotational kinematics.

5.   describe Newton's law of gravitation and how it relates to the acceleration due to gravity, and apply the general formulation of gravitational potential energy.

6.   state and explain Kepler's laws of planetary motion and describe the orbits and motions of satellites.

II.   Key Terms

Upon completion of this chapter, you should be able to define and/or explain the following key terms:

angular displacement	centripetal acceleration
radian (rad)	centripetal force
average angular speed	average angular acceleration
angular velocity	tangential acceleration
average	universal law of gravitation
instantaneous	universal gravitational constant, G
tangential speed	gravitational potential energy
period (T)	Kepler's first law (the law of orbits)
frequency (f)	Kepler's second law (the law of area)
hertz (Hz)	Kepler's third law (the law of periods)
uniform circular motion	escape speed

The definitions and/or explanations of the most important key terms can be found in the following section: **III. Chapter Summary and Discussion.**

III.   Chapter Summary and Discussion

1.   Angular Measure (Section 7.1)

Circular motion is conveniently described using the polar coordinates (r, θ) because r is a constant and only θ varies. The relationships between rectangular coordinates and polar coordinates are as follows:

$$x = r \cos \theta \quad \text{and} \quad y = r \sin \theta, \text{ where } \theta \text{ is measured ccw from the } +x \text{ axis}$$

Angular distance ($\Delta \theta = \theta - \theta_0 = \theta$ if $\theta_0 = 0$) is measured in degrees or radians. A **radian** (rad) is defined as the angle that subtends an arc length that is equal to the radius: 1 rad = 57.3° or 2π rad = 360°. Similarly, $s = r\, \theta$,   where s is the arc length, r is the radius, and θ is the angle in radians.

Example 7.1   When you are watching the NASCAR Daytona 500, the 5.5-meter long race car subtends an angle of 0.31°. What is the distance from the race car to you?

Solution: Given: $\theta = 0.31° = (0.31°) \times \dfrac{\pi \text{ rad}}{180°} = 5.41 \times 10^{-3}$ rad, $s = 5.5$ m.

Find: r.

From $s = r\, \theta$, we have $r = \dfrac{s}{\theta} = \dfrac{5.5 \text{ m}}{5.41 \times 10^{-3} \text{ rad}} = 1.0 \times 10^4 3\text{m} = 1.0$ km.

Note: The angle θ must be in radians.

---

2.   Angular Speed and Velocity (Section 7.2)

Angular speed and **angular velocity** are defined analogously to their linear motion counterparts. The direction of angular velocity is determined by the right-hand rule: when the fingers of the right hand are curled in the direction of the rotation, the extended thumb points in the direction of the velocity vector. Therefore, there are only two possible directions for angular velocity, perpendicularly in or out of a plane, for a given observer. For convenience, they may be expressed with plus (ccw) and minus (cw) signs. They both have the units rad/s. Like their linear conterparts, angular speed and angular velocity could both be average and instantaneous. Instantaneous angular speed is the magnitude of the instantaneous angular velocity. Often, the word instantaneous is omitted.

Average angular speed is defined by $\overline{\omega} = \dfrac{\Delta \theta}{\Delta t} = \dfrac{\theta - \theta_0}{t - t_0}$.

Sometimes angular speed and velocity are given in units of rpm (revolutions per minute). You should first convert them to rad/s before trying to solve the problems. For example:

$$(33 \text{ rmp}) = \frac{33 \text{ rev}}{\text{min}} \times \frac{2\pi \text{ rad}}{1 \text{ rev}} \times \frac{1 \text{ min}}{60 \text{ s}} = 3.5 \text{ rad/s}.$$

Tangential (linear) speed and angular speed are related to each other through $v = r\,\omega$, where r is the radius. (Generally, linear quantity = radius × angular quantity, e.g., $s = r\theta$.)

In circular motion, the time it takes for an object to go through one revolution is called the **period** (T) and the number of revolutions in one second is called **frequency** (f). They are related through $f = \frac{1}{T}$. The SI unit of frequency is 1/s or hertz (Hz).

Example 7.2   A bicycle wheel rotates uniformly through 2.0 revolutions in 4.0 s.

      (a) What is the average angular speed of the wheel?

      (b) What is the tangential speed of a point 0.10 m from the center of the wheel?

      (c) What is the period?

      (d) What is the frequency?

Solution: Given: $\theta = 2.0 \text{ rev} = (2.0 \text{ rev}) \times \dfrac{2\pi \text{ rad}}{1 \text{ rev}} = 4\pi \text{ rad}, \quad t = 4.0 \text{ s}, \quad r = 0.10 \text{ m}.$

        Find: (a) $\overline{\omega}$ (b) v (c) T (d) f.

(a) $\overline{\omega} = \dfrac{\Delta\theta}{\Delta t} = \dfrac{4\pi \text{ rad}}{4.0 \text{ s}} = \pi \text{ rad/s} = 3.1 \text{ rad/s}.$

(b) Since the rotation is uniform, $\omega = \overline{\omega}$.   So $v = r\omega = (0.10 \text{ m})(3.1 \text{ rad/s}) = 0.31 \text{ m/s}.$

(c) Period is the time for one revolution.   So $T = \dfrac{4.0 \text{ s}}{2.0} = 2.0 \text{ s}.$

(d) $f = \dfrac{1}{T} = \dfrac{1}{2.0 \text{ s}} = 0.50 \text{ Hz}.$

---

3.   Uniform Circular Motion and Centripetal Acceleration (Section 7.3)

For **uniform circular motion**, the speed (tangential and angular) is a constant, but there is an acceleration. The linear tangential velocity vector changes direction as the object moves along the circle. Since velocity has both magnitude and direction, a change in direction results in a change in velocity, therefore, an acceleration (see the discussion in Chapter 2). This acceleration is called **centripetal acceleration** (center-seeking) because it is always directed toward the center of the circle. The magnitude of centripetal acceleration is given by $a_c = \dfrac{v^2}{r} = r\omega^2$.

From Newton's second law, net force = mass × acceleration, we conclude that there must be a net force associated with centripetal acceleration. In the case of uniform circular motion, this force is called **centripetal force** and always directed toward the center of the circle since we know the net force on an object is in the same direction as acceleration. Its magnitude is $F_c = ma_c = \dfrac{m v^2}{r} = mr\omega^2$.

When analyzing circular motions, always think about centripetal acceleration and centripetal force. (**Note**: centripetal force is not a separate or extra force. It is a net force toward the center of the circle.) A centripetal force is always required for objects to stay in circular path. Without the necessary centripetal force, an object will not stay in a circular orbit and instead fly out along a tangent line due to its inertia. Can you identify the centripetal force when a car makes a turn on a flat road and when clothes move around in a dryer? Or how about an Earth satellite orbiting the Earth?

Example 7.3   A car of mass 1500 kg is negotiating a flat circular curve of radius 50 m with a speed of 20 m/s.

      (a) What is the source of centripetal force on the car?

      (b) What is the magnitude of the centripetal acceleration of the car?

      (b) What is the magnitude of the centripetal force on the car?

Solution: Given: $m = 1500$ kg, $r = 50$ m, $v = 20$ m/s.

              Find: (b) a_c (b) F_c.

(a) Since the car is on a flat curve, the only forces acting on it are the vertically downward weight, the vertically upward normal force by the surface, and a static friction force. Also because the center of the circle is a horizontal distance 50 m away from the car, so the centripetal force cannot be provided by the vertical forces. Therefore we conclude that the static friction force is the source of this centripetal force. this conclusion is also supported vy your everyday experience. For example, why new tires "grip" better and icy road makes driving dangerous when you negotiate a curve?

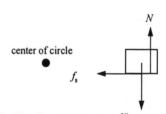

(b) $a_c = \dfrac{v^2}{r} = \dfrac{(20\ \text{m/s})^2}{50\ \text{m}} = 8.0\ \text{m/s}^2$.

(c) $F_c = ma_c = (15{=}00\ \text{kg})(8.0\ \text{m/s}^2) = 1.2 \times 10^4$ N.

---

4. Angular Acceleration (Section 7.4)

Average **angular acceleration** is defined analogously as its linear counterpart.

$$\overline{\alpha} = \frac{\Delta \omega}{\Delta t} = \frac{\omega - \omega_0}{t - t_0}.$$ The unit of angular acceleration is rad/s².

An object having angular acceleration has also tangential acceleration, that is, its tangential speed changes. They are related by $a_t = r\alpha$ (again, linear quantity = radius × angular quantity, as in $s = r\theta$ and $v = r\omega$). If an object is in nonuniform circular motion, it will have both centripetal acceleration (due to circular motion) and tangential acceleration (due to the change in tangential speed). the total acceleration is given by $\mathbf{a} = a_t \hat{\mathbf{t}} + a_c \hat{\mathbf{r}}$, where $\hat{\mathbf{t}}$ is the unit vector in the tangential direction and $\hat{\mathbf{r}}$ is the unit vector pointing toward the center of the circle.

Note: There is *always* centripetal acceleration no matter whether the circular motion is uniform or nonuniform. It is the tangential acceleration that is zero in uniform circular motion.

As you can see, circular motion and linear motion are analogous (at least mathematically). As a matter of fact, we can arrive at all the equations for angular motion from the kinematic equations in Chapter 2 if we just use the following substitutions: $\theta \rightarrow x$, $\omega \rightarrow v$, and $\alpha \rightarrow a$ (or add the word angular to every linear quantity except time). The angular kinematic equations are (their linear counterparts are in parentheses):

$$\theta = \overline{\omega}\, t \qquad\qquad (x = \overline{v}\, t),$$

$$\overline{\omega} = \frac{\omega + \omega_0}{2} \qquad\qquad (\overline{v} = \frac{v + v_0}{2}),$$

$$\omega = \omega_0 + \alpha t \qquad\qquad (v = v_0 + a t),$$

$$\theta = \omega_0 t + \tfrac{1}{2}\alpha t^2 \qquad\qquad (x = v_0 t + \tfrac{1}{2}a t^2),$$

$$\omega^2 = \omega_0^2 + 2\alpha\theta \qquad\qquad (v^2 = v_0^2 + 2 a x).$$

Example 7.4   A wheel is rotating with a constant angular acceleration of 3.5 rad/s². If the initial angular velocity is 2.0 rad/s and is speeding up, find

(a) the angle the wheel rotates through in 2.0 s,

(b) the angular speed at $t = 2.0$ s.

Solution: Given: $\alpha = 3.5$ rad/s², $\omega_0 = 0$, $t = 2.0$ s.

Find: (a) θ (b) ω.

(a) $\theta = \omega_0 t + \tfrac{1}{2}at^2 = (2.0\ \text{rad/s})(2.0\ \text{s}) + \tfrac{1}{2}(3.5\ \text{rad/s}^2)(2.0\ \text{s})^2 = 11$ rad.

(b) $\omega = \omega_0 + \alpha t = 2.0$ rad/s $+ (3.5\ \text{rad/s})(2.0\ \text{s}) = 9.0$ rad/s.

Example 7.5   The power on a medical centrifuge rotating at 12 000 rpm is cut off. If the magnitude of the maximum deceleration of the centrifuge is 50 rad/s^2, how many revolutions does it rotate before coming to rest?

Solution: Given: $\omega_0 = 12\,000 \text{ rpm} = \dfrac{12\,000 \text{ rev}}{\text{min}} \times \dfrac{2\pi \text{ rad}}{1 \text{ rev}} \times \dfrac{1 \text{ min}}{60 \text{ s}} = 1.26 \times 10^3 \text{ rad/s},$

$\omega = 0$ (coming to rest), $\alpha = -50 \text{ rad/s}^2$ (deceleration).

Find: θ (in terms of revolutions).

The unknown (how many turns) is related to the angle, θ, since 2π rad is one revolution or turn.

From $\omega^2 = \omega_0^2 + 2\alpha\theta$, we have

$$\theta = \frac{\omega^2 - \omega_0^2}{2\,\alpha} = \frac{0 - (1.26 \times 10^3 \text{ rad/s})^2}{2(-50 \text{ rad/s}^2)} = (1.59 \times 10^4 \text{ rad}) \times \frac{1 \text{ rev}}{2\pi \text{ rad}} = 2.5 \times 10^3 \text{ rev (turns)}.$$

5.   Newton's Law of Gravitation (Section 7.5)

Newton's **universal law of gravitation** states that the mutual gravitational attraction between any two point masses is directly proportional to the product of their masses and inversely proportional to the square of the distance between them. The magnitude is given by $F = \dfrac{Gm_1 m_2}{r^2}$, where $G = 6.67 \times 10^{-11}$ N·m^2/kg^2, is the **universal gravitational constant**.

Note: The magnitude of the force depends on $\dfrac{1}{r^2}$ (inverse square). For example, if the distance between the masses is doubled, the force between them would decrease to $\dfrac{1}{2^2} = \dfrac{1}{4}$ (not $\dfrac{1}{2}$).

According to Newton's second law, $F = ma$, the acceleration due to gravity by a mass M acting on another object m is then equal to $a_g = \dfrac{F}{m} = \dfrac{GM}{r^2}$. At the Earth's surface: $a_g = g = \dfrac{GM_E}{R_E^2}$, where M_E and R_E are the mass of radius of the Earth. At an altitude h above the Earth: $a_g = \dfrac{GM_E}{(R_E + h)^2}$.

The **gravitational potential energy** between two masses is given by $U = -\dfrac{Gm_1 m_2}{r}$, **note**: it is just r (not r^2) in the denominator. The minus sign arises from the choice of zero reference point (point where $U = 0$), which is $r = \infty$.

Example 7.6   The hydrogen atom consists of a proton of mass 1.67×10^{-27} kg and an orbiting electron of mass 9.11×10^{-31} kg. In one of its orbits, the electron is 5.3×10^{-11} m from the proton and in another orbit, it is 10.6×10^{-11} m from the proton.

(a) What are the mutual attractive forces when the electron is in these orbits, respectively?

(b) If the electron jumps from the large orbit to the small one, what is the change in potential energy?

Solution: Given: $m_1 = 1.67 \times 10^{-27}$ kg, $m_2 = 9.11 \times 10^{-31}$ kg, $r_1 = 5.3 \times 10^{-11}$ m, $r_2 = 10.6 \times 10^{-11}$ m.

Find: (a) F_1 and F_2 (b) ΔU.

(a) From Newton's law of gravitation,

$$F_1 = \frac{Gm_1 m_2}{r_1^2} = \frac{(6.67 \times 10^{-11} \text{ N·m}^2/\text{kg}^2)(1.67 \times 10^{-27} \text{ kg})(9.11 \times 10^{-31} \text{ kg})}{(5.3 \times 10^{-11} \text{ m})^2} = 3.6 \times 10^{-47} \text{ N}.$$

Since $r_2 = 2r_1$ and $F \propto \dfrac{1}{r^2}$, $\dfrac{F_2}{F_1} = \dfrac{r_1^2}{r_2^2} = \dfrac{1}{2^2} = \dfrac{1}{4}$.

So $F_2 = \dfrac{1}{4} F_1 = \dfrac{1}{4} (3.6 \times 10^{-47} \text{ N}) = 9.0 \times 10^{-48}$ N.

(c) $\Delta U = U_1 - U_2 = -\dfrac{Gm_1 m_2}{r_1} - \left(-\dfrac{Gm_1 m_2}{r_2} \right) = Gm_1 m_2 \left(\dfrac{1}{r_2} - \dfrac{1}{r_1} \right)$

$$= (6.67 \times 10^{-11} \text{ N·m}^2/\text{kg}^2)(1.67 \times 10^{-27} \text{ kg})(9.11 \times 10^{-31} \text{ kg}) \left(\frac{1}{10.6 \times 10^{-11} \text{ m}} - \frac{1}{5.3 \times 10^{-11} \text{ m}} \right)$$

$$= -9.6 \times 10^{-58} \text{ J}.$$

So the atom loses potential energy. You will learn in Chapter 27 that a loss of energy is accompanied by the emission of light.

Example 7.7   Calculate the acceleration due to gravity at the surface of the Moon.

Solution: Given: $R_M = 1750$ km $= 1.75 \times 10^6$ m, $M_M = 7.4 \times 10^{22}$ kg (from back inside cover).

Find: a_g.

$$a_g = \frac{GM}{r^2} = \frac{GM_M}{R_M^2} = \frac{(6.67 \times 10^{-11} \text{ N·m}^2/\text{kg}^2)(7.4 \times 10^{22} \text{ kg})}{(1.75 \times 10^6 \text{ m})^2} = 1.6 \text{ m/s}^2.$$

This is about $\dfrac{1}{6}$ of the value at the surface of the Earth, $g = 9.80$ m/s^2.

6. Kepler's Laws and Earth Satellites (Section 7.6)

(1) The law of orbits: Planets move in elliptical orbits with the Sun at one of the focal points.

(2) The law of areas: A line from the Sun to a planet sweeps out equal area in equal time interval.

(3) The law of periods: The square of the period of a planet is directly proportional to the cube of the average distance of the planet from the Sun, that is, $T^2 \propto r^3$. Or $T^2 = K r^3$, where $K = 2.97 \times 10^{-19} \text{ s}^2/\text{m}^3$ for the Sun.

The **escape speed** is the initial speed needed to escape from the surface of a planet or moon. For Earth satellites, this speed is $v_{esc} = \sqrt{\dfrac{2GM_E}{R_E}} = \sqrt{2gR_E}$, since $g = \dfrac{GM_E}{R_E}$.

Example 7.8 The planet Saturn is 1.43×10^{12} m from the Sun. How long does it take for Jupiter to orbit once about the Sun?

Solution: Given: $r = 1.43 \times 10^{12}$ m.

Find: T.

From Kepler's law of periods, $T^2 = (2.97 \times 10^{-19} \text{ s}^2/\text{m}^3)r^3$, we have

$T = \sqrt{(2.97 \times 10^{-19} \text{ s}^2/\text{m}^3)r^3} = \sqrt{(2.97 \times 10^{-19} \text{ s}^2/\text{m}^3)(1.43 \times 10^{12} \text{ m})^3} = 9.32 \times 10^8 \text{ s} \approx 30 \text{ years.}$

Example 7.9 If a satellite were launched from the surface of the Moon, at what initial speed would it need to begin in order for it to escape the gravitational attraction of the Moon?

Solution: With the known data:

$v_{esc} = \sqrt{\dfrac{2GM_M}{R_M}} = \sqrt{\dfrac{2(6.67 \times 10^{-11} \text{ N·m}^2/\text{kg}^2)(7.4 \times 10^{22} \text{ kg})}{1.75 \times 10^6 \text{ m}}} = 2.4 \times 10^3 \text{ m/s.}$

IV. Mathematical Summary

Arc Length	$s = r\theta$	(7.3)	Relates arc length to radius and angle (in radians).
Angular Kinematic Equation 1	$\theta = \overline{\omega}\, t$	(7.5)	Relates angular displacement with average angular velocity and time.
Angular Kinematic Equation 2	$\overline{\omega} = \dfrac{\omega + \omega_0}{2}$	(2, Table 7.2)	Defines average angular velocity for circular motion with constant angular acceleration.

Angular Kinematic Equation 3	$\omega = \omega_0 + \alpha t$ (7.12)	Relates final angular velocity with initial angular velocity, angular acceleration, and time (constant angular acceleration only).
Angular Kinematic Equation 4	$\theta = \omega_0 t + \frac{1}{2}\alpha t^2$ (4, Table 7.2)	Relates angular displacement with initial angular velocity, angular acceleration, and time (constant angular acceleration only).
Angular Kinematic Equation 5	$\omega^2 = \omega_0^2 + 2\alpha\theta$ (5, Table 7.2)	Relates final angular velocity with initial angular velocity, angular acceleration, and angular displacement (constant angular acceleration only).
Tangential and Angular Speeds	$v = r\omega$ (7.6)	Relates tangential speed with radius and angular speed.
Angular Speed	$\omega = \frac{2\pi}{T} = 2\pi f$ (7.8)	Defines angular speed for uniform circular motion.
Frequency and Period	$f = \frac{1}{T}$ (7.7)	Relates frequency with period.
Centripetal Acceleration	$a_c = \frac{v^2}{r} = r\omega^2$ (7.10)	Defines centripetal acceleration.
Tangential and Angular Accelerations	$a_t = r\alpha$ (7.13)	Relates tangential acceleration with radius and angular acceleration.
Centripetal Force and Acceleration	$F_c = m\, a_c = \frac{mv^2}{r}$ (7.11)	Defines centripetal force.
Newton's Law of Gravitation	$F = \frac{Gm_1 m_2}{r^2}$ (7.15) $G = 6.67 \times 10^{-11}\ \text{N·m}^2/\text{kg}^2$	Calculates the mutual attractive forces between two masses.
Acceleration due to Gravity at an Altitude h	$a_g = \frac{GM_E}{(R_E + h)^2}$ (7.18)	Calculates the acceleration due to gravity at an altitude of h above the Earth's surface.
Gravitational Potential Energy	$U = -\frac{Gm_1 m_2}{r}$ (7.19)	Defines gravitational potential energy for a two mass system ($U_\infty = 0$).
Kepler's Law of Periods	$T^2 = Kr^3$ (7.23) $K = 2.97 \times 10^{-19}\ \text{s}^2/\text{m}^3$ for Sun	Relates period with radius for objects orbiting the Sun.
Escape Speed	$v_{esc} = \sqrt{\dfrac{2GM_E}{R_E}}$ $= \sqrt{2gR_E}$ (7.24)	Gives the escape speed of the Earth.
Energy of Orbiting Satellites Orbiting the Earth	$E = -\dfrac{GmM_E}{2r}$ (7.28) $K = |E|$ (7.29)	Calculates the total energy and kinetic energy of objects orbiting the Earth.

V. Solutions of Selected Exercises and Paired/Trio Exercises

8. From $s = r\theta$, we have $r = \dfrac{s}{\theta} = \dfrac{5.0 \text{ m}}{2.0°(\pi \text{ rad}/180°)} = \boxed{1.4 \times 10^2 \text{ m}}$.

10. In 3 months, the Earth travels $\dfrac{3}{12} = \dfrac{1}{4}$ of a circle or $\dfrac{1}{4} \times (2\pi \text{ rad}) = \dfrac{\pi}{2} \text{ rad}$.

    So $s = r\theta = (1.5 \times 10^8 \text{ km}) \times \dfrac{\pi \text{ rad}}{2} = \boxed{2.4 \times 10^8 \text{ km}}$.

12. In 30 min, the hour hand travels $\pi/12 =$ rad, the minute hand π rad, and the second hand $30(2\pi) = 60\pi$ rad.

    Hour hand: $s = r\theta = (0.25 \text{ m})(\pi/12 \text{ rad}) = \boxed{0.065 \text{ m}}$.

    Minute hand: $s = (0.30 \text{ m})(\pi \text{ rad}) = \boxed{0.94 \text{ m}}$.

    Second hand: $s = (0.35 \text{ m})(60\pi \text{ rad}) = \boxed{66 \text{ m}}$.

15. (a) $\theta = \dfrac{s}{r} = \dfrac{3500 \text{ km}}{3.8 \times 10^5 \text{ km}} = \boxed{9.2 \times 10^{-3} \text{ rad} = 0.53°}$.

    (b) $\theta = \dfrac{2(6.4 \times 10^3 \text{ km})}{3.8 \times 10^5 \text{ km}} = \boxed{3.4 \times 10^{-2} \text{ rad} = 1.9°}$.

23. The $\boxed{\text{point farthest from the center}}$ has the greatest tangential speed and $\boxed{\text{the point closest to the center}}$ has the smallest tangential speed, because the tangential speed is directly proportional to the radius.

26. $\omega = \dfrac{\Delta\theta}{\Delta t} = \dfrac{2.5(2\pi \text{ rad})}{(3.0 \text{ min})(60 \text{ s/min})} = \boxed{0.087 \text{ rad/s}}$.

30. $\omega = \dfrac{v_t}{r} = \dfrac{3.0 \text{ m/s}}{0.20 \text{ m}} = 15 \text{ rad/s}$.

    From $\omega = \dfrac{\Delta\theta}{\Delta t}$, we have $\Delta t = \dfrac{\Delta\theta}{\omega} = \dfrac{2\pi \text{ rad}}{15 \text{ rad/s}} = \boxed{0.42 \text{ s}}$.

32. (a) $\boxed{3.5 \times 10^{-2} \text{ rad/s}}$.   (b) $\boxed{17 \text{ m/s}}$.

39.   There is insufficient centripetal force (provided by friction and adhesive forces) on the water drops so the water drops fly out along a tangent and the clothes get dry.

44.   120 km/h = 33.33 m/s. $a_c = \dfrac{v^2}{r} = \dfrac{(33.33 \text{ m/s})^2}{(1.00 \times 10^3 \text{ m})} = \boxed{1.11 \text{ m/s}^2}$.

48.   (a) $v = \dfrac{d}{t} = \dfrac{2\pi r}{t} = \dfrac{2\pi(1.50 \text{ m})}{1.20 \text{ s}} = \boxed{7.85 \text{ m/s}}$.

       (b) $F_c = ma_c = m\dfrac{v^2}{r} = \dfrac{(0.250 \text{ kg})(7.85 \text{ m/s})^2}{1.50 \text{ m}} = \boxed{10.3 \text{ N}}$.

       (c) $\boxed{\text{No}}$, the string cannot be exactly horizontal. There must be

       something upward (a component of the tension) to balance the downward gravitational force.

51.   Static friction force provides centripetal force. $f_s = \mu_s N = \mu_s mg = F_c = m\dfrac{v^2}{r}$,

       so $v = \sqrt{\mu_s gr} = \sqrt{0.50(9.80 \text{ m/s}^2)(20 \text{ m})} = \boxed{9.9 \text{ m/s}}$.

52.   (a) 700 km/h = 194 m/s. At the bottom, the centripetal force is provided by

       the difference $N - mg$.   So $F_c = N - mg = m\dfrac{v^2}{r}$,

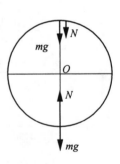

       $N = mg + m\dfrac{v^2}{r} = mg + m\dfrac{(194 \text{ m/s})^2}{2.0 \times 10^3 \text{ m}} = mg + m(18.8 \text{ m/s}^2)$

       $= mg + 1.9mg = \boxed{2.9mg}$.

       (b) At the top, the centripetal force is provided by $N + mg$.

       $N = m\dfrac{v^2}{r} - mg = m\dfrac{(194 \text{ m/s})^2}{2.0 \times 10^3 \text{ m}} - mg = m(18.8 \text{ m/s}^2) - mg = 1.93mg - mg = \boxed{0.93mg}$.

59. $\boxed{\text{No}}$, this is not possible. Any car in circular motion always has centripetal acceleration.

60.   700 rpm = 73.3 rad/s,   3000 rpm = 314 rad/s.

       $\alpha = \dfrac{\Delta\omega}{\Delta t} = \dfrac{314 \text{ rad/s} - 73.3 \text{ rad/s}}{3.0 \text{ s}} = \boxed{80 \text{ rad/s}^2}$.

62.   (a) $\boxed{4.28 \text{ rad}}$.   (b) $\boxed{2.14 \text{ ft}}$.

64. (a) Given: $\omega_0 = 0$, $\omega = \dfrac{v}{r} = \dfrac{2.20 \text{ m/s}}{17.5 \text{ m}} = 0.126 \text{ rad/s}$, $t = 15.0$ s.   Find: α.

$\alpha = \dfrac{\omega - \omega_0}{t} = \dfrac{0.126 \text{ rad/s} - 0}{15.0 \text{ s}} = \boxed{8.40 \times 10^{-3} \text{ rad/s}^2}$.

(b) After reaching the constant operating speed, $\alpha = 0$ and so $a_t = r\alpha = \boxed{0}$.

67. (a) Given: $\omega_0 = 0$, $\alpha = 4.5 \times 10^{-3} \text{ rad/s}^2$, $\theta = 1$ rev $= 2\pi$ rad.   Find: t.

from $\theta = \omega_0 t + \frac{1}{2}\alpha t^2 = 0 + \frac{1}{2}\alpha t^2$,

we have $t = \sqrt{\dfrac{2\theta}{\alpha}} = \sqrt{\dfrac{2(2\pi \text{ rad})}{4.5 \times 10^{-3} \text{ rad/s}^2}} = \boxed{53 \text{ s}}$.

(b) After half a lap, $\omega^2 = \omega_0^2 + 2\alpha\theta = 0 + 2(4.5 \times 10^{-3} \text{ rad/s}^2)(\pi \text{ rad}) = 0.0283 \text{ rad}^2/\text{s}^2$,

so $\omega = 0.168$ rad/s.

The centripetal acceleration is $a_c = r\omega^2 = (0.30 \times 10^3 \text{ m})(0.168 \text{ rad/s})^2 = 8.5 \text{ m/s}^2$,

the tangential acceleration is $a_t = r\alpha = (0.30 \times 10^3 \text{ m})(4.5 \times 10^{-3} \text{ rad/s}^2) = 1.4 \text{ m/s}^2$.

So the total acceleration is $\mathbf{a} = \boxed{(8.5 \text{ m/s}^2)\,\hat{\mathbf{r}} + (1.4 \text{ m/s}^2)\,\hat{\mathbf{t}}}$.

73. $\boxed{\text{Yes}}$, if we also know the radius of the Earth. The acceleration due to gravity near the surface of the Earth

can be written as $a_g = \dfrac{GM_E}{R_E^2}$. By simply measuring a_g, you can determine $M_E = \dfrac{a_g R_E^2}{G}$.

76. $F = \dfrac{GM_E M_M}{r_{E\text{-}M}^2} = \dfrac{(6.67 \times 10^{-11} \text{ N·m}^2/\text{kg}^2)(5.98 \times 10^{24} \text{ kg})(7.4 \times 10^{22} \text{ kg})}{(3.8 \times 10^8 \text{ m})^2} = \boxed{2.0 \times 10^{20} \text{ N}}$.

79. $F_1 = F_3 = \dfrac{Gm^2}{d^2} = \dfrac{(6.67 \times 10^{-11} \text{ N·m}^2/\text{kg}^2)(2.5 \text{ kg})^2}{(1.0 \text{ m})^2} = 4.17 \times 10^{-10} \text{ N}$,

The diagonal distance is $\sqrt{(1.0 \text{ m})^2 + (1.0 \text{ m})^2} = \sqrt{2}$ m,

so $F_3 = \dfrac{(6.67 \times 10^{-11} \text{ N·m}^2/\text{kg}^2)(2.5 \text{ kg})^2}{(\sqrt{2} \text{ m})^2} = 2.08 \times 10^{-10} \text{ N}$.

From symmetry the net force is

$F = \sqrt{(4.17 \times 10^{-10} \text{ N})^2 + (4.17 \times 10^{-10} \text{ m})^2} + 2.08 \times 10^{-10} \text{ N} = \boxed{8.0 \times 10^{-10} \text{ N, toward opposite corner}}$.

80. $a_g = \dfrac{GM_E}{(R_E + h)^2} = \dfrac{(6.67 \times 10^{-11} \text{ N·m}^2/\text{kg}^2)(5.98 \times 10^{24} \text{ kg})}{(6.38 \times 10^6 \text{ m} + 8.80 \times 10^3 \text{ m})^2} = \boxed{9.77 \text{ m/s}^2}$.

82.   When the Earth's gravitational force equals the lunar gravitational force,

$$F_E = \frac{GM_Em}{x^2} = F_M = \frac{GM_Mm}{(3.8 \times 10^8 \text{ m} - x)^2}.$$

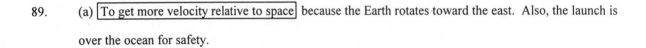

Taking the square root on both sides gives

$$\frac{\sqrt{M_E}}{x} = \frac{\sqrt{M_M}}{3.8 \times 10^8 \text{ m} - x},$$

or $\sqrt{7.4 \times 10^{22} \text{ kg}}\ x = \sqrt{5.98 \times 10^{24} \text{ kg}}\ (3.8 \times 10^8 \text{ m} - x).$

Solving, $x = \boxed{3.4 \times 10^8 \text{ m from Earth}}.$

$\boxed{\text{No}}$, there are still other gravitational forces from the other planets and the Sun.

89.   (a) $\boxed{\text{To get more velocity relative to space}}$ because the Earth rotates toward the east. Also, the launch is over the ocean for safety.

(b) The $\boxed{\text{tangential speed of the Earth is higher in Florida}}$ because Florida is closer to the equator than California. Also, California launches are polar (not eastward) for safety.

92.   For the Earth orbiting the Sun, $K = \frac{4\pi^2}{GM_S}$.

Replace the mass of the Sun with the mass of the Earth for satellites orbiting the Earth.

$$K = \frac{4\pi^2}{GM_E} = \frac{4\pi^2}{(6.67 \times 10^{-11} \text{ N·m}^2/\text{kg}^2)(5.98 \times 10^{24} \text{ kg})} = 9.90 \times 10^{-14} \text{ s}^2/\text{m}.$$

$T = 1\,\text{day} = 24(3600 \text{ s}) = 86\,400 \text{ s}$ (synchronous satellite).

From $T^2 = Kr^3 = K(R_E + h)^3$, we have

$$h = \sqrt[3]{\frac{T^2}{K}} - R_E = \sqrt[3]{\frac{(86\,400 \text{ s})^2}{9.90 \times 10^{-14} \text{ s}^2/\text{m}}} - 6.38 \times 10^6 \text{ m} = \boxed{3.6 \times 10^7 \text{ m}}.$$

96.   The Moon's gravitational attraction on the near side is greater on the water than on the Earth and produces one bulge for one tide; the attraction is greater on the Earth than on the water on the far side and so the Earth moves toward the Moon and leaves the water behind for another bulge.

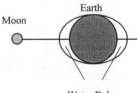

97. The centripetal force is provided by the combination of

$T - mg\cos\theta = m\dfrac{v^2}{r}$, where $\cos\theta = \dfrac{h}{l}$ and v could be found from energy

conservation. When the girl is h below the original position,

we have $\frac{1}{2}m(0)^2 + mgh = \frac{1}{2}mv^2 + mg(0)$, so $v^2 = 2gh$.

Therefore $T = mg\cos\theta + m\dfrac{v^2}{r} = \dfrac{mgh}{l} + \dfrac{m(2gh)}{l} = \dfrac{3mgh}{l}$.

(a) $h = 0$, so $T = \boxed{0}$.

(b) $h = 12\ m - 5.0\ \text{m} = 7.0\ \text{m}$, so $T = \dfrac{3(60\ \text{kg})(9.80\ \text{m/s}^2)(7.0\ \text{m})}{10\ \text{m}} = \boxed{1.2 \times 10^3\ \text{N}}$.

(c) $h = 12\ \text{m} - 2.0\ \text{m} = 10\ \text{m}$, so $T = \dfrac{3(60\ \text{kg})(9.80\ \text{m/s}^2)(10\ \text{m})}{10\ \text{m}} = \boxed{1.8 \times 10^3\ \text{N}}$.

103. For the Earth, $T = 1$ year and $r = 1$ AU.

From Kepler's third law, $T^2 = Kr^3$., we have $(1\ \text{y})^2 = K(1\ \text{AU})^3$.

So $K = \boxed{1\ \text{y}^2/\text{AU}^3}$.

VI. Practice Quiz

1. Convert 5.0 rad/s to revolutions per minute.

   (a) 0.53 rpm   (b) 16 rpm   (c) 31 rpm   (d) 48 rpm   (e) 150 rpm

2. The cutting cord on a gas-powered weed cutter is 0.16 m in length. If the motor rotates at the rate of 20 rev/s, what is the approximate tangential velocity of the end of the cord?

   (a) 20 m/s   (b) 25 m/s   (c) 35 m/s   (d) 65 m/s   (e) 630 m/s

3. A 0.300-kg mass, attached to the end of a 0.75-meter string, is whirled around in a smooth level table. If the maximum tension that the string can withstand is 250 N, then what maximum tangential speed can the mass have if the string is not to break?

   (a) 22.4 m/s   (b) 25 m/s   (c) 19.4 m/s   (d) 32.7 m/s   (e) 275 m/s

4. How many revolutions does a 0.100-meter radius wheel rotate after starting from rest and accelerating at a constant angular acceleration of 2.00 rad/s^2 over a 5.00 s interval?

   (a) 157 rev   (b) 50.0 rev   (c) 3.98 rev   (d) 12.5 rev   (e) 25.0 rev

5. The gravitation attractive force between two masses is F. If the masses are moved to twice of their initial distance, what is the gravitational attractive force?

(a) $F/4$ (b) $F/2$ (c) F (d) $2F$ (e) $4F$

6. A hypothetical planet has a mass of half that of the Earth and a radius of twice that of the Earth. What is the acceleration due to gravity on the planet in terms of g?

(a) g (b) $g/2$ (c) $g/4$ (d) $g/8$ (e) $g/16$

7. A centrifuge, rotating at 1.2×10^3 rad/s, slows to a stop while turning through 2.0×10^3 revolutions. What is the magnitude of the angular acceleration of the centrifuge?

(a) 0.30 rad/s^2 (b) 1.7 rad/s^2 (c) 57 rad/s^2 (d) 1.1×10^2 rad/s^2 (e) 3.6×10^2 rad/s^2

8. An object moves with a constant speed of 20 m/s on a circular track of radius 100 m. What is the magnitude of the centripetal acceleration of the object?

(a) zero   (b) 0.20 rad/s^2 (c) 0.40 m/s^2 (d) 2.0 m/s^2 (e) 4.0 rad/s^2

9. The minute hand in a clock has a length of 0.30 m. What distance does the tip of the minute hand sweep through in 15 minutes?

(a) 0.12 m   (b) 0.47 m   (c) 0.36 m   (d) 0.24 m   (e) 27 m

10. At what angular speed must a circular, rotating spacestation ($r = 1000$ m) rotate to produce an artificial gravity of 9.80 m/s^2?

(a) 9.8×10^{-3} rad/s   (b) 9.9×10^{-2} rad/s   (c) 0.98 rad/s   (d) 10 rad/s   (e) 1.0×10^2 rad/s

Answers to Practice Quiz:

1. d 2. a 3. b 4. c 5. a 6. d 7. c 8. c 9. b 10. b

CHAPTER 8

Rotational Motion and Equilibrium

I. Chapter Objectives

Upon completion of this chapter, you should be able to:

1. distinguish between pure translational and pure rotational motions of a rigid body, and state the condition(s) for rolling without slipping.

2. define torque, apply the conditions for mechanical equilibrium, and describe the relationship between the location of the center of gravity and stability.

3. describe the moment of inertia of a rigid body and apply the rotational form of Newton's second law to physical situations.

4. discuss, explain, and use the rotational forms of work, kinetic energy, and power.

5. define angular momentum and apply the conservation of angular momentum to physical situations.

II. Key Terms

Upon completion of this chapter, you should be able to define and/or explain the following key terms:

rigid body	center of gravity
translational motion	stable equilibrium
rotational motion	unstable equilibrium
instantaneous axis of rotation	moment of inertia
moment (lever arm)	parallel axis theorem
torque	rotational work
translational equilibrium	rotational power
concurrent forces	rotational kinetic energy
rotational equilibrium	angular momentum
mechanical equilibrium	conservation of angular momentum
static equilibrium	

The definitions and/or explanations of the most important key terms can be found in the following section:
III. Chapter Summary and Discussion.

III. Chapter Summary and Discussion

1. Rigid Bodies, Translation, and Rotation (Section 8.1)

A **rigid body** is an object or system of particles in which the interparticle distances (distances between particles) are fixed and remain constant. In pure **translational motion**, every particle of an object or system of particles has the same instantaneous velocity, which means there is no rotation. In pure **rotational motion**, all particles of an object or system of particles have the same instantaneous angular velocity.

Generally, rigid body motion is a combination of translational and rotational motions. If an object is rolling without slipping, the following conditions must be satisfied:

$$s = r\theta, \quad \text{or} \quad v_{CM} = r\omega, \quad \text{or} \quad a_{CM} = r\alpha,$$ where the subscripts CM stand for center of mass.

Note: When $v_{CM} = r\omega$, the object is rolling without slipping. If $v_{CM} > r\omega$, the object is sliding. The object rolls with slipping when $v_{CM} < r\omega$.

2. Torque, Equilibrium, and Stability (Section 8.2)

The magnitude of the **torque** (τ) produced by a force about an axis of rotation is given by $\tau = r_\perp F$. The distance $r_\perp = r\sin\theta$ is called the **moment arm** or **lever arm**, which is the *perpendicular* distance from the axis of rotation to the line of the action of the force. The SI unit of torque is m·N (the same as for work N·m or J, but which we write in reverse for distinction). By convention, counterclockwise torque is positive and clockwise torque is negative.

Note: The moment arm is the *perpendicular* distance from the axis of rotation to the *line of the action* of the force, which is usually not the same as simply the distance from the axis to the force.

An object is in **translational equilibrium** if the net force on it is zero, and an object is in **rotational equilibrium** if the net torque is zero. If an object is in both translational equilibrium and rotational equilibrium, it is said to be in **mechanical equilibrium**. So the conditions for an object in mechanical equilibrium are $\Sigma \mathbf{F}_i = 0$ and $\Sigma \tau_i = 0$. A special case of mechanical equilibrium is **static equilibrium** in which objects are and remain at rest.

Note: When applying $\Sigma \tau_i = 0$ to rotational or static equilibrium situations, you can choose anywhere on the object as the axis of rotation because the object is not rotating about any point. However, some choices of axes of rotation can greatly simplify the solutions of such problems. Generally, the axis of rotation should be chosen through a point through which some unknown force(s) act so to make the moment of arm(s) of the force(s) zero, therefore the torque(s) zero.

Example 8.1   The bolts on a car wheel require tightening to a torque of 90 m·N. If a 20-centimeter long wrench is used, what is the magnitude of the force required

(a) when the force is perpendicular to the wrench,

(b) when the force is $\theta = 35°$ to the wrench as shown.

Solution: Given: $r = 0.20$ m, $\tau = 90$ m·N,   (a) $\theta = 90°$,   (b) $\theta = 35°$.

Find: F in (a) and (b).

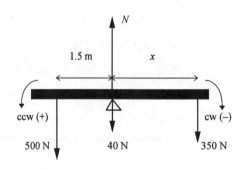

(a) When the force is applied at a 90° angle as shown, the lever arm is simply equal to the length of the wrench.

$r_\perp = r \sin\theta = (0.20 \text{ m}) \sin 90° = 0.20$ m.

From $\tau = r_\perp F$, we have $F = \dfrac{\tau}{r_\perp} = \dfrac{90 \text{ m·N}}{0.20 \text{ m}} = 4.5 \times 10^2$ N.

(b) When the force is applied at a non-90° angle, the lever arm *is not* equal to the length of the wrench. The perpendicular distance from the axis to the line of the action of the force is

$r_\perp = (0.20 \text{ m}) \sin 35° = 0.115$ m. $F = \dfrac{90 \text{ m·N}}{0.115 \text{ m}} = 7.8 \times 10^2$ N.

Why does it need more force in (b) than in (a)?

Example 8.2   A uniform board of weight 40 N supports two children weighing 500 N and 350 N, respectively. If the support is at the center of the board and the 500-newton child is 1.5 m from the center, what is the position of the 350-newton child?

Solution:

Since the board is uniform, its center of mass and center of gravity are at the center of the board. We choose the support as the axis of rotation since there is an unknown normal force N by the support on the board (although we can find it easily with $\Sigma \mathbf{F_i} = 0$ but this way we can save a step). Once an axis of rotation is chosen, we can clearly see, from the free-body diagram of the board, that the weight of the 500-newton child produces a ccw (counterclockwise) torque, the weight of the 350-newton child causes a cw (clockwise) torque, and the weight of the board and the normal force have no contribution to the total torque because their moment arms are zero (force through the axis).

Note: Whether a torque is ccw or cw depends on the location of the axis of rotation. For example, if the axis is to the left of the 500-newton force it generates a cw torque. So make sure you chose an axis before you apply the condition for rotational equilibrium.

From the condition for rotational equilibrium $\Sigma \tau_i = 0$,

we have $\Sigma \tau_i = (500 \text{ N})(1.5 \text{ m}) - (350 \text{ N}) x + N(0) = 0.$ Solving, $x = 2.1$ m.

Example 8.3   A 10-meter long uniform beam of weight 100 N is supported by two ropes at the ends as shown. If a 400-newton person sits at 2.0 m from one end of the beam, what are the tensions in the ropes?

Solution:

Since the beam is uniform, its center of mass and center of gravity are at its geometrical center, which is 5.0 m from one end. We first draw the free-body diagram of the beam. Where should we choose the axis of rotation? What if we chose the center of the beam as the axis? If we do so, we will have two unknowns (T_1 and T_2) in our $\Sigma \tau_i$ = 0 equation. We cannot solve for two unknowns within just one equation. (Although we can get another equation from $\Sigma F_i = 0$. Then we have to solve simultaneous equations.) We can simplify the solution of this problem by choosing our axis at either end (either at T_1 or T_2). Let's try T_2 first, i.e., we chose the axis of rotation at the right end of the beam.

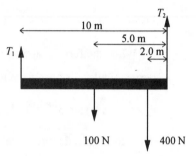

About the right end, T_1 has a cw torque, the 100-newton and 400-newton forces have ccw torques, and T_2 has zero torque since its lever arm is zero.

From the condition for rotational equilibrium, $\Sigma \tau_i = 0$, we have

$\Sigma \tau_i = -T_1 (10 \text{ m}) + (100 \text{ N})(5.0 \text{ m}) + (400 \text{ N})(2.0 \text{ m}) + T_2 (0) = 0.$ Solving, $T_1 = 130$ N.

To find T_2, we can choose the left end as the axis of rotation and repeat our calculation. Or we can use the condition for translational equilibrium, $\Sigma F_i = 0$.

$\Sigma F_i = T_1 + T_2 - 100 \text{ N} - 400 \text{ N} = 0.$ So $T_2 = 500 \text{ N} - T_1 = 500 \text{ N} - 130 \text{ N} = 370$ N.

If an object is in **stable equilibrium**, any small displacement results in a restoring force or torque, which tends to return the object to its original equilibrium. For an object in **unstable equilibrium**, any small displacement from equilibrium results in a force or torque that tends to take the object farther away from its equilibrium position. An object will be in stable equilibrium as long as its center of gravity lies above and inside its original base of support., i.e., line of action of the weight force of the center of gravity intersects the original base of support.

3.   Rotational Dynamics (Section 8.3)

The torque on a particle due to a constant force is $\tau = mr^2\alpha$. Similarly for a rotating body about a fixed axis, $\tau = \Sigma (m_i r_i^2)\alpha = I\alpha$, where $I = \Sigma m_i r_i^2$ is called the **moment of inertia** (rotational analog of mass). For a single particle, the moments of inertia is simply equal to $I = mr^2$, where r is the distance from the mass to the axis. However, for many continuous objects (objects treated as an infinite number of particles, each with a very small mass) the moments of inertia have to be determined with calculus. The results are listed in Figure 8.16 on page 269 in the textbook.

The moment of inertia about an axis parallel to an axis through its center of mass is given by the **parallel axis theorem**: $I = I_{CM} + Md^2$, where I_{CM} is the moment of inertia about an axis through the center of mass, M is the mass of the object, and d is the distance between the parallel axes.

In Chapter 7, we discussed the fact that the kinematic equations for angular motion are identical to those for linear motion if the linear quantities are replaced by their respective angular counterparts (x by θ, v by ω, and a by α). In dynamics, the rotational counterparts of force and mass are torque and moment of inertia. Therefore we can write the **rotational form of Newton's second law** as $\tau = I\alpha$ (for linear motion, it is $\mathbf{F} = m\mathbf{a}$).

Example 8.4   The moment of inertia of a solid disk rotating about an axis through its center is $\frac{1}{2}MR^2$, where M is the mass of the disk and R is the radius of the disk. What is the moment of inertia of the disk about an axis through the contact point when the disk is rolling on a horizontal surface?

Solution:

When the disk is rolling, the axis of rotation is though the point where the disk makes contact with the horizontal surface. The distance from this axis to an axis through the center of the disk is R, the radius of the disk. From the parallel axis theorem, we have $I = I_{CM} + Md^2 = \frac{1}{2}MR^2 + MR^2 = \frac{3}{2}MR^2$.

Example 8.5   A solid cylinder of mass 10 kg is pivoted about a frictionless axis through its center O. A rope wrapped around the outer radius $R_1 = 1.0$ m, exerts a force of $F_1 = 5.0$ N to the right. A second rope wrapped around another section of radius $R_2 = 0.50$ m exerts a force of $F_2 = 6.0$ N downward.

(a) What is the angular acceleration of the disk?

(b) If the disk starts from rest, how many radians does it rotate through in the first 5.0 s?

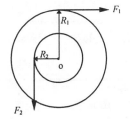

Solution: Given: $M = 10$ kg, $F_1 = 5.0$ N, $R_1 = 1.0$ m, $F_2 = 6.0$ N, $R_2 = 0.50$ m, $t = 5.0$ s.

Find: (a) α (b) θ.

(a) About an axis through O, the torque by F_1 is cw and the torque by F_2 is ccw. From Figure 8.16 in the textbook, the moment of inertia of a disk is given by $I = \frac{1}{2} MR_2^2 = \frac{1}{2}(10 \text{ kg})(1.0 \text{ m})^2 = 5.0$ kg·m².

The net torque on the disk is $\tau_{net} = -F_1 R_1 + F_2 R_2 = -(5.0 \text{ N})(1.0 \text{ m}) + (6.0 \text{ N})(0.50 \text{ m}) = -2.0$ m·N.

From Newton's second law, $\tau_{net} = I\alpha$, we have $\alpha = \dfrac{\tau_{net}}{I} = \dfrac{-2.0 \text{ m·N}}{5.0 \text{ kg·m}^2} = -0.40$ rad/s². The negative sign indicates that the disk will accelerate clockwise.

(b) From rotational kinematics, $\theta = \omega_0 t + \frac{1}{2} \alpha t^2 = 0 + \frac{1}{2}(-0.40 \text{ rad/s}^2)(5.0 \text{ s})^2 = -5.0$ rad.

Again, the negative sign indicates that the angular displacement is counterclockwise.

---

4.   Rotational Work and Kinetic Energy (Section 8.4)

As was discussed earlier, the rotational counterparts of the linear quantities x, v, a, F, and m are θ, ω, α, τ, and I. The equations for work and energy in linear motion are still valid if we replace the linear quantities by their respective rotational counterparts. The following are the equations for rotational work and energy with the original linear equations being given in parentheses (a detailed comparison can be found in Table 8.1 on page 275 in the textbook).

Work:	$W = \tau\theta$	$(W = Fd = Fx)$
Kinetic Energy:	$K = \frac{1}{2} I\omega^2$	$(K = \frac{1}{2} mv^2)$
Power:	$P = \tau\omega$	$(P = Fv)$
Work-Energy Theorem:	$W = \Delta K = \frac{1}{2} I\omega^2 - \frac{1}{2} I\omega_0^2$	$(W = \Delta K = \frac{1}{2} mv^2 - \frac{1}{2} mv_0^2)$

The kinetic energy of a **rolling body** (without slipping) relative to an axis through the contact point is the sum of the rotational kinetic energy about an axis through the center of mass and the translational kinetic energy of the center of mass and, i.e., $K = \frac{1}{2} I_{CM}\, \omega^2 + \frac{1}{2} Mv_{CM}^2$.

Example 8.6   A solid sphere of mass 1.0 kg and radius 0.010 m rolls with a speed of 10 m/s. How high up an inclined plane can it roll before coming to rest?

Solution: Given: $M = 1.0$ kg, $R = 0.010$ m, $v_{CM} = 10$ m/s.

Find: h.

We use the conservation of mechanical energy to solve this problem. First we need to choose the bottom of the inclined plane as the reference level for zero gravitational potential energy ($U_0 = 0$ at $h_0 = 0$). The mechanical energy at the bottom should be equal to the mechanical energy at the top. When the sphere is at the bottom of the plane, its potential energy is zero, and it has both translational and rotational kinetic energies. When the sphere reaches the maximum height, both its linear and angular speeds are zero so it has only potential energy. We also need to use $v_{CM} = R\omega$ and $I_{CM} = \frac{2}{5}MR^2$ in our solution.

At the bottom the mechanical energy is $\frac{1}{2}I_{CM}\omega^2 + \frac{1}{2}Mv_{CM}^2 = \frac{1}{2}\left(\frac{2}{5}MR^2\right)\frac{v_{CM}^2}{R^2} + \frac{1}{2}Mv_{CM}^2 = \frac{7}{10}Mv_{CM}^2$.

At the highest point on the incline, the mechanical energy is Mgh.

From the conservation of mechanical energy, we have $\frac{7}{10}Mv_{CM}^2 = Mgh$.

Solving, $h = \frac{7v_{CM}^2}{10\,g} = \frac{7(10 \text{ m/s})^2}{10(9.80 \text{ m/s}^2)} = 7.1$ m.

Example 8.7   A cylindrical hoop of mass 10 kg and radius 0.20 m is accelerated by a motor from rest to an angular speed of 20 rad/s during a 0.40 s interval.

(a) How much work is required?

(b) What is the power output of the motor?

Solution: Given: $M = 10$ kg, $R = 0.20$ m, $\omega_o = 0$, $\omega = 20$ rad/s, $t = 0.40$ s.

Find: (a) W (b) P.

(a) From Fig. 8.16 in the textbook, the moment of inertia of a cylindrical hoop is given by $I = MR^2 = (10 \text{ kg})(0.20 \text{ m})^2 = 0.40$ kg·m^2. From the work-energy theorem, we have

$W = \Delta K = \frac{1}{2}I\omega^2 - \frac{1}{2}I\omega_o^2 = \frac{1}{2}I\omega^2 = \frac{1}{2}(0.40 \text{ kg·m}^2)(20 \text{ rad/s})^2 = 80$ J.

(b) $P = \dfrac{W}{t} = \dfrac{80 \text{ J}}{0.40 \text{ s}} = 200$ W.

---

5.   Angular Momentum (Section 8.5)

The definition of angular momentum is analogous to linear momentum. It is defined as the product of moment of inertia and angular velocity. $\mathbf{L} = I\omega$ (it is $p = mv$ for linear momentum). The SI unit of angular momentum is kg·m^2/s. The vector direction of angular momentum is determined by the right hand rule as applied to angular velocity in Chapter 7. The rotational form of Newton's second law can also be written in terms of angular momentum: $\tau = \dfrac{\Delta \mathbf{L}}{\Delta t}$.

In the absence of an external, unbalanced torque, the total angular momentum of a system is conserved (remains constant). Mathematically, this can be written as $\mathbf{L} = \mathbf{L}_o$ or $I\omega = I_o\,\omega_o$.

Angular momentum has many practical applications in our everyday lives such as the gyroscope used in navigation, the twin-rotor helicopters, and even the tight spiral of a football.

Example 8.8   A figure skater rotating at 4.00 rad/s with arms extended has a moment of inertia of 2.25 kg·m². If she pulls her arms in so the moment of inertia decreases to 1.80 kg·m², what is the magnitude of her final angular speed?

Solution: Given: $I_0 = 2.25$ kg·m² $I = 1.80$ kg·m²

$\omega_0 = 4.00$ rad/s

Find: ω.

From the conservation of angular momentum, $I\omega = I_0\,\omega_0$, we have

$$\omega = \frac{I_0\,\omega_0}{I} = \frac{(2.25 \text{ kg·m}^2)(4.00 \text{ rad/s})}{1.80 \text{ kg·m}^2} = 5.00 \text{ rad/s}.$$

The figure skater rotates faster after pulling the arms in.

Why does the moment of inertia decrease when the arms are pulled in?

IV.   Mathematical Summary

Condition for Rolling without Slipping	$v_{CM} = r\omega$ (or $s = r\theta$ or $a_{CM} = r\alpha$)	(8.1)	Defines the condition for rolling without slipping.
Torque (magnitude)	$\tau = r_\perp F = rF\sin\theta$	(8.2)	Defines the magnitude of torque.
Conditions for Equilibrium	$\Sigma \mathbf{F}_i = 0$ and $\Sigma \tau_i = 0$	(8.3)	Define the conditions for translational, rotational, and mechanical equilibria.
Torque on a particle (magnitude)	$\tau = mr^2\alpha$	(8.4)	Calculates the magnitude of torque on a particle.
Moment of Inertia	$I = \Sigma m_i r_i^2$	(8.6)	Defines the moment of inertia.
Newton's Second Law (magnitude)	$\tau = I\alpha$	(8.7)	Gives the rotational form of Newton's second law.
Parallel Axis Theorem	$I = I_{CM} + Md^2$	(8.8)	Calculates the moment of inertia about an axis parallel to one through the center of mass.
Rotational Work	$W = \tau\theta$	(8.9)	Defines rotational work.
Rotational Power	$P = \tau\omega$	(8.10)	Defines rotational power.
Work-Energy Theorem	$W = \Delta K = \frac{1}{2}I\omega^2 - \frac{1}{2}I\omega_0^2$	(8.11)	Relates rotational work with change in rotational kinetic energy.

Rotational Kinetic Energy	$K = \frac{1}{2}I\omega^2$ (8.12)	Defines rotational kinetic energy.
Kinetic Energy of a Rolling Object	$K = \frac{1}{2}I_{CM}\,\omega^2 + \frac{1}{2}Mv_{CM}^2$ (8.13)	Calculates the total kinetic energy of a rolling object.
Angular Momentum of a Particle (magnitude)	$L = mr_\perp^2\,\omega$ (8.14)	Defines the angular momentum of a particle in circular motion.
Angular Momentum of a Rigid Body	$\mathbf{L} = I\omega$ (8.16)	Defines the angular momentum of a rigid body.
Torque in Terms of Angular Momentum	$\tau = \dfrac{\Delta \mathbf{L}}{\Delta t}$ (8.17)	Expresses Newton's second law in terms of angular momentum.
Conservation of Angular Momentum	$I\omega = I_o\,\omega_o$ (8.18)	Gives the law of the conservation of angular momentum when the net external torque is zero.

V. Solutions of Selected Exercises and Paired/Trio Exercises

6. $s = r\theta = (0.065\text{ m})(4\text{ rev})(2\pi\text{ rad/rev}) = \boxed{1.6\text{ m}}$.

10.   From $v_{CM} = r\omega$, we have $\omega = \dfrac{v_{CM}}{r} = \dfrac{0.25\text{ m/s}}{0.15\text{ m}} = \boxed{1.7\text{ rad/s}}$.

13. $\alpha = \dfrac{a_t}{r} = \dfrac{0.018\text{ m/s}^2}{0.10\text{ m}} = 0.18\text{ rad/s}^2$. From $\omega^2 = \omega_o^2 + 2\alpha\theta$,

     we have $\theta = \dfrac{\omega^2 - \omega_o^2}{2\alpha} = \dfrac{(1.25\text{ rad/s})^2 - (0.50\text{ rad/s})^2}{2(0.18\text{ rad/s}^2)} = 36.5\text{ rad} = \boxed{0.58\text{ rotations}}$.

18.   When the force is applied perpendicular to the length of the wrench, minimum force is required and the lever arm equals the length of the wrench. At $\theta = 90°$, $r_\perp = 0.15\text{ m}$.

     From $\tau = Fr_\perp$, we have $F = \dfrac{\tau}{r_\perp} = \dfrac{25\text{ m·N}}{0.15\text{ m}} = \boxed{1.7 \times 10^2\text{ N}}$.

19.   In this case, the lever arm is *not* equal to the length of the wrench.

     $r_\perp = (0.15\text{ m})\sin 30°$.

     $F = \dfrac{\tau}{r_\perp} = \dfrac{25\text{ m·N}}{(0.15\text{ m})\sin 30°} = \boxed{3.3 \times 10^2\text{ N}}$.

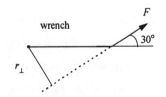

20. $\boxed{5.6 \times 10^2\text{ N}}$.

24.   (a) Apply $\Sigma\tau = 0$,   we have $(0.100 \text{ kg})g(0.500 \text{ m} - 0.250 \text{ m}) - (0.0750 \text{ kg})g(x - 0.500 \text{ m}) = 0$.

Solving, $x = 0.833 \text{ m} = \boxed{83.3 \text{ cm}}$.

(b) $(0.100 \text{ kg})g(0.500 \text{ m} - 0.250 \text{ m}) - m(0.900 \text{ m} - 0.500 \text{ m}) = 0$,

so $0.0625 \text{ kg} = \boxed{62.5 \text{ g}}$.

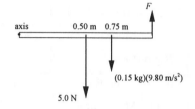

29.   Apply $\Sigma\tau = 0$,   we have

$-(5.0 \text{ N})(0.50 \text{ m}) - (0.15 \text{ kg})(9.80 \text{ m/s}^2)(0.75 \text{ m}) + F(1.0 \text{ m}) = 0$.

Solving, $F = \boxed{3.6 \text{ N}}$.

32.   (a) The center of gravity (CG) of the first book is at the center. So for the last book not to fall, its CG can not displace more than 12.5 cm relative to the CG of the first book (within the base). The CG of each successive book on the top is moved 1.5 cm relative to that of the one below. Therefore the number of books which can be on top of the first one is $\dfrac{12.5 \text{ cm}}{1.5 \text{ cm}} = 8.33$.

Thus we can stack a total of $1 + 8 = \boxed{9}$ books, including the first one.

(b) The total height is $9(5.0) \text{ cm} = 45 \text{ cm}$. So the CM is at $\dfrac{45 \text{ cm}}{2} = \boxed{22.5 \text{ cm}}$.

37.   (a) Assume the distance from the center of gravity (CG) to the point where the wheel touches the ground is d. We choose the axis of rotation through the CG.   Apply $\Sigma\tau = 0$, we have

$f_s d \cos\theta - Nd \sin\theta = 0$,   or $f_s d \cos\theta = Nd \sin\theta$,   so $\tan\theta = f_s / N$.

(b) $f_s = \mu_s N = N \tan\theta$,   or $\mu_s = \tan\theta = \tan 11° = \boxed{0.19}$.

(c) $\Sigma F_y = N - mg = 0$,   or $N = mg$.

So $f_s = \mu_s N = \mu_s mg = F_c = m\dfrac{v^2}{r}$,   and $v = \sqrt{\mu_s gr} = \sqrt{0.19(9.80 \text{ m/s}^2)(6.5 \text{ m})} = \boxed{3.5 \text{ m/s}}$.

43.   This is the rotational analog of "pulling the table cloth" in Exercise 4.9. It takes a certain amount of torque to accelerate the paper tower and the paper can only exert certain amount of force, and therefore torque. When the paper is pulled quickly (a great force is required to accelerate the roll), the force the paper can provide is not great enough to accelerate the paper roll. However, if the paper is pulled slowly, the paper is strong enough to accelerate the roll because the force required is smaller. The amount of paper on the roll affects the results. The more paper the roll has, the greater the moment of inertia, the greater the force required to accelerate the roll, and therefore the easier to tear.

46. From Fig. 8.16 on page 269 in the textbook, $I = \frac{1}{2}mr^2 = \frac{1}{2}(0.15 \text{ kg})(0.075 \text{ m})^2 = 4.219 \times 10^{-4} \text{ kg·m}^2$.

From Newton's second law $\tau = I\alpha$, we have $\alpha = \dfrac{\tau}{I} = \dfrac{6.4 \text{ m·N}}{4.219 \times 10^{-4} \text{ kg·m}^2} = \boxed{3.4 \times 10^4 \text{ rad/s}^2}$.

50. We first calculate the angular acceleration from kinematics.

$$\alpha = \frac{\Delta\omega}{\Delta t} = \frac{2.0 \text{ rad/s} - 0}{12 \text{ s}} = 0.167 \text{ rad/s}^2.$$

Then $\tau = I\alpha = \frac{1}{2}mr^2\alpha = \frac{1}{2}(2000 \text{ kg})(30 \text{ m})^2 (0.167 \text{ rad/s}^2) = \boxed{1.5 \times 10^5 \text{ m·N}}$.

52. $\boxed{0.45 \text{ m·N}}$.

55. Apply Newton's second law and note $a = r\alpha$.

m_2: $m_2 g - T_2 = m_2 a$,   Eq. (1)

pulley: $T_2 R - T_1 R - \tau_f = I\alpha = \frac{1}{2}MR^2\alpha = \frac{1}{2}MRa$,

       or $T_2 - T_1 - \dfrac{\tau_f}{R} = \frac{1}{2}Ma$,   Eq. (2)

m_1: $T_1 - m_1 g = m_1 a$.   Eq. (3)

Eq. (1) + Eq. (2) + Eq. (3) gives $(m_2 - m_1)g - \dfrac{\tau_f}{R} = (m_1 + m_2 + 0.5M)a$,

so $a = \dfrac{(m_2 - m_1)g - \dfrac{\tau_f}{R}}{m_1 + m_2 + 0.5M} = \dfrac{(0.80 \text{ kg} - 0.40 \text{ kg})(9.80 \text{ m/s}^2) - \dfrac{0.35 \text{ m·N}}{0.15 \text{ m}}}{0.40 \text{ kg} + 0.80 \text{ kg} + 0.5(0.20 \text{ kg})} = \boxed{1.2 \text{ m/s}^2}$.

59. (a) From Fig. 8.16 on page 269 in the textbook, the moment of inertia of a meterstick about its end is $I = \frac{1}{3}ML^2$. The torque is generated by the weight of the stick through its center of mass.

So $\tau = RF = \frac{1}{2}LMg = I\alpha = \frac{1}{3}ML^2\alpha$,   or $\alpha = \dfrac{3g}{2L}$.

Therefore $a = r\alpha = L\dfrac{3g}{2L} = \boxed{1.5g}$.

(b) $a = g = r\dfrac{3g}{2L}$,   or $r = \dfrac{2L}{3} = 0.67 \text{ m} = \boxed{67 \text{ cm position}}$.

62. The hoop rotates about an instantaneous axis of rotation through the point of contact (point O). From the parallel-axis theorem, the moment of inertia about this axis is

$I = I_{\text{CM}} + Md^2 = MR^2 + MR^2 = 2MR^2$.

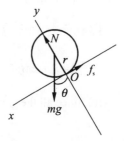

The torque by gravity is $\tau = MgR\sin\theta$.

So $a = R\alpha = R\dfrac{\tau}{I} = R\dfrac{MgR\sin\theta}{2MR^2} = \dfrac{g\sin\theta}{2}$

$$= \dfrac{(9.80 \text{ m/s}^2)\sin 15°}{2} = \boxed{1.3 \text{ m/s}^2}.$$

68. . From the work energy theorem:

$$W = \tau\theta = \tfrac{1}{2}I\omega^2 - \tfrac{1}{2}I\omega_0^2 = \tfrac{1}{2}I\omega^2 - \tfrac{1}{2}I(0)^2 = \tfrac{1}{2}I\omega^2 = \tfrac{1}{2}\tfrac{1}{2}MR^2\omega^2 = \tfrac{1}{4}MR^2\omega^2.$$

So $\omega = \sqrt{\dfrac{4\tau\theta}{MR^2}} = \sqrt{\dfrac{4(10 \text{ m·N})(2.0)(2\pi \text{ rad})}{(10 \text{ kg})(0.20 \text{ m})^2}} = \boxed{35 \text{ rad/s}}.$

72. The center of mass (CM) lowers by an amount of $h = 0.50$ m.

From energy conservation: $\tfrac{1}{2}I(0)^2 + mgh = \tfrac{1}{2}I\omega^2 + mg(0)$,

so $\omega = \sqrt{\dfrac{2mgh}{I}} = \sqrt{\dfrac{2mgh}{\tfrac{1}{3}mL^2}} = \sqrt{\dfrac{6gh}{L^2}} = \dfrac{6(9.80 \text{ m/s}^2)(0.50 \text{ m})}{(1.0 \text{ m})^2} = \boxed{5.4 \text{ rad/s}}.$

74. $\boxed{3.4 \text{ m/s}}$.

76. $K = \tfrac{1}{2}mv^2 + \tfrac{1}{2}I\omega^2 = \tfrac{1}{2}mv^2 + \tfrac{1}{2}I\dfrac{v^2}{R^2}$. So $K_h = \tfrac{1}{2}mv_h^2 + \tfrac{1}{2}mR^2\dfrac{v_h^2}{R^2} = mv_h^2$,

$K_c = \tfrac{1}{2}mv_c^2 + \tfrac{1}{2}(\tfrac{1}{2}mR^2)\dfrac{v_c^2}{R^2} = \tfrac{3}{4}mv_c^2$, and $K_s = \tfrac{1}{2}mv_s^2 + \tfrac{1}{2}(\tfrac{2}{5}mR^2)\dfrac{v_s^2}{R^2} = \tfrac{7}{10}mv_s^2$.

Since they are all released from the same height, the K's are the same. That means the v is the greatest for the sphere and smallest for the hoop. Therefore the sphere gets to the bottom first and the hoop last.

$v_s = \sqrt{\dfrac{10K}{7m}}$, $v_c = \sqrt{\dfrac{4K}{3m}}$, and $v_h = \sqrt{\dfrac{K}{m}}$.

79. (a) The centripetal force is provided solely by gravity at the minimum speed. In this case the centripetal

force at the top of the track is just equal to the weight of the ball. $Mg = F_c = M\dfrac{v^2}{R}$, so $v = \boxed{\sqrt{gR}}$.

(b) From energy conservation:

$$\tfrac{1}{2}M(0)^2 + \tfrac{1}{2}I(0)^2 + Mgh = \tfrac{1}{2}Mv^2 + \tfrac{1}{2}I\omega^2 + Mg(2R) = \tfrac{1}{2}Mv^2 + \tfrac{1}{2}\tfrac{2}{5}MR^2\dfrac{v^2}{R^2} + Mg(2R)$$

$$= \tfrac{7}{10}MgR + 2MgR = \tfrac{27}{10}MgR, \quad \text{so} \quad h = \boxed{2.7R}.$$

(c) Since all the gravity is "used up" as centripetal force, the rider feels $\boxed{\text{weightless}}$.

83.   The arms and legs are put onto these positions to decrease the moment of inertia. This decrease in moment of inertia increases the rotational speed.

88.   Orbital: The Earth can be considered as a particle to the sun.

$$\omega_o = \frac{2\pi \text{ rad}}{(365)(24)(3600 \text{ s})} = 1.99 \times 10^{-7} \text{ rad/s}, \quad \text{and} \quad I_o = MR^2,$$

so $L_o = I_o \, \omega_o = (6.0 \times 10^{24} \text{ kg})(1.5 \times 10^{11} \text{ m})^2(1.99 \times 10^{-7} \text{ rad/s}) = 2.69 \times 10^{40} \text{ kg·m}^2\text{/s}.$

Spin: The Earth has to be considered as a sphere.

$$\omega_s = \frac{2\pi \text{ rad}}{(24)(3600 \text{ s})} = 7.27 \times 10^{-5} \text{ rad/s}, \quad \text{and} \quad I_s = \tfrac{2}{5}Mr^2,$$

so $L_s = \tfrac{2}{5}(6.0 \times 10^{24} \text{ kg})(6.4 \times 10^6 \text{ m})^2 (7.27 \times 10^{-5} \text{ rad/s}) = 7.15 \times 10^{33} \text{ kg·m}^2\text{/s}.$

Therefore $\dfrac{L_o}{L_s} = \dfrac{2.69 \times 10^{40}}{7.15 \times 10^{33}} = \boxed{3.8 \times 10^6}.$

They are not in the same direction because the Earth's axis is tilted.

92.   From angular momentum conservation: $I\omega = I_o \, \omega_o,$ we have

$$\omega = \frac{I_o \, \omega_o}{I} = \frac{(100 \text{ kg·m}^2)(2.0 \text{ rps})}{75 \text{ kg·m}^2} = \boxed{2.7 \text{ rps}}.$$

96.   We chose the axis of rotation at the support on the right. Apply $\Sigma \tau = 0$, we have $Mg(5.0 \text{ m}) + mg(7.0 \text{ m}) - F_L(10 \text{ m}) = 0,$ so

$(10\,000 \text{ kg})(9.80 \text{ m/s}^2)(5.0 \text{ m}) + (2000 \text{ kg})(9.80 \text{ m/s}^2)(7.0 \text{ m}) = F_L(10 \text{ m})$

Solving, $F_L = 6.27 \times 10^4 \text{ N} = \boxed{6.3 \times 10^4 \text{ N}}.$

Apply $\Sigma F_y = F_L + F_R - mg - Mg = 0$, we have

$F_R = (m + M)g - F_L = (10\,000 \text{ kg} + 2000 \text{ kg})(9.80 \text{ m/s}^2) - 6.27 \times 10^4 \text{ N}$

$= \boxed{5.5 \times 10^4 \text{ N}}.$

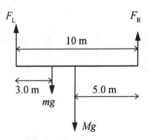

99.   (a) For the mass: $\Sigma F = mg - T = ma$ Eq. (1)

For the pulley: $\Sigma \tau = rT = I\alpha = \tfrac{1}{2}Mr^2 \dfrac{a}{r} = \tfrac{1}{2}Mra,$ or $T = \tfrac{1}{2}Ma$ Eq. (2)

Solving for $a = \dfrac{mg}{m + \tfrac{1}{2}M} = \dfrac{(5.0 \text{ kg})(9.80 \text{ m/s}^2)}{5.0 \text{ kg} + \tfrac{1}{2}(10 \text{ kg})} = \boxed{4.9 \text{ m/s}^2}.$

(b) $\alpha = \dfrac{a}{r} = \dfrac{4.9 \text{ m/s}^2}{0.50 \text{ m}} = \boxed{9.8 \text{ rad/s}^2}.$

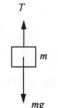

103.   (a) The first brick is not displaced.

So for each brick, the maximum displacement is $\dfrac{20 \text{ cm}}{8} = \boxed{2.5 \text{ cm}}$.

(b) The height of the center of mass is $\dfrac{9(8.0 \text{ cm})}{2} = \boxed{36 \text{ cm}}$.

108.   On the verge of the unstable equilibrium, the weight goes through the lower

corner of the trailer.   So $\tan\theta = \dfrac{L/2}{h} = \dfrac{3.66 \text{ m}}{2(3.58 \text{ m})} = 0.511$,

therefore $\theta = \boxed{27°}$.

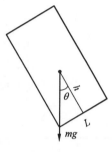

VI.   Practice Quiz

1.   A meter stick is supported by a knife edge at the 50 cm mark and has masses of 0.40 kg and 0.60 kg

hanging at the 20 cm and 80 cm marks, respectively. Where (at what mark) should a third mass of 0.30 kg

be hung to keep the stick balanced?

(a) 20 cm   (b) 25 cm   (c) 30 cm   (d) 50 cm   (e) 70 cm

2.   A constant torque of 20 m·N is applied to a flywheel. If the wheel starts from rest and has moment of

inertia of 12 kg·m^2,   what is its angular speed after it rotated through 5.0 revolutions?

(a) 4.1 rad/s   (b) 7.2 rad/s   (c) 10 rad/s   (d) 14 rad/s   (e) 17 rad/s

3.   The Earth moves about the Sun in an elliptical orbit. As the Earth moves farther from the Sun, then which

of the following best describes the orbiting speed of the Earth about the Sun?

(a) increases   (b) decreases   (c) conserves   (d) remains constant   (e) none of the above

4.   A wheel with moment of inertia of 2.00 kg·m^2 accelerates with an angular acceleration of 5.0 rad/s^2. What

is the net torque on the wheel?

(a) 0.40 m·N   (b) 2.5 m·N   (c) 3.0 m·N   (d) 7.0 m·N   (e) 10 m·N

5.   A wheel of radius 0.20 m rolls without slipping. If the wheel rotates through 3.0 revolutions, how far does

it roll?

(a) 0.60 m   (b) 1.2 m   (c) 1.9 m   (d) 3.8 m   (e) 7.6 m

6.   The rotational analog of mass in linear motion is

(a) moment of inertia.   (b) kinetic energy.   (c) moment arm.   (d) work.   (e) torque.

7.   A sphere of mass 10 kg and radius 0.010 m is released from the top of a 1.0-meter high inclined plane. What is the speed of the sphere when it reaches the bottom of the inclined plane?

(a) 3.7 m/s   (b) 4.4 m/s   (c) 5.6 m/s   (d) 6.3 m/s   (e) 7.2 m/s

8.   If the net torque on a rigid object is zero, that object will

(a) rotate with a constant angular velocity.   (b) rotate with a constant angular acceleration.

(c) not rotate.   (d) either (a) or (b).

(e) either (a) or (c).

9.   A 4.00-meter long rod is hinged at one end. The rod is initially held in the horizontal position, and then released as the free end is allowed to fall. What is the angular acceleration as it is released? (The moment of inertia of a rod about one end is $ML^2/3$.)

(a) 2.45 rad/s^2 (b) 3.68 rad/s^2 (c) 4.90 rad/s^2 (d) 6.75 rad/s^2 (e) 7.35 rad/s^2

10.   A 500-newton person stands on a uniform board of weight 100 N and length 8.0 m. The board is supported at each end. If the support force at the right end is three times that at the left end, how far from the right end is the person?

(a) 4.0 m   (b) 2.0 m   (c) 1.6 m   (d) 1.2 m   (e) 6.4 m

Answers to Practice Quiz:

1. c 2. c 3. b 4. e 5. d 6. a 7. a 8. e 9. b 10. c

CHAPTER 9

Solids and Fluids

I.   Chapter Objectives

Upon completion of this chapter, you should be able to:

1.   distinguish between stress and strain, and use elastic moduli to compute dimensional changes.

2.   explain the pressure-depth relationship and state Pascal's principle and describe how it is used in practical applications.

3.   relate the buoyant force and Archimedes' principle and tell whether an object will float in a fluid based on relative densities.

4.   identify the simplifications used in describing ideal fluid flow and use the continuity equation and Bernoulli's equation to explain common effects.

*5.   describe the source of surface tension and its effect and discuss fluid viscosity.

II.   Key Terms

Upon completion of this chapter, you should be able to define and/or explain the following key terms:

fluid	absolute pressure
stress	gauge pressure
strain	buoyant force
elastic modulus	Archimedes' principle
elastic limit	specific gravity
Young's modulus	ideal fluid
shear modulus	streamlines
bulk modulus	equations of continuity
compressibility	flow rate equation
pressure	Bernoulli's equation
pascal (Pa)	*surface tension
atmosphere (atm)	*viscosity
Pascal's principle	*Poiseuille's law

The definitions and/or explanations of the most important key terms can be found in the following section: **III. Chapter Summary and Discussion.**

III.   Chapter Summary and Discussion

1.   Solids and Elastic Moduli (Section 9.1)

All materials are elastic to some degree and can be deformed. **Stress** is the quantity that describes the force causing the deformation and **strain** is a relative measure of how much deformation a given stress produces. Quantitatively, stress is the applied force per unit cross-sectional area, and strain is the ratio of the change in dimensions to the original dimensions.

$$\text{Stress} = \frac{F}{A}, \quad \text{tensile strain (in length)} = \frac{\Delta L}{L_o}, \quad \text{shear strain (in area)} = \frac{x}{h}, \quad \text{volume strain} = \frac{\Delta V}{V_o}.$$

In general, stress is proportional to strain up to the elastic limit. The constant of proportionality, which depends on the nature of the material, is called the **elastic modulus** and is defined as the ratio of stress to strain,

$$\text{elastic modulus} = \frac{\text{stress}}{\text{strain}}.$$

There are three types of elastic moduli, the Young's modulus Y, the shear modulus S, and the bulk modulus B. $Y = \dfrac{F/A}{\Delta L/L_o}, \quad S = \dfrac{F/A}{x/h} \approx \dfrac{F/A}{\phi}, \quad B = \dfrac{F/A}{-\Delta V/V_o} = -\dfrac{\Delta p}{\Delta V/V_o}.$ The SI units of moduli are N/m^2 or pascal (Pa). The compressibility is the inverse of the bulk modulus, $\quad k = \dfrac{1}{B}.$

Example 9.1   A steel wire 2.0 m in length and 2.0 mm in diameter supports a 10-kilogram mass.

(a) What is the stress in the wire?

(b) What is the elongation of the wire?

Solution: Given: $F = w = mg = (10 \text{ kg})(9.80 \text{ m/s}^2) = 98 \text{ N}, \quad L_o = 2.0 \text{ m},$

$r = d/2 = 1.0 \text{ mm} = 1.0 \times 10^{-3} \text{ m}, \quad Y = 20 \times 10^{10} \text{ N/m}^2 \text{ (from Table 9.1)}$

Find: (a) stress $= F/A$ (b) ΔL.

(a) stress $= \dfrac{F}{A} = \dfrac{F}{\pi r^2} = \dfrac{98 \text{ N}}{\pi(1.0 \times 10^{-3} \text{ m})^2} = 3.1 \times 10^7 \text{ N/m}^2.$

(b) From the definition of Young's modulus, $Y = \dfrac{F/A}{\Delta L/L_o} = \dfrac{L_o F/A}{\Delta L},$

we have $\Delta L = \dfrac{L_o F/A}{Y} = \dfrac{(2.0 \text{ m})(3.1 \times 10^7 \text{ N/m}^2)}{(20 \times 10^{10} \text{ N/m}^2)} = 3.1 \times 10^{-4} \text{ m} = 0.31 \text{ mm}.$

Example 9.2   A shear force of 2.0×10^3 N is applied to one face of an aluminum cube with sides of 15 cm. What is the resulting relative displacement?

Solution: Given: $F = 2.0 \times 10^3$ N, $h = 0.15$ m, $S = 2.5 \times 10^{10}$ N/m^2 (from Table 9.1)

Find: x.

The area of a side is $A = h^2$.   From $S = \dfrac{F/A}{x/h}$,

We have $x = \dfrac{F/A}{S/h} = \dfrac{Fh}{AS} = \dfrac{Fh}{h^2 S} = \dfrac{F}{hS} = \dfrac{2.0 \times 10^3 \text{ N}}{(0.15 \text{ m})(2.5 \times 10^{10} \text{ N/m}^2)} = 5.3 \times 10^{-7}$ m.

2.   Fluids: Pressure and Pascal's Principle (Section 9.2)

A **fluid** is a substance that flows and cannot support a shear. Both liquids and gases are fluids. **Pressure** is defined as the force per unit area, $P = \dfrac{F}{A}$, and has units of N/m^2 or pascal (Pa).

Absolute pressure on an object submerged at a depth h below the surface of a fluid is given by the **pressure-depth equation**, $p = p_o + \rho g h$, where p_o is the pressure on the fluid surface and ρ is the density of the fluid. When this equation is applied to fluid on the surface of the Earth, $p_o = p_a = 1.01 \times 10^5$ Pa, where p_a is called the atmospheric pressure and is measured with a barometer. **Gauge pressure** (static pressure) is the difference between the absolute pressure and the atmospheric pressure, i.e., $p_{gauge} = p - p_a = \rho g h$. When a gauge is used to measure pressure, it measures the gauge pressure.

Pascal's principle states that any external pressure applied to an *enclosed* fluid is transmitted undiminished to every point in the fluid and to the walls of its container.

Hydraulic press and lift are good practical applications of this principle. In a press or lift, there are usually two pistons, one with a larger area than the other. If a small force, F_1, is applied to the smaller piston of cross sectional area A_1, the external pressure is then $p_1 = \dfrac{F_1}{A_1}$. According to Pascal's principle, this pressure will be transmitted to the larger piston of cross sectional area A_2, and $p_2 = p_1$ or $\dfrac{F_2}{A_2} = \dfrac{F_1}{A_1}$. Since $A_2 > A_1$, $F_2 > F_1$.

Example 9.3   (a) What is the absolute pressure at a location 5.00 m below the surface of a lake?

(b) What is the gauge pressure there?

Solution: Given: $p_o = p_a = 1.01 \times 10^5$ Pa, $h = 5.00$ m, $\rho = 1.00 \times 10^3$ kg/m³ (water from Table 9.2)

Find: (a) p (b) $p - p_a$.

(a) $p = p_o + \rho g h = 1.01 \times 10^5$ Pa + $(1.00 \times 10^3$ kg/m³$)(9.80$ m/s²$)(5.00$ m$) = 1.50 \times 10^5$ Pa.

(b) $p_{gauge} = p - p_a = 1.50 \times 10^5$ Pa $- 1.01 \times 10^5$ Pa $= 4.9 \times 10^4$ Pa.

Example 9.4   In a hydraulic garage lift, the small piston has a radius of 5.0 cm and the large piston has a radius of 15 cm. What force must be applied on the small piston in order to lift a car weighing 20 000 N on the large piston?

Solution: Given: $r_1 = 0.050$ m, $r_2 = 0.15$ m, $F_2 = 20\,000$ N

Find: F_1.

According to Pascal's principle, the external pressure exerted at the small piston is transmitted undiminished to the large piston. Therefore $p_1 = \dfrac{F_1}{A_1} = p_2 = \dfrac{F_2}{A_2}$.

So $F_1 = \left(\dfrac{A_1}{A_2}\right) F_2 = \left(\dfrac{\pi r_1^2}{\pi r_2^2}\right) F_2 = \left(\dfrac{r_1}{r_2}\right)^2 F_2 = \left(\dfrac{0.050}{0.15}\right)^2 (20\,000$ N$) \approx 2200$ N.

Note: What a wonderful device! The radius of the large piston is only three times the radius of the small piston, $(0.15$ m$)/(0.050$ m$) = 3$. However, the force required on the small piston is only 1/9 of the force on the large piston, $(20\,000$ N$)/(2200$ N$) = 1/9$. Why?

---

3.   Buoyancy and Archimedes' Principle (Section 9.3)

An object that is either partially or completely submerged in a fluid will experience an upward buoyant force exerted by the fluid. **Archimedes' principle** states that the buoyant force is equal to the weight of the fluid displaced., i.e., if an object displaces 3.0 N of fluid, it will receive a 3.0 N buoyant force. Mathematically, the principle is expressed as $F_b = w_f = m_f g = (\rho_f V_f)g = \rho_f V_f g$, where ρ_f is the density of the fluid, V_f is the volume of fluid displaced (which is the same as the volume fraction of the object submerged *in* the fluid).

Archimedes' principle is very useful in determining the volume of an object and therefore its density. That was what Archimedes allegedly used to determine that the crown of his king was not made of pure gold.

From Archimedes' principle, we can draw some simple conclusions about density and floating:

(1) An object will float in a fluid if the average density of the object is less than the density of the fluid.

(2) An object will sink in a fluid if the average density of the object is greater than the density of the fluid.

(3) An object will be in equilibrium at any submerged depth in a fluid if the average densities of the object and fluid are equal.

The **specific gravity** (sp. gr.) of a substance is the ratio of the density of the substance to the density of water, and so is a pure number (dimensionless). For example, aluminum has a density of 2700 kg/m^3 and water has a density of 1000 kg/m^3. So the sp. gr. of aluminum is simply $(2700$ kg/m$^3)/(1000$ kg/m$^3) = 2.7$.

Example 9.5   A boat approximating a rectangular box measures 5.0 m long, 1.0 wide and 0.50 m high. Is it safe to transport a 3000-kilogram machine part with this boat across a lake? (Neglect the mass of the boat itself.)

Solution: Given: $L = 5.0$ m, $W = 1.0$ m, $H = 0.50$ m, $\rho_{water} = 1.0 \times 10^3$ kg/m^3, $m = 3000$ kg.

Find: Will the boat float or sink.

The average density of the boat (with the machine part) is

$$\rho = \frac{m}{V} = \frac{3000 \text{ kg}}{(5.0 \text{ m})(1.0 \text{ m})(0.50 \text{ m})} = 1.2 \times 10^4 \text{ kg/m}^3 > \rho_{water} = 1.0 \times 10^4 \text{ kg/m}^3.$$

Since the average density of the boats is greater than that of water, the boat would sink. So it is not safe to transport the machine part with this boat.

Example 9.6   A bargain hunter bought a pure (24-karat) "gold" crown at a flea market. After getting it home, she suspends the crown on a scale and finds its weight to be 7.84 N. She then weighs the crown while it is immersed in water and finds that the scale reads 6.86 N. Is the crown made of pure gold? (Similar to Archimedes' problem!)

Solution: Given: $w = 7.84$ N, w (in water) $= 6.86$ N, $\rho_{water} = 1.0 \times 10^3$ kg/m^3.

Find: ρ.

There are many different physical quantities we can use to identify matter based on the unique physical quantities of different elements. For example, pure gold and aluminum each have their own unique density. So if we can determine the density of the crown, we can determine whether it is made of pure gold. When the crown is completely submerged in water, the volume of water displaced is equal to the volume of the crown. The "apparent" weight of the crown decreases because the buoyant force is canceling out part of the crown's weight. The difference between the true weight and the apparent weight of the crown is therefore the buoyant force, $F_B = 7.84$ N $- 6.86$ N $= 0.98$ N.

We first find the volume of the crown with Archimedes' principle.

From $F_b = \rho_f V_f g$,   we have $V_f = \dfrac{F_b}{\rho_f g} = \dfrac{0.98 \text{ N}}{(1.0 \times 10^3 \text{ kg/m}^3)(9.80 \text{ m/s}^2)} = 1.00 \times 10^{-4}$ m^3.

Now we can find the density of the crown.

$$\rho = \frac{m}{V} = \frac{w/g}{V} = \frac{w}{gV} = \frac{7.84 \text{ N}}{(9.80 \text{ m/s}^2)(1.00 \times 10^{-4} \text{ m}^3)} = 8.0 \times 10^3 \text{ kg/m}^3.$$

From Table 9.2, this crown is not made of pure gold because the density of gold is 19.3×10^3 kg/m^3.

---

4. Fluid Dynamics and Bernoulli's Equation (Section 9.4)

Ideal fluid flow is a steady, irrotational, nonviscous, and incompressible flow. From the conservation of mass, the **equation of continuity** for fluid flow can be derived: $\rho A v$ = constant, where ρ is the density of the fluid, A is the cross-sectional area, and v is the speed. If the fluid is incompressible (ρ = constant), then Av = constant, or $A_1 v_1 = A_2 v_2$, which is called the **flow rate equation**, since the quantity, Av, measures flow rate (volume per unit time).

Bernoulli's equation is the direct result of the application of conservation of energy to fluid. It is written as $p + \frac{1}{2}\rho v^2 + \rho g y$ = constant. If the flow is at a constant height, y = constant, then $p + \frac{1}{2}\rho v^2$ = constant. In verbal statement, this is "the higher the speed, the lower the pressure, or vice versa." Since speed is related to kinetic energy and pressure is associated with fluid height and so potential energy, Bernoulli's equation simply restates that kinetic energy plus potential energy is a constant. This equation is very successful in explaining physical phenomena such as the curve ball in a baseball game.

Example 9.7   An ideal fluid flows at 4.0 m/s in a horizontal circular pipe. If the pipe narrows to half of its original radius, what is the flow speed in narrow section?

Solution: Given: $v_1 = 12$ m/s, $r_2 = r_1/2$
Find: v_2.

From the flow rate equation, $A_1 v_1 = A_2 v_2$, we have

$$v_2 = \frac{A_1 v_1}{A_2} = \frac{\pi r_1^2 v_1}{\pi r_2^2} = \frac{r_1^2 v_1}{(r_1/2)^2} = \frac{v_1}{1/4} = 4\, v_1 = 4(4.0 \text{ m/s}) = 16 \text{ m/s}.$$

Note: The pipe narrows to half of its original radius and the speed increases by a factor of 4. Why?

Example 9.8   In Example 9.7, if the fluid is water and the pressure at the narrow section is 1.8×10^5 Pa, what is the pressure at the wide section?

Solution: Given: $\rho = 1.0 \times 10^3$ kg/m^3, $p_2 = 1.8 \times 10^5$ Pa, $v_1 = 4.0$ m/s, $v_2 = 16$ m/s (Example 9.7).

Find: p_1.

Since the fluid is flowing horizontally, y is constant, so Bernoulli's equation can be written as

$p_1 + \frac{1}{2}\rho v_1^2 = p_2 + \frac{1}{2}\rho v_2^2$.

So $p_1 = p_2 + \frac{1}{2}\rho(v_2^2 - v_1^2) = 1.8 \times 10^5$ Pa $+ \frac{1}{2}(1.0 \times 10^3$ kg/m$^3)[(16$ m/s$)^2 - (4.0$ m/s$)^2] = 3.0 \times 10^5$ Pa.

Note: the speed is lower but the pressure is higher in the wide section.

---

*5. Surface Tension; Viscosity and Poiseuille's Law (Section 9.5)

The **surface tension** of a liquid is caused by the inward pull on the surface molecules which causes the surface of the liquid to contract and be stretched. **Coefficient of viscosity** (or **viscosity**), η, is a measure of the fluid's internal resistance to flow.

The average **flow rate**, $Q = Av$ (volume/time, m^3/s), depends on the characteristics of the fluid and the pipe as well as on the pressure difference between the ends of the pipe. **Poiseuille's law** gives a relationship for the flow rate: $Q = \dfrac{\pi r^4 \Delta p}{8 \eta L}$, where r is the radius of the pipe, L its length, Δp the pressure difference between the ends of the pipe, and η the viscosity of the fluid. An interesting result is that the flow rate is proportional to r^4, that is, doubling the radius of a pipe can increase the flow rate by $2^4 =$ ~ 16 times.

IV. Mathematical Summary

Stress	$\text{stress} = \dfrac{F}{A}$ (9.1)	Defines stress.
Strain (tensile)	$\text{strain} = \dfrac{\Delta L}{L_0} = \dfrac{L - L_0}{L_0}$ (9.2)	Defines tensile strain.
Young's Modulus	$Y = \dfrac{F/A}{\Delta L/L_0}$ (9.4)	Relates Young's modulus with stress and strain.
Shear Modulus	$S = \dfrac{F/A}{x/h} \approx \dfrac{F/A}{\phi}$ (9.5)	Relates shear modulus with stress and shear strain.
Bulk Modulus	$B = \dfrac{F/A}{-\Delta V/V_0} = -\dfrac{\Delta p}{\Delta V/V_0}$ (9.6)	Relates bulk modulus with stress and volume strain.
Compressibility	$k = \dfrac{1}{B}$ (9.7)	Defines compressibility.

Pressure	$p = \dfrac{F}{A}$ (9.8a)	Defines pressure in terms of force and area.
Pressure-Depth Equation	$p = p_o + \rho g h$ (9.10)	Expresses pressure as a function of depth in fluid.
Archimedes' Principle	$F_b = m_f g = \rho_f g V_f$ (9.14)	Calculates buoyant force.
Equation of Continuity	$\rho_1 A_1 v_1 = \rho_2 A_2 v_2$ or $\rho A v = $ constant (9.16)	Relates flow characteristics for ideal fluid.
Flow Rate Equation	$A_1 v_1 = A_2 v_2$ or $A v = $ constant (9.17)	Rewrites equation of continuity for an incompressible fluid.
Bernoulli's Equation	$p_1 + \frac{1}{2}\rho v_1^2 + \rho g y_1$ $= p_2 + \frac{1}{2}\rho v_2^2 + \rho g y_2$ (9.18)	Applies the conservation of energy to a fluid.
*Poiseuille's Law	$Q = \dfrac{\pi r^4 \Delta p}{8 \eta L}$ (9.19)	Relates the flow rate to fluid characteristics.

V.   Solutions of Selected Exercises and Paired/Trio Exercises

9.   Stress $= \dfrac{F}{A} = \dfrac{mg}{\pi r^2} = \dfrac{(6.0\ \text{kg})(9.80\ \text{m/s}^2)}{\pi(0.50 \times 10^{-3}\ \text{m})^2} = 7.49 \times 10^7\ \text{N/m}^2$,

strain $= \dfrac{\Delta L}{L_o} = \dfrac{1.4 \times 10^{-3}\ \text{m}}{2.0\ \text{m}} = 7.0 \times 10^{-4}$.

So the Young's modulus is $Y = \dfrac{\text{stress}}{\text{strain}} = \dfrac{7.49 \times 10^7\ \text{N/m}^2}{7.0 \times 10^{-4}} = \boxed{1.1 \times 10^{11}\ \text{N/m}^2}$.

10.   From $Y = \dfrac{F/A}{\Delta L / L_o} = \dfrac{F L_o}{A \Delta L}$,   we have

$\Delta L = \dfrac{F L_o}{YA} = \dfrac{(5.0\ \text{kg})(9.80\ \text{m/s}^2)(2.0\ \text{m})}{(7.0 \times 10^{10}\ \text{N/m}^2)(\pi)(1.0 \times 10^{-3}\ \text{m})^2} = 4.5 \times 10^{-4}\ \text{m} = \boxed{0.45\ \text{mm}}$.

15.   The shear strain is given by

$\phi = \dfrac{\text{shear stress}}{S} = \dfrac{F/A}{S} = \dfrac{F}{SA} = \dfrac{500\ \text{N}}{(2.5 \times 10^{10}\ \text{N/m}^2)(0.10\ \text{m})^2} = 2.0 \times 10^{-6}$

Also $\phi = \dfrac{x}{h}$,   so $x = (2.0 \times 10^{-6})(0.10\ \text{m}) = \boxed{2.0 \times 10^{-7}\ \text{m}}$.

19.   (a) Since the bulk modulus is $B = -\dfrac{\Delta p}{\Delta V/V_0}$, the smaller the B, the greater the compressibility.

So $\boxed{\text{ethyl alcohol}}$ has the greatest compressibility.

(b) $\dfrac{\Delta p_w}{\Delta p_e} = \dfrac{B_w}{B_e} = \dfrac{2.2 \times 10^9 \text{ N/m}^2}{1.0 \times 10^9 \text{ N/m}^2} = \boxed{2.2 \text{ times}}$.

21.   We first need to find the tensions on the cables. The system is symmetrical.

In the vertical direction: $2T \sin 15° = mg$,   so $T = \dfrac{(45 \text{ kg})(9.80 \text{ m/s}^2)}{2 \sin 15°} = 852 \text{ N}$.

$\dfrac{\Delta L}{L_0} = \dfrac{\text{stress}}{Y} = \dfrac{T/A}{Y} = \dfrac{T}{YA} = \dfrac{852 \text{ N}}{(20 \times 10^{10} \text{ N/m}^2)(\pi)(0.50 \times 10^{-2} \text{ m})^2} = \boxed{5.4 \times 10^{-5}}$.

27.   Bicycle tires have a much smaller contact area with the ground so they need a higher pressure to balance the weight of the bicycle and the rider.

30.   (a) The pressure due to water is $p_w = \rho g h = (1000 \text{ kg/m}^3)(9.80 \text{ m/s}^2)(15 \text{ m}) = \boxed{1.5 \times 10^5 \text{ Pa}}$.

(b) The total pressure is $p = p_0 + p_w = 1.01 \times 10^5 \text{ Pa} + 1.5 \times 10^5 \text{ Pa} = \boxed{2.5 \times 10^5 \text{ Pa}}$.

36.   The pressure difference due to the 35 m high air is

$\Delta p = \rho g h = (1.29 \text{ kg/m}^3)(9.80 \text{ m/s}^2)(35 \text{ m}) = 442 \text{ Pa}$.

So the fractional decrease is $\dfrac{\Delta p}{p_a} = \dfrac{442 \text{ Pa}}{1.013 \times 10^5 \text{ Pa}} = 4.37 \times 10^{-3} = \boxed{0.44\%}$.

42.   (a) From Pascal's principle, $p_{\text{input}} = p_{\text{output}} = \dfrac{F_{\text{output}}}{A_{\text{output}}} = \dfrac{1.5 \times 10^6 \text{ N}}{0.20 \text{ m}^2} = \boxed{7.5 \times 10^6 \text{ Pa}}$.

(b) $F_{\text{input}} = P_{\text{input}} A_{\text{input}} = (7.5 \times 10^6 \text{ Pa})(\pi)(0.025 \text{ m})^2 = \boxed{1.5 \times 10^4 \text{ N}}$.

44. $\boxed{1.2 \times 10^6 \text{ Pa}}$.

49.   There is no change. As the ice melts, the volume of the newly converted water decreases; however, the ice which was initially above the water surface is now under the water. This compensated for the decrease in volume. It does not matter whether the ice is hollow or not. Both can be proved mathematically.

52.   The buoyant force is $F_b = \rho V g = (1000 \text{ kg/m}^3)(0.085 \text{ m})^3(9.80 \text{ m/s}^2) = 6.02 \text{ N}$.

The weight is $W = mg = (0.65 \text{ kg})(9.80 \text{ m/s}^2) = 6.37 \text{ N}$.

Since $W > F_b$, $\boxed{\text{no}}$, the object sinks.

56. We first need to find the volume of the crown with Archimedes' principle.

$$F_b = 8.0 \text{ N} - 4.0 \text{ N} = 4.0 \text{ N} = \rho_f g V,$$

so $V = \dfrac{F_b}{\rho_f g} = \dfrac{4.0 \text{ N}}{(1000 \text{ kg /m}^3)(9.80 \text{ m/s}^2)} = 4.08 \times 10^{-4} \text{ m}^3.$

The density $\rho = \dfrac{m}{V} = \dfrac{w/g}{V} = \dfrac{8.0 \text{ N}}{(9.80 \text{ m/s}^2)(4.08 \times 10^{-4} \text{ m}^3)} = \boxed{2.0 \times 10^3 \text{ kg/m}^3}.$

58. $\boxed{1.50 \times 10^{-5} \text{ m}^3}$ and $\boxed{6.00 \times 10^3 \text{ kg/m}^3}$.

60. For the open cube to float, $w = mg = F_b = \rho_f g V_f = \rho_f g L^3,$ where V_f is the volume of the open cube.

So $L = \sqrt[3]{\dfrac{m}{\rho_f}} = \sqrt[3]{\dfrac{\rho V}{\rho_f}} = \sqrt[3]{\dfrac{(7.8 \times 10^3 \text{ kg/m}^3)(1.0 \text{ m})^3}{1000 \text{ kg/m}^3}} = \boxed{2.0 \text{ m}},$

where V is the volume of the solid cube.

63. We first need to find the volume from Archimedes' principle.

The buoyant force is equal to $F_b = 9.8 \text{ N} - 9.1 \text{ N} = 0.7 \text{ N}.$

From $F_b = \rho_f g V,$ we have $V = \dfrac{F_b}{\rho_f g} = \dfrac{0.7 \text{ N}}{(1000 \text{ kg/m}^3)(9.80 \text{ m/s}^2)} = 7.0 \times 10^{-5} \text{ m}^3.$

The density $\rho = \dfrac{m}{V} = \dfrac{w/g}{V} = \dfrac{9.8 \text{ N}}{(7.0 \times 10^{-5} \text{ m}^3)(9.80 \text{ m/s}^2)} = 14.3 \times 10^3 \text{ kg/m}^3 < \rho_g = 19.3 \times 10^3 \text{ kg/m}^3.$

$\boxed{\text{No}}$, the bar is not gold.

69. The aspirator has suction because the water flowing by decreases the pressure at the end of the tube away from the flask. This creates a pressure difference, and suction is created in the tube. The egg is kept aloft by the pressure of the air coming out of the end of the tube. As the egg moves to one side, there is a change in the flow speed around the egg that creates an inward pressure that makes the egg move back to midstream.

70. From the equation of continuity, $A_1 v_1 = A_2 v_2,$ we have

$$v_2 = \dfrac{A_1}{A_2} v_1 = \dfrac{\pi(0.20 \text{ m})^2}{\pi(0.35 \text{ m})^2} \times (3.0 \text{ m/s}) = \boxed{0.98 \text{ m/s}}.$$

75. (a) We use Bernoulli's principle, $p_1 + \frac{1}{2}\rho v_1^2 + \rho g y_1 = p_2 + \frac{1}{2}\rho v_2^2 + \rho g y_2.$

Here $p_1 = p_2 = p_a,$ and $v_1 \approx 0$ (at the top of the water surface),

so $v_2 = \sqrt{2g\Delta y}.$

For the 40-cm high (upper) hole, $v_u = \sqrt{2(9.80 \text{ m/s}^2)(0.05 \text{ m})} = \boxed{0.99 \text{ m/s}}$;

for the 30-cm high (middle) hole, $v_m = \sqrt{2(9.80 \text{ m/s}^2)(0.15 \text{ m})} = \boxed{1.7 \text{ m/s}}$;

for the 20-cm high (bottom) hole, $v_b = \sqrt{2(9.80 \text{ m/s}^2)(0.25 \text{ m})} = \boxed{2.2 \text{ m/s}}$;

for the 10-cm high (bottom) hole, $v_b = \sqrt{2(9.80 \text{ m/s}^2)(0.35 \text{ m})} = \boxed{2.6 \text{ m/s}}$.

(b) They are all horizontal projectile motions. First find the time of flight from the vertical motion.

$$y = \tfrac{1}{2}gt^2, \quad \text{so} \quad t = \sqrt{\frac{2y}{g}}. \quad \text{Therefore the range is } x = v_x t = v_x \sqrt{\frac{2y}{g}}.$$

$$x_1 = (0.99 \text{ m/s})\sqrt{\frac{2(0.40 \text{ m})}{9.80 \text{ m/s}^2}} = 0.28 \text{ m};$$

$$x_2 = (1.7 \text{ m/s})\sqrt{\frac{2(0.30 \text{ m})}{9.80 \text{ m/s}^2}} = 0.42 \text{ m};$$

$$x_3 = (2.2 \text{ m/s})\sqrt{\frac{2(0.20 \text{ m})}{9.80 \text{ m/s}^2}} = 0.44 \text{ m};$$

$$x_4 = (2.6 \text{ m/s})\sqrt{\frac{2(0.10 \text{ m})}{9.80 \text{ m/s}^2}} = 0.37 \text{ m}.$$

Hence the $\boxed{20\text{-centimeter hole}}$ has the greatest range.

83. From Poiseuille's law, $Q = \dfrac{\pi r^4 \Delta p}{8 \eta L}$, we have

$$\Delta p = \frac{8 \eta L Q}{\pi r^4} = \frac{8(1.00 \times 10^{-3} \text{ Pl})(6.0 \text{ m})(40 \text{ L/min})(10^{-3} \text{ m}^3/\text{L})(1 \text{ min}/60 \text{ s})}{\pi (0.015 \text{ m})^4} = \boxed{2.0 \times 10^2 \text{ Pa}}.$$

87. (a) The weight of the wood must be equal to the buoyant force.

$$w = mg = F_b = \rho_f V g, \quad \text{so} \quad V = \frac{m}{\rho_f} = \frac{\rho V}{\rho_f} = \frac{(700 \text{ kg/m}^3)(0.30 \text{ m})^3}{1000 \text{ kg/m}^3} = 0.0189 \text{ m}^3.$$

Therefore the distance from the top of the wood to the water surface is

$$0.30 \text{ m} - \frac{0.0189 \text{ m}^3}{(0.30 \text{ m})(0.30 \text{ m})} = \boxed{0.09 \text{ m}}.$$

(b) The 0.09 m above the water surface can support the mass on top of the wood.

The mass is $\dfrac{(1000 \text{ kg/m}^3)(0.30 \text{ m})^2(0.09 \text{ m})g}{g} = \boxed{8.1 \text{ kg}}$.

91. The pressure by the oil is equal to the pressure by mercury.

$$p_o = \rho_o g h_o = \rho_m g h_m, \quad \text{so} \quad \rho_o = \frac{\rho_m h_m}{h_o} = \frac{(13.6 \times 10^3 \text{ kg/m}^3)(5.0 \text{ cm})}{80 \text{ cm}} = \boxed{8.5 \times 10^2 \text{ kg/m}^3}.$$

99.   (a) When it is in the water, the force applied is

$$F = mg - F_b = \rho Vg - \rho_f gV$$

$$= (\rho - \rho_f)gV = (7.8 \times 10^3 \text{ kg/m}^3 - 1.0 \times 10^3 \text{ kg/m}^3)(9.80 \text{ m/s}^2)(0.25 \text{ m})(0.20 \text{ m})(10 \text{ m})$$

$$= \boxed{3.3 \times 10^4 \text{ N}}.$$

(b) When it is out of water, the force applied is

$$F = mg = \rho Vg = (7.8 \times 10^3 \text{ kg/m}^3)(9.80 \text{ m/s}^2)(0.25 \text{ m})(0.20 \text{ m})(10 \text{ m}) = \boxed{3.8 \times 10^4 \text{ N}}.$$

VI.   Practice Quiz

1.   What is the gauge pressure at the bottom of a 5.0-meter deep swimming pool?

(a) 4.9×10^4 Pa   (b) 7.2×10^4 Pa   (c) 1.01×10^5 Pa   (d) 1.5×10^5 Pa   (e) 2.51×10^5 Pa

2.   A 4.00-kilogram cylinder made of solid iron is supported by a string while submerged in water. What is the tension in the string? (The density of iron is 7.86×10^3 kg/m^3.)

(a) 0 N   (b) 2.50 N   (c) 19.6 N   (d) 23.7 N   (e) 34.2 N

3.   When you turn on a shower head quickly, you notice that the shower curtain moves inward. This is because

(a) the air inside the curtain moves faster and the pressure is higher.

(b) the air inside the curtain moves faster and the pressure is lower.

(c) the air inside the curtain moves slower and the pressure is lower.

(d) the air inside the curtain moves slower and the pressure is higher.

(e) none of the above applies.

4.   An ideal fluid flows through a pipe made of two sections with diameters of 2.0 and 6.0 cm, respectively. What is the speed ratio of the flow through the 2.0 cm section to that through the 6.0 cm section?

(a) 1/9   (b) 1/3   (c) 1   (d) 3   (e) 9

5.   A 25 000-newton truck on a hydraulic lift rests on a cylinder with a piston of diameter 0.40 m. What is the pressure applied to the liquid in order to lift the car?

(a) 2.0×10^4 N/m^2 (b) 4.0×10^4 N/m^2 (c) 5.0×10^4 N/m^2 (d) 1.6×10^5 N/m^2 (e) 2.0×10^5 N/m^2

6.   Which of the following is associated with the law of conservation of energy for fluids?

(a) Archimedes' principle   (b) Bernoulli's principle   (c) Pascal's principle

(d) Poiseuille's law   (e) equation of continuity

7.   The same object is hung from identical wires made of aluminum, brass, copper, and steel. Which wire will stretch the least?

(a) aluminum   (b) brass   (c) copper   (d) steel   (e) all the same.

8.   Water flows with a speed of 8.0 m/s in a horizontal pipe 2.0 cm in diameter. The water then enters a horizontal pipe 4.0 cm in diameter. What is the difference in pressure between the two segments?

(a) 1.0×10^4 N/m^2 (b) 2.0×10^4 N/m^2 (c) 3.0×10^4 N/m^2 (d) 4.0×10^4 N/m^2 (e) 5.0×10^4 N/m^2

9.   In a gasoline spill, a gasoline layer of 0.50 cm thick was found above a surface. Estimate the gauge pressure of the gasoline.

(a) 0.33 N/m^2 (b) 3.3 N/m^2 (c) 33 N/m^2 (d) 3.3×10^2 N/m^2 (e) 3.3×10^3 N/m^2

10.   If the density of gold is 19.3×10^3 kg/m^3, what buoyant force does a 0.60-kilogram gold crown experience when it is immersed in water?

(a) 3.0×10^{-5} N   (b) 3.0×10^{-4} N   (c) 3.0×10^{-2} N   (d) 0.30 N   (e) 3.0 N

Answers to Practice Quiz:

1. a 2. e 3. b 4. e 5. e 6. b 7. d 8. c 9. c 10. d

CHAPTER 10

Temperature

I.   Chapter Objectives

Upon completion of this chapter, you should be able to:

1.   distinguish between temperature and heat.

2.   explain how a temperature scale is constructed and convert temperatures from one scale to another.

3.   describe the ideal gas law and explain how it is used to determine absolute zero.

4.   calculate the thermal expansions of solids and liquids.

5.   relate kinetic theory and temperature and explain the process of diffusion.

*6.   understand the difference between monatomic and diatomic gases, the meaning of the equipartition theorem, and the expression for the internal energy of a diatomic gas.

II.   Key Terms

Upon completion of this chapter, you should be able to define and/or explain the following key terms:

temperature	Kelvin temperature scale
heat	kelvin
internal energy	triple point of water
thermometer	thermal coefficient of linear expansion
thermal expansion	thermal coefficient of area expansion
Fahrenheit temperature scale	thermal coefficient of volume expansion
Celsius temperature scale	kinetic theory of gases
ideal (perfect) gas law	diffusion
mole	osmosis
Avogadro's number	*degree of freedom
absolute zero	*equipartition theorem

The definitions and/or explanations of the most important key terms can be found in the following section:
III. Chapter Summary and Discussion.

III. Chapter Summary and Discussion

1. Temperature and Heat (Section 10.1)

Temperature is a relative measure, or indication, of hotness and coldness. The result of the kinetic theory of gases states that temperature is a measure of the average kinetic energy of the molecules.

Heat is the net energy transferred from one object to another because of a temperature difference. The total energy (kinetic plus potential) of all molecules of a body or system is the **internal energy**. When heat is transferred out of or into a system while there is no other physical process present, the internal energy of the system will change.

When heat is transferred between two objects, whether or not they are physically touching, they are said to be in **thermal contact**. When there is no longer a net heat transfer between objects in thermal contact, they are at the same temperature and are said to be in **thermal equilibrium**.

2. The Celsius and Fahrenheit Temperature Scales (Section 10.2)

The two most common temperature scales are the **Celsius temperature scale** and the **Fahrenheit temperature scale**. On the Celsius scale, water freezes at 0°C and boils at 100°C; while on the Fahrenheit scale, water freezes at 32°F and boils at 212°F. The conversions between the two scales are

$$T_F = \tfrac{9}{5} T_C + 32, \text{ or } T_C = \tfrac{5}{9} (T_F - 32).$$

Note: To convert T_C to T_F, you need to multiply by $\tfrac{9}{5}$ and then add 32. However, to convert T_F to T_C, you need to subtract 32 and then multiply by $\tfrac{5}{9}$.

Example 10.1   What is the temperature 50.0°F on the Celsius scale?

Solution: Given: $T_F = 50.0°F$.   Find: T_C.

$T_C = \tfrac{5}{9} (T_F - 32) = \tfrac{5}{9} (50 - 32) = 10°C.$

Example 10.2   The temperature changes from 35°F during the night to 75°F during the day. What is the temperature change on the Celsius scale?

Solution: Given: $\Delta T_F = 75°F - 35°F = 40$ F°. [Note: the unit for Fahrenheit temperature change is F° (Fahrenheit degree), not °F (degrees Fahrenheit).]

Find: ΔT_C.

Although we can convert both 35°F and 75°F to their Celsius temperatures and then calculate the change, we can take advantage of the conversion equations to calculate the temperature change ΔT.

From $T_C = \frac{5}{9}(T_F - 32)$, we can see that for every Fahrenheit degree increase, the Celsius temperature increases by $\frac{5}{9}$ of a degree, or $\Delta T_C = \frac{5}{9}\Delta T_F$.

So $\Delta T_C = \frac{5}{9}\Delta T_F = \frac{5}{9}(40) = 22$ C°. (Note: the unit for Celsius temperature change is C°, not °C.)

---

3.   Gas Law and Absolute Temperature (Section 10.3)

An ideal gas is a low density and low pressure gas. The **ideal** (or perfect) **gas law** relates the pressure, volume and the temperature of the gas, $pV = Nk_B T$, where N is the number of molecules and k_B is the Boltzmann's constant which has a value of 1.38×10^{-23} J/K. It describes real, low-density gases fairly well. Two other forms of the ideal gas law are $pV = nRT$ (macroscopic form) and $\frac{p_1 V_1}{T_1} = \frac{p_2 V_2}{T_2}$, where n is the number of moles of molecules and R is the universal gas constant which has values of 8.31 J/(mol·K). One mole of substance is defined as containing $N_A = 6.02 \times 10^{23}$ molecules (Avogadro's number).

A gas can be used to measure temperature as a function of time at a constant volume. Extrapolation to zero pressure defines absolute zero temperature. **Absolute zero** is the foundation for the **Kelvin temperature scale**, which uses absolute zero and the **triple point of water** as fixed points. The conversion between Kelvin and Celsius temperatures is $T_K = T_C + 273.15$, or more commonly to three significant figures, $T_K = T_C + 273$.

Note: The temperature in the ideal gas law *must* be the Kelvin temperature.

Example 10.3   What is −40°F on the Kelvin scale?

Solution: Given: $T_F = -40°C$.   Find: T_K.

Since there is no direct conversion given between Fahrenheit and Kelvin temperatures, we first convert Fahrenheit to Celsius.

$T_C = \frac{5}{9}(T_F - 32) = \frac{5}{9}(-40 - 32) = \frac{5}{9}(-72) = -40°C$.

(Wow! The Celsius and Fahrenheit have the same nominal reading at this temperature.)

Next, $T_K = T_C + 273 = -40 + 273 = 233$ K.

Example 10.4   A gas has a volume of 0.20 m³, a temperature of 30°C, and a pressure of 1.0 atm (one atmosphere of pressure). It is heated to 60°C and compressed to a volume of 0.15 m³. Find the new pressure in atmospheres.

Solution: Given: $T_1 = 30°C$, $T_2 = 60°C$,

$V_1 = 0.20$ m³, $V_2 = 0.15$ m²,

$p_1 = 1.0$ atm.

Find: p_2 (in atm).

The temperatures are given on the Celsius scale, so we first convert them to Kelvin temperatures.

$T_1 = 30°C = 30 + 273 = 303$ K   and $T_2 = 60°C = 60 + 273 = 333$ K.

From $\dfrac{p_1 V_1}{T_1} = \dfrac{p_2 V_2}{T_2}$, we have $p_2 = \dfrac{V_1}{V_2} \times \dfrac{T_2}{T_1} \times p_1 = \dfrac{0.20 \text{ m}^3}{0.15 \text{ m}^2} \times \dfrac{333 \text{ K}}{303 \text{ K}} \times (1.0 \text{ atm}) = 1.5$ atm.

Note: Leaving the pressure in units of atm gives the new pressure in atm because of the ratio form of the ideal gas law. Also, the temperatures used in the ideal gas law *must* be the Kelvin temperatures. What would happen if you used a temperature of $T_1 = 0°C$ in the example?

Example 10.5   An ideal gas in a container of volume of 1000 cm³ (one liter) at 20.0°C has a pressure of 1.00×10^4 N/m². Determine the number of gas molecules and the number of moles of gas in the container.

Solution: Given: $V = 1000$ cm³, $T = 20.0°C$, $p = 1.00 \times 10^4$ N/m².

Find: N and n.

The volume is not given in m³ (the standard SI unit), so we first need to convert the volume to m³ and temperature to Kelvin. Although 1 m = 100 cm, 1 m³ ≠ 100 cm³.

Actually 1 m³ = (1 m)³ = (100 cm)³ = 10⁶ cm³.

$V = (1000 \text{ cm}^3) \times \dfrac{1 \text{ m}^3}{10^6 \text{ cm}^3} = 1.00 \times 10^{-3}$ m³.

$T = T_C + 273 = 20.0 + 273 = 293$ K.

From $pV = Nk_B T$, $N = \dfrac{pV}{k_B T} = \dfrac{(1.00 \times 10^4 \text{ N/m}^2)(1.00 \times 10^{-3} \text{ m}^3)}{(1.38 \times 10^{-23} \text{ J/K})(293 \text{ K})} = 2.47 \times 10^{21}$ molecules.

$n = \dfrac{N}{N_A} = \dfrac{2.47 \times 10^{21} \text{ molecules}}{6.02 \times 10^{23} \text{ molecues/mol}} = 4.10 \times 10^{-3}$ mol.

4.   Thermal Expansion (Section 10.4)

The **thermal expansion** of a material is characterized by its coefficient of expansion. For solids, the **thermal coefficient of linear expansion,** α, applies to one-dimensional length changes; the **thermal coefficient of area expansion** (approximately equal to 2α) applies to two-dimensional area changes; and the **thermal coefficient of volume expansion** (approximately equal to 3α) applies to three-dimensional volume changes. For fluids with no definite shape, only volume expansion is applicable and a special thermal coefficient of volume expansion β is used. The equations we use to calculate the thermal expansions are:

Linear: $\dfrac{\Delta L}{L_0} = \alpha \Delta T.$ Area: $\dfrac{\Delta A}{A_0} = 2\alpha \Delta T.$

Volume: $\dfrac{\Delta V}{V_0} = 3\alpha \Delta T$ (for solids)   Volume: $\dfrac{\Delta V}{V_0} = \beta \Delta T$ (for fluids).

Example 10.6   You are installing some outdoor copper electric wire to a backyard fish pond on a hot $40°C$ summer day. The temperature could be as low as $-20°C$ in your area during a cold winter night. How much extra wire (minimum) do you have to include to allow for thermal expansion if the distance from the electric service to the pond is 100 m?

Solution: Given: $L_0 = 100$ m, $T_i = 40°C$, $T_F = -20°C$, $\alpha = 17 \times 10^{-6}$ /C° (from Table 10.1).
       Find: ΔL.

If you cut the wire exactly 100 m in the summer, then it shrinks to a length smaller than 100 m in the winter and the wire will snap. So we want to make sure that the wire is at least 100 m long during the winter and calculate the corresponding length in the summer.

$\Delta T = T_f - T_i = -20°C - (40°C) = -60$ C°.

From $\dfrac{\Delta L}{L_0} = \alpha \Delta T$, we have $\Delta L = \alpha \Delta T L_0 = (17 \times 10^{-6}$ /C°$)(-60$ C°$)(100$ m$) = -0.10$ m.

Here the negative sign simply means that is a compression (negative expansion).

Example 10.7   A 500-milliliter glass beaker of water is filled to the rim at a temperature of $0°C$. How much water will overflow if the water is heated to a temperature of $95°C$? (Ignore the expansion of the beaker, why?)

Solution: Given: $V_0 = 500$ mL, $T_i = 0°C$, $T_f = 95°C$, $\beta = 2.1 \times 10^{-4}$ /C° (from Table 10.1).
       Find: ΔV.

$\Delta T = T_f - T_i = 95°C - 0°C = 95\ C°$.

The amount of water that overflows is simply equal to the change in volume of the water.

From $\dfrac{\Delta V}{V_0} = \beta \Delta T$, we have $\Delta V = \beta \Delta T V_0 = (2.1 \times 10^{-4}\ /C°)(95\ C°)(500\ \text{mL}) = 10\ \text{mL}$.

Note: In this Example, we can ignore the expansion of the beaker because glass has a much smaller coefficient of volume expansion (approximately $10 \times 10^{-6}\ /C°$).

5. The Kinetic Theory of Gases (Section 10.5)

The **kinetic theory of gases** uses statistical methods to derive the ideal gas law from mechanical principles. Some of the important conclusions of this theory are:

(1) Temperature is a measure of the average kinetic energy of the molecules, $\frac{1}{2} m v_{rms}^2 = \frac{3}{2} k_B T$, where v_{rms} is the root-mean-square speed of the molecules and m is the mass of the molecule.

(2) The ideal gas law can be expressed in terms of the root-mean-square speed of the molecules, $pV = \frac{1}{3} N m v_{rms}^2$.

(3) The total internal energy of an ideal monatomic gas is given by $U = \frac{3}{2} N k_B T = \frac{3}{2} n R T$.

Diffusion is a process of random molecular mixing in which particular molecules move from a region of higher concentration to one of lower concentration. **Osmosis** is the diffusion of a liquid across a permeable membrane because of a concentration gradient (difference).

Example 10.8 Calculate the rms (root–mean–square) speed of a hydrogen molecule and an oxygen molecule at a temperature of 300 K. (The masses of hydrogen and oxygen molecules are 3.3×10^{-27} kg and 5.3×10^{-26} kg, respectively.)

Solution: Given: $T = 300\ \text{K}, \quad m_H = 3.3 \times 10^{-27}\ \text{kg}, \quad m_O = 5.3 \times 10^{-26}\ \text{kg}$.

Find: v_{rms} for H_2 and O_2.

From $\frac{1}{2} m v_{rms}^2 = \frac{3}{2} k_B T$, we have $v_{rms} = \sqrt{\dfrac{3 k_B T}{m}}$.

For H_2: $v_{rms} = \sqrt{\dfrac{3(1.38 \times 10^{-23}\ \text{J/K})(300\ \text{K})}{3.3 \times 10^{-27}\ \text{kg}}} = 1.9 \times 10^3$ m/s (about 4000 mi/h).

For O_2: $v_{rms} = \sqrt{\dfrac{3(1.38 \times 10^{-23}\ \text{J/K})(300\ \text{K})}{5.3 \times 10^{-26}\ \text{kg}}} = 4.8 \times 10^2$ m/s.

Why does more massive molecule gas gave smaller v_{rms}?

Example 10.9   If the temperature of a gas increases from 20°C to 40°C, by what factor does the rms (root–mean–square) speed increases?

Solution: Given: $T_1 = 20°C = 20 + 273 = 293$ K, $T_2 = 40°C = 40 + 273 = 313$ K.

(The temperatures *must* be in Kelvins!)

Find: $\dfrac{(v_{rms})_2}{(v_{rms})_1}$.

From $\frac{1}{2} m v_{rms}^2 = \frac{3}{2} k_B T$,   we have $v_{rms} = \sqrt{\dfrac{3 k_B T}{m}}$.

So $\dfrac{(v_{rms})_2}{(v_{rms})_1} = \sqrt{\dfrac{T_2}{T_1}} = \sqrt{\dfrac{313 \text{ K}}{293 \text{ K}}} = 1.03$,   or it increases by only 3%.

Note: Since k_B and m are constant, they are divided out in the ratio $\dfrac{(v_{rms})_2}{(v_{rms})_1}$. The Celsius temperature doubles in this Example, but the rms speed does not double (it increases by only 3%), nor does the Kelvin temperature.

*6.   Kinetic Theory, Diatomic Gases, and the Equipartition Theorem
(Section 10.6)

For monatomic gases, the total internal energy, U, consists solely of translational kinetic energy because a monatomic molecule cannot rotate. However, diatomic molecules can rotate so the kinetic energy associated with these motions should also be included. The average translational kinetic energy of any molecule is always equal to $\frac{1}{2} m v_{rms}^2 = \frac{3}{2} k_B T$.

For a monatomic molecule, there are three independent ways of possessing kinetic energy: with x, y, or z linear motion. For a diatomic molecule, there are three independent ways of possessing translational kinetic energy and two independent ways of possessing rotational kinetic energy. Each independent way a molecule has for possessing energy is called a **degree of freedom**.

The **equipartition theorem** states: On average, the total internal energy (U) of an ideal gas is divided equally among each degree of freedom its molecules possess. Furthermore, each degree of freedom contributes $\frac{1}{2} N k_B T$ (or $\frac{1}{2} nRT$) to the total internal energy.

According to this theorem, a monatomic gas possesses $3(\frac{1}{2} Nk_B T) = \frac{3}{2} Nk_B T$ (or $\frac{3}{2} nRT$) of internal energy because it has three degrees of freedom, and a diatomic gas possesses $5(\frac{1}{2} Nk_B T) = \frac{5}{2} Nk_B T$ (or $\frac{5}{2} nRT$) of internal energy because it has five degrees of freedom. Among the $\frac{5}{2} Nk_B T$ for a diatomic gas, $\frac{3}{2} Nk_B T$ are from translational motion and $\frac{2}{2} Nk_B T$ are from rotational motion. (A diatomic molecule has a vibrational degree of freedom as well. However, the contribution from vibration at room temperature is negligible to total internal energy.)

IV. Mathematical Summary

Celsius—Fahrenheit Conversion	$T_F = \frac{9}{5} T_C + 32$ (10.1) $T_C = \frac{5}{9}(T_F - 32)$ (10.2)	Converts between Celsius temperature scale and Fahrenheit temperature scale.
Ideal (or perfect) Gas Law (always absolute temperature)	$pV = Nk_B T$ (10.5) or $\dfrac{p_1 V_1}{T_1} = \dfrac{p_2 V_2}{T_2}$ (10.6) or $pV = nRT$ (10.7)	Relates pressure, volume, and absolute temperature. $k_B = 1.38 \times 10^{-23}$ J/K $R = 8.31$ J/(mol·K) Avogadro's number: $N_A = 6.02 \times 10^{23}$ molecules/mole
Kelvin—Celsius Conversion	$T_K = T_C + 273.15$ (10.8) or $T_K = T_C + 273$	Converts between Celsius temperature scale and Kelvin temperature scale.
Thermal Expansion of Solids: Linear	$\dfrac{\Delta L}{L_o} = \alpha \Delta T$ (10.9) or $L = L_o(1 + \alpha \Delta T)$ (10.10)	Calculates the fractional change in length in terms of the coefficient of linear expansion and change in temperature.
Thermal Expansion of Solids: Area	$\dfrac{\Delta A}{A_o} = 2\alpha \Delta T$ (10.11) or $A = A_o(1 + 2\alpha \Delta T)$	Calculates the fractional change in area in terms of the coefficient of linear expansion and change in temperature.
Thermal Expansion of Solids: Volume	$\dfrac{\Delta V}{V_o} = 3\alpha \Delta T$ (10.12) or $V = V_o(1 + 3\alpha \Delta T)$	Calculates the fractional change in volume in terms of the coefficient of linear expansion and change in temperature.
Thermal Volume Expansion of Fluids	$\dfrac{\Delta V}{V_o} = \beta \Delta T$ (10.13)	Calculates the fractional change in volume in terms of the coefficient of volume expansion and change in temperature.
Results of Kinetic Theory of Gases	$pV = \frac{1}{3} Nmv_{rms}^2$ (10.14) $\frac{1}{2} mv_{rms}^2 = \frac{3}{2} k_B T$ (10.15) $U = \frac{3}{2} Nk_B T = \frac{3}{2} nRT$ (10.16) $U = \frac{5}{2} Nk_B T = \frac{5}{2} nRT$ (10.17)	Gives the results of the kinetic theory of gases. Relate absolute temperature to kinetic energy. For ideal monatomic gas only. For diatomic gas only.

V. Solutions of Selected Exercises and Paired/Trio Exercises

8. $T_F = \frac{9}{5} T_C + 32 = \frac{9}{5}(39.4) + 32 = \boxed{103°F}$.

10.   (a) $\boxed{245°F}$ is lower.   (b) $\boxed{375°F}$ is lower.

17.   (a) If pressure is held constant, the volume will decrease as temperature decreases, according to the ideal gas law. So the density $\boxed{\text{increases}}$.

       (b) If the volume is held constant, the density is also $\boxed{\text{constant}}$.

22.   (a) $T_K = T_C + 273.15 = 0 + 273.15 = \boxed{273 \text{ K}}$.   (b) $T_K = 100 + 273.15 = \boxed{373 \text{ K}}$.

       (c) $T_K = 20 + 273.15 = \boxed{293 \text{ K}}$.   (d) $T_K = -35 + 273.15 = \boxed{238 \text{ K}}$.

29.   From $\dfrac{p_1 V_1}{T_1} = \dfrac{p_2 V_2}{T_2}$ and since $T_1 = T_2$,

       we have $p_2 = \dfrac{V_1}{V_2} \times p_1 = \dfrac{0.10 \text{ m}^3}{0.12 \text{ m}^3} \times (1.4 \times 10^5 \text{ Pa}) = \boxed{1.2 \times 10^5 \text{ Pa}}$.

30.   From $pV = nRT$, we have

$$V = \frac{nRT}{p} = \frac{(1 \text{ mol})[8.31 \text{ /(mol·K)}](273 \text{ K})}{1.01 \times 10^5 \text{ Pa}} = \boxed{0.0224 \text{ m}^3 = 22.4 \text{ L}}.$$

32.   (a) $\boxed{2.68 \times 10^{22}}$.

       (b) $\boxed{2.68 \times 10^{19}}$.

34. $T_1 = 92°F = \frac{5}{9}(92 - 32) \text{ °C} = 33.3°C = 306.3 \text{ K}$, $T_2 = 32°F = 0°C = 273 \text{ K}$.

       From $\dfrac{p_1 V_1}{T_1} = \dfrac{p_2 V_2}{T_2}$, we have

$$V_2 = \frac{p_1 V_1 T_2}{T_1 p_2} = \frac{(20.0 \text{ lb/in}^2)(0.20 \text{ m}^3)(273 \text{ K})}{(306.3 \text{ K})(14.7 \text{ lb/in}^2)} = \boxed{0.24 \text{ m}^3}.$$

       Here we do not have to use pascal as the unit for pressure and can use lb/in^2 because pressure is involved in a ratio.

37.　(a) The gas $\boxed{\text{expands}}$ according to ideal gas law.

(b) From $\dfrac{p_0 V_0}{T_0} = \dfrac{pV}{T}$, we have $\dfrac{V}{V_0} = \dfrac{Tp_0}{T_0 p} = \dfrac{T}{T_0} = \dfrac{313 \text{ K}}{283 \text{ K}} = 1.106$.

So the fractional change is $\dfrac{V - V_0}{V_0} = \dfrac{V}{V_0} - 1 = 0.106 = \boxed{10.6\%}$.

43.　Water expands when cooled from 4°C to 2°C as it exhibits abnormal expansion between 0°C and 4°C.

48.　Only the higher temperature needs to be considered because the slabs will not touch under the lower

temperature $\Delta L = \alpha L_0 \Delta T = (12 \times 10^{-6} \text{ C}^{\circ-1})(10.0 \text{ m})[45°C - (20°C)] = 3.0 \times 10^{-3} \text{ m} = \boxed{3.0 \text{ mm}}$.

52.　(a) $\boxed{\text{Larger}}$ due to expansion.

(b) $A_0 = \pi r^2 = \pi (4.00 \text{ cm})^2 = 50.27 \text{ cm}^2$.

$\Delta A = 2\alpha A_0 \Delta T = 2(24 \times 10^{-6} \text{ C}^{\circ-1})(50.27 \text{ cm}^2)(150°C - 20°C) = 0.314 \text{ cm}^2$.

So the final area is $50.27 \text{ cm}^2 + 0.314 \text{ cm}^2 = \boxed{50.6 \text{ cm}^2}$.

57.　From the definition of density $\rho = \dfrac{m}{V}$, we have $\dfrac{\rho}{\rho_0} = \dfrac{V_0}{V} = \dfrac{V_0}{V_0(1 + \beta \Delta T)} = \dfrac{1}{1 + \beta \Delta T}$.

$\rho = \dfrac{1}{1 + \beta \Delta T} \rho_0 = \dfrac{1}{1 + (1.8 \times 10^{-4} \text{ C}^{\circ-1})(100°C - 0°C)} \times (13.6 \times 10^3 \text{ kg/m}^3) = \boxed{13.4 \times 10^3 \text{ kg/m}^3}$.

60.　At 20°C, $V_{ob} = 1000 \text{ cm}^3$ and $V_{om} = 990 \text{ cm}^3$.

At a temperature T, both the beaker and the mercury have the same volume, $V_b = V_m$.

$V = V_0 + \Delta V = V_0 (1 + 3\alpha\Delta T) = V_0 (1 + \beta\Delta T)$, ☞ $V_{ob}(1 + 3\alpha_b \Delta T) = V_{om}(1 + \beta\Delta T)$.

$\Delta T = \dfrac{V_{ob} - V_{om}}{V_{om}\beta_m - 3V_{ob}\alpha_b} = \dfrac{1000 \text{ cm}^3 - 990 \text{ cm}^3}{(990 \text{ cm}^3)(1.8 \times 10^{-4} \text{ C}^{\circ-1}) - 3(1000 \text{ cm}^3)(3.3 \times 10^{-6} \text{ C}^{\circ-1})} = 59.4 \text{ C}°$.

So the temperature is $T = 20°C + 59.4 \text{ C}° = \boxed{79.4°C}$.

66.　From $U = \frac{3}{2} nRT$, we have $\Delta U = \frac{3}{2} nR\Delta T = \frac{3}{2} (2.0 \text{ mol})[8.31 \text{ J/(mol·K)}](50°C - 20°C) = \boxed{7.5 \times 10^2 \text{ J}}$.

74.　For monatomic gas, $U_1 = \frac{3}{2} nRT$. For diatomic gas, $U_2 = \frac{5}{2} nRT$.

So $U_2 = \frac{5}{3} U_1 = \frac{5}{3} (5.0 \times 10^3 \text{ J}) = \boxed{8.3 \times 10^3 \text{ J}}$.

77. $\Delta V_{gas} = \beta V_0 \Delta T = (9.5 \times 10^{-4} \text{ C}^{\circ -1})(25 \text{ gal})(30°C - 10°C) = 0.48 \text{ gal.}$

$\Delta V_{tank} = 3\alpha V_0 \Delta T = 3(12 \times 10^{-6} \text{ C}^{\circ -1})(25 \text{ gal})(30°C - 10°C) = 0.018 \text{ gal.}$

So the spilled volume is $\Delta V_{gas} - \Delta V_{tank} = 0.48 \text{ gal} - 0.018 \text{ gal} = \boxed{0.46 \text{ gal}}.$

83. From $\Delta L = \alpha L_0 \Delta T$, we have $\dfrac{\Delta L}{L_0} = \alpha \Delta T.$ So $\dfrac{F}{A} = Y \dfrac{\Delta L}{L_0} = Y\alpha\Delta T.$

Therefore $\Delta T = \dfrac{F/A}{Y\alpha} = \dfrac{8.0 \times 10^7 \text{ N/m}^2}{(20 \times 10^{10} \text{ N/m}^2)(12 \times 10^{-6} \text{ C}^{\circ -1})} = \boxed{33 \text{ C}^\circ}.$

89. From $pV = Nk_B T$, we have $N = \dfrac{pV}{k_B T} = \dfrac{(20 \text{ Pa})(0.20 \text{ m}^3)}{(1.38 \times 10^{-23} \text{ J/K})(293 \text{ K})} = \boxed{9.9 \times 10^{20}}.$

95. From $\dfrac{p_1 V_1}{T_1} = \dfrac{p_2 V_2}{T_2}$ and since $T_2 = T_1$,

we have $p_2 = \dfrac{V_1}{V_2} p_1 = 2(1 \text{ atm}) = 2 \text{ atm} = 2(1.013 \times 10^5 \text{ Pa}) = \boxed{2.026 \times 10^5 \text{ Pa}}.$

VI.   Practice Quiz

1. Nitrogen condenses into a liquid at approximately 77 K. What temperature, in degrees Fahrenheit, does this correspond to?

(a) –353°F   (b) –321°F   (c) –196°F   (d) –171°F   (e) –139°F

2. An air bubble of volume 1.0 cm^3 was released under water. As it rises to the surface of the water, its volume expands. What will be its new volume if its original temperature and pressure are 5.0°C and 1.2 atm, and its final temperature and pressure are 20°C and 1.0 atm?

(a) 1.1 cm^3 (b) 1.3 cm^3 (c) 3.3 m^3 (d) 4.0 cm^3 (e) 4.8 cm^3

3. A fixed container holds oxygen and hydrogen gases at the same temperature. Which one of the following statements is correct?

(a) The oxygen molecules have the greater kinetic energy.

(b) The hydrogen molecules have the greater kinetic energy.

(c) The oxygen molecules have the greater speed.

(d) The hydrogen molecules have the greater speed.

(e) They have the same speed and kinetic energy.

4. The absolute (Kelvin) temperature of an ideal gas is directly proportional to which of the following properties, when taken as an average, of the molecules of the gas?

(a) speed (b) momentum (c) mass (d) kinetic energy (e) potential energy

5. An aluminum beam is 15.0 m long at a temperature of −15°C. What is its expansion when the temperature is 35°C?

(a) 1.8×10^{-2} m (b) 9.0×10^{-3} m (c) 7.2×10^{-3} m (d) 1.2×10^{-3} m (e) 3.6×10^{-4} m.

6. An ideal gas sample has a pressure of 2.5 atm, a volume of 1.0 L at a temperature of 30°C. How many moles of gas are in the sample?

(a) 9.9×10^{-4} (b) 1.0×10^{-2} (c) 0.10 (d) 1.1 (e) 2.5

7. A brass cube, 10 cm on a side, is heated with a temperature change of 200 C°. By what percentage does its volume change?

(a) 5.7×10^{-3} % (b) 0.57 % (c) 1.1 % (d) 1.1×10^{-3} % (e) 0.10 %

8. When water warms from 0°C to 4°C, the density of the water

(a) increases (b) decreases (c) remains constant (d) becomes zero (e) none of the above

9. A molecule has a rms speed of 500 m/s at 20°C. What is its rms speed at 80°C?

(a) 500 m/s (b) 1,000 m/s (c) 2,000 m/s (d) 550 m/s (e) 600 m/s

10. If one mole of a monatomic gas has a total internal energy of 3.7×10^{3} J, what is the total internal energy of one mole of diatomic gas at the same temperature?

(a) zero (b) 2.2×10^{3} J (c) 6.2×10^{-3} J (d) 1.1×10^{4} J (e) 1.9×10^{4} J

Answers to Practice Quiz:

1. b 2. b 3. d 4. d 5. a 6. c 7. c 8. a 9. d 10. c

CHAPTER 11

Heat

I. Chapter Objectives

Upon completion of this chapter, you should be able to:

1. distinguish the various units of heat and define the mechanical equivalent of heat.

2. describe specific heat and explain how the specific heats of materials are obtained from calorimetry.

3. compare and contrast the three common phases of matter and relate latent heat to phase changes.

4. describe the three methods of heat transfer and give practical and/or environmental examples of each.

II. Key Terms

Upon completion of this chapter, you should be able to define and/or explain the following key terms:

heat	latent heat
kilocalorie (kcal)	latent heat of fusion
calorie (cal)	latent heat of vaporization
British thermal unit (Btu)	evaporation
mechanical equivalent of heat	conduction
specific heat	thermal conductors
solid phase	thermal insulators
melting point	thermal conductivity
freezing point	convection
liquid phase	radiation
gaseous (vapor) phase	infrared radiation
boiling point	Stefan's law
condensation point	emissivity
sublimation	black body

The definitions and/or explanations of the most important key terms can be found in the following section:

III. Chapter Summary and Discussion.

III. Chapter Summary and Discussion

1. Units of Heat (Section 11.1)

Heat is the net energy transferred from one object to another because of a temperature difference (Chapter 10). The SI unit of heat is joule (J).

Other units of heat commonly used are the **calorie** (cal) and the **kilocalorie** (kcal, or food Calorie, note the capital C for the "large" food calorie). A kilocalorie is the amount of heat required to raise the temperature of 1 kg of water by 1 C° (from 14.5°C to 15.5°C).

A **British thermal unit (Btu)** is the amount heat needed to raise the temperature of 1 lb of water by 1 F° (from 63°F to 64°F). This definition has its limitations because it is based on the weight of water and weight can change depending on the acceleration due to gravity.

The relationship between unit of heat and the standard SI unit (J) is called the **mechanical equivalent of heat**: 1 kcal = 4186 J = 4.186 kJ, or 1 cal = 4.186 J.

2. Specific Heat (Section 11.2)

Specific heat is the amount of heat required to change the temperature of 1 kg of substance by 1 C°. Among the substances known to us, water has a relatively large specific heat (more heat required per kg per C°) value of 1 cal/(g·C°), or 1 kcal/(kg·C°), or 4186 J/(kg·C°). The 4186 J/(kg·C°) value simply means that it takes 4186 J of heat to change t1 kg of =water by a temperature difference of 1 C°. The high specific heat of water explains why the climate doesn't vary much with season in places near a large body of water.

The heat needed to change the temperature of a mass m by ΔT is then $Q = cm\Delta T$, where Q is heat and c is the specific heat. When an object gains heat, $(\Delta T > 0)$, $Q > 0$. However, if an object loses heat $(\Delta T < 0)$, $Q < 0$.

Within an isolated system, the heat lost by an object will be gained by other objects within the system. According to the conservation of energy, $\Sigma Q = 0$ or the absolute value of heat lost = heat gained. (The absolute value of heat lost has to be used because heat lost is a negative quantity.) This is the basis of **calorimetry** (the method of mixing).

Note: Since heat lost is a negative quantity, when applying calorimetry, you should write $-Q_{lost} = Q_{gained}$, (that is, equate magnitude).

Example 11.1   A 2.0-kilogram steel block is originally at 10°C. If 140 kJ of heat energy are added to the block, what is its final temperature?

Solution: Given: $m = 2.0$ kg, $c = 460$ J/(kg·C°) (Table 11.1), $Q = 140$ kJ $= 140 \times 10^3$ J, $T_i = 10$°C.
              Find: T_f.

From $Q = cm\Delta T$, we have $\Delta T = \dfrac{Q}{cm} = \dfrac{140 \times 10^3 \text{ J}}{[460 \text{ J/(kg·C°)}](2.0 \text{ kg})} = 152$ C°.

So $T_f = T_i + \Delta T = 10$°C $+ 152$ C° $= 162$°C.

Example 11.2   A 0.600-kilogram piece of a pure metal is heated to 100°C and placed in an aluminum can of mass 0.200 kg which contains 0.500 kg of water initially at 17.3°C. The final equilibrium temperature of the mixture is 20.2°C. What kind of metal is it?

Solution: Use the following subscripts:
              u = unknown metal,   w = water,   a = aluminum,   i = initial,   and   f = final.
              Given: $m_u = 0.600$ kg, m_a 0.200 kg, $m_w = 0.500$ kg, $t_{iu} = 100$°C, $t_{fu} = 20.2$°C,
                    $t_{ia} = t_{iw} = 17.3$°C, $T_{fa} = T_{fw} = 20.2$°C, $c_w = 4186$ J/(kg·C°), $c_a = 920$ J/(kg·C°).
              Find: c_u.

Here we want to identify the metal. If we can find its specific heat (remember it is specific to the substance), we can identify the metal by looking up Table 11.1.

In the mixing process, the metal loses heat, and aluminum and the water gain heat.

The heat lost by metal is $Q_{lost} = c_u \, m_u \, \Delta T_u = c_u$ (0.600 kg)(20.2°C $-$ 100°C) $= -(47.9$ kg·C°)c_u.

The heat gained by aluminum and water is

$Q_{gained} = c_a \, m_a \, \Delta T_a + c_w \, m_w \, \Delta T_w = [920$ J/(kg·C°)](0.200 kg)(20.2°C $-$ 17.3°C)

    $+ [4186$ J/(kg·C°)](0.500 kg)(20.2°C $-$ 17.3°C) $= 6.60 \times 10^3$ J.

From calorimetry, $\Sigma Q = 0$, or the absolute value of heat lost = heat gained ($-Q_{lost} = Q_{gained}$),

we have $-[-(47.9$ kg·C°)$c_u] = 6.60 \times 10^3$ J.   Solving, $c_u = 138$ J/(kg·C°).

Looking up Table 11.1, we conclude that the metal is lead.

---

3. Phase Changes and Latent Heat (Section 11.3)

Matter normally exists in three phases: solid, liquid, and gas. **Latent heat** is the heat involved in a phase change such as from solid to liquid or from liquid to gas, and does not go into changing the temperature, but in breaking or forming molecular bonds.

The latent heat for a solid-liquid phase change is called the **latent heat of fusion** (L_f), which is the heat required to change 1 kg of solid to liquid while the temperature remains constant at the **freezing point**. The latent heat for a liquid-gas phase change is called the **latent heat of vaporization** (L_v), which is the heat required to change 1 kg of liquid to gas while the temperature remains constant at the **boiling point**. Therefore, the heat involved for a mass m is given by $Q = mL$, where L is the latent heat.

Example 11.3   How much heat energy is needed to change 10 kg of ice at $-20°C$ to steam at $120°C$?

Solution: Use the following subscripts: i = ice,   w = water,   and   s = steam.

Given: $m = 10$ kg, $c_i = 2100$ J/(kg·C°), $c_w = 4186$ J/(kg·C°), $c_s = 2010$ J/(kg·C°),
    $L_f = 3.33 \times 10^5$ J/kg, $L_v = 22.6 \times 10^5$ J/kg.

Find: Q_{tot} (total heat).

Ice at $-20°C$ first needs to be raised to $0°C$, and then melted to $0°C$ liquid water. The water at $0°C$ needs to be raised to $100°C$, where it is vaporized to steam. Finally, the temperature of steam is raised to $120°C$.

The heat required to raise the temperature of ice from $-20°C$ to $0°C$ is

$$Q_1 = c_i\, m_i\, \Delta T_i = [2100 \text{ J/(kg·C°)}](10 \text{ kg})[0°C - (-20°C)] = 4.2 \times 10^5 \text{ J.}$$

The heat required to melt ice at $0°C$ to water at $0°C$ is

$$Q_2 = mL_f = (10 \text{ kg})(3.33 \times 10^5 \text{ J/kg}) = 3.33 \times 10^6 \text{ J.}$$

The heat required to raise the temperature of water from $0°C$ to $100°C$ is

$$Q_3 = c_w\, m_w\, \Delta T_w = [4186 \text{ J/(kg·C°)}](10 \text{ kg})(100°C - 0°C) = 4.186 \times 10^6 \text{ J.}$$

The heat required to vaporize water at $100°C$ to steam at $100°C$ is

$$Q_4 = mL_v = (10 \text{ kg})(22.6 \times 10^5 \text{ J/kg}) = 22.6 \times 10^6 \text{ J.}$$

The heat required to raise the temperature of steam from $100°C$ to $120°C$ is

$$Q_5 = c_s\, m_s\, \Delta T_s = [2010 \text{ J/(kg·C°)}](10 \text{ kg})(120°C - 100°C) = 4.02 \times 10^5 \text{ J.}$$

So the total heat required is $Q_{tot} = \Sigma Q_i = Q_1 + Q_2 + Q_3 + Q_4 + Q_5 = 3.1 \times 10^7$ J.

Note: Q_4 (liquid to gas) is by far the largest heat required in the above five processes. That is why our bodies cool after perspiration evaporates.

---

4.   Heat Transfer (Section 11.4)

There are three mechanisms of heat transfer: **conduction**, **convection**, and **radiation**.

The **thermal conductivity** (k) characterizes the heat-conducting ability of a material. Its SI unit is J/(m·s·C°). Materials with large k values are called **thermal conductors** while materials with small k values are called **thermal insulators**. **Conduction** is the transfer of energy through a process of molecular collision. The heat conduction (flow) rate is given by $\dfrac{\Delta Q}{\Delta t} = \dfrac{kA\Delta T}{d}$, where A is the cross-sectional area, d the thickness, and ΔT is the temperature difference.

Convection involves mass transfer (conduction does not). For example, when cold water is in contact with a hot object, such as the bottom of a pot on a stove, the object transfers heat to the water by conduction, and the warmer water carries the heat away with it by convection as it rises (warmer water is less dense than cold water at the top).

Radiation is the transfer of energy by electromagnetic waves and requires no transfer medium (conduction and convection require a transfer medium). The rate at which an object radiates energy is given by $P = \sigma A e T^4$, where P is the power radiated in watts (W) or joules/s (J/s), e is the **emissivity** which characterizes the radiating property of the material and has dimensionless value between 0 and 1, A is the surface area of the object, T is the absolute (Kelvin) temperature of the object, and $\sigma = 5.67 \times 10^{-8}$ W/(m²·K⁴) is called the *Stefan-Boltzmann constant*. A good emitter of radiation is also a good absorber. A **black body** is a perfect emitter and absorber ($e = 1$).

If an object is at temperature T and its surroundings are at a temperature of T_s, then the net rate of energy loss or gain due to radiation alone is given by $P_{net} = \sigma A e(T_s^4 - T^4)$. The temperatures used in the radiation equations *must* be in Kelvin.

Example 11.4   A window glass 0.50-centimeter thick has dimensions of 2.0 m by 1.0 m. If the outside temperature is −10°C and the inside temperature 20°C,
(a) what is the rate of heat conduction through the window?
(b) How much heat flows through the window in 1.0 h due to conduction only?

Solution: Given: $d = 0.50$ cm, $A = LW = (2.0$ m$)(1.0$ m$) = 2.0$ m², $\Delta T = 20°C − (−10°C) = 30$ C°,
$k = 0.84$ J/(m·s·C°) (Table 11.3).

Find: (a) $\Delta Q/\Delta t$ (b) ΔQ.

(a) $\dfrac{\Delta Q}{\Delta t} = \dfrac{kA\,\Delta T}{d} = \dfrac{[0.84\ \text{J/(m·s·C°)}](2.0\ \text{m}^2)(30\ \text{C°})}{0.50 \times 10^{-2}\ \text{m}} = 1.0 \times 10^4\ \text{J/s} = 1.0 \times 10^4\ \text{W} = 10\ \text{kW}.$

(b) $\Delta Q = \dfrac{\Delta Q}{\Delta t}\,\Delta t = (1.0 \times 10^4\ \text{J/s})(1.0\ \text{h})(3600\ \text{s/h}) = 3.6 \times 10^7\ \text{J}.$

Example 11.5   A radiator has an emissivity of 0.70 and its exposed area is 1.50 m². The temperature of the radiator is 100°C and the surrounding temperature is 20°C. What is the net heat flow rate from the body?

Solution: Given: $e = 0.70,$ $A = 1.50\ \text{m}^2,$ $T = 100°\text{C} = 273 + 100 = 373\ \text{K},$

   $T_\text{s} = 20°\text{C} = 273 + 20 = 293\ \text{K}.$

   Find: $P_\text{net}.$

$P_\text{net} = \sigma A e(T_\text{s}^4 - T^4) = [5.67 \times 10^{-8}\ \text{W/(m}^2\cdot\text{K}^4)](1.50\ \text{m}^2)(0.70)[(293\ \text{K})^4 - (373\ \text{K})^4]$

$= -7.1 \times 10^2\ \text{J/s} = -7.1 \times 10^2\ \text{W}.$

The negative sign indicates that the radiator is losing energy.

IV.   Mathematical Summary

Mechanical Equivalent of Heat	1 kcal = 4186 J = 4.186 kJ 1 cal = 4.186 J		Gives the conversion between calorie and joules.
Specific Heat (specific heat capacity)	$Q = mc\Delta T$	(11.1)	Relates heat with mass, specific heat, and change in temperature.
Latent Heat	$L = mL$	(11.2)	Relates heat with latent heat and mass. (for water) $L_\text{f} = 3.3 \times 10^5$ J/kg (80 kcal/kg) $L_\text{v} = 22.6 \times 10^5$ J/kg (540 kcal/kg)
Thermal Conduction	$\dfrac{\Delta Q}{\Delta t} = \dfrac{kA\,\Delta T}{d}$	(11.3)	Calculates the heat conduction rate.
Stefan's Law	$P = \dfrac{\Delta Q}{\Delta t} = \sigma A e T^4$	(11.5)	Computes the heat power.
Net Radiant Power (Loss or Gain)	$P_\text{net} = \sigma A e(T_\text{s}^4 - T^4)$	(11.6)	Computes the net heat power.

V.   Solutions of Selected Exercises and Paired/Trio Exercises

7.   (a) The work done in each lift is $W = Fd = (20 \text{ kg})(9.80 \text{ m/s}^2)(1.0 \text{ m}) = 196 \text{ J}$.

The energy input is $E = 2800 \text{ Cal} = 2800 \times 10^3 \text{ cal} = (2800 \times 10^3 \text{ cal})(4.186 \text{ J/cal}) = 1.17 \times 10^7 \text{ J}$.

So the number of lifts required is $\dfrac{1.17 \times 10^7 \text{ J}}{196 \text{ J}} = \boxed{60\,000 \text{ times}}$.

(b) $t = 60\,000(5.0 \text{ s}) = 3.0 \times 10^5 \text{ s} = \boxed{83 \text{ h}}$.

10. $\boxed{\text{No}}$. A negative specific heat corresponds to an increase in temperature when heat is removed.

16.   From $Q = mc\Delta T$, we have $\Delta T = \dfrac{Q}{cm} = \dfrac{(200 \text{ J})}{[920 \text{ J/(kg·C°)}](5.0 \times 10^{-3} \text{ kg})} = 43°\text{C}$.

So the final temperature is $43°\text{C} + 20°\text{C} = \boxed{63°\text{C}}$.

18.   The heat gained by the cup is $Q_c = c_c\, m_c\, \Delta T_c = c_c\,(0.250 \text{ kg})(80°\text{C} - 20°\text{C}) = (15 \text{ kg·C°})c_c$,

The heat lost by the coffee is $Q_{cof} = [4186 \text{ J/(kg·C°)}](0.250 \text{ kg})(80°\text{C} - 100°\text{C}) = -2.093 \times 10^4 \text{ J}$.

From calorimetry, $-Q_{lost} = Q_{gained}$, we have $2.093 \times 10^4 \text{ J} = (15 \text{ kg·C°})c_c$.

Solving, $c_c = \boxed{1.40 \times 10^3 \text{ J/(kg·C°)}}$.

20. $\boxed{1.27 \text{ kg}}$

24.   The metal loses heat and the water and the cup gain heat.

The heat lost by the metal is $Q_m = cm\Delta T = c(0.150 \text{ kg})(30.5°\text{C} - 400°\text{C}) = -(55.43 \text{ kg·C°})c$.

The heat gained by the water and the cup is

$Q_G = [4186 \text{ J/(kg·C°)}](0.400 \text{ kg})(30.5°\text{C} - 10.0°\text{C}) + [920 \text{ J/(kg.C°)}](0.200 \text{ kg})(30.5°\text{C} - 10.0°\text{C})$

$= 3.810 \times 10^4 \text{ J}$.

From calorimetry, $-Q_{lost} = Q_{gained}$, we have $(55.43 \text{ kg·C°})c = 3.810 \times 10^4 \text{ J}$.

Solving, $c = \boxed{687 \text{ J/(kg·C°)}}$.

27.   Assume the final temperature is T. $Q = cm\Delta T$.

The heat lost by aluminum is $Q_a = [920 \text{ J/(kg·C°)}](0.100 \text{ kg})(T - 90.0°\text{C}) = (92 \text{ J/C°})(T - 90.0°\text{C})$,

The heat gained by water is $Q_w = [4186 \text{ J/(kg·C°)}](1.00 \text{ kg})(T - 20°\text{C}) = (4186 \text{ J/C°})(T - 20°\text{C})$.

From calorimetry, $-Q_{lost} = Q_{gained}$, we have $-92(T - 90.0°\text{C}) = 4186(T - 20°\text{C})$.

Solving, $T = \boxed{21.5°\text{C}}$.

33. This is due to the |high value of the latent heat of vaporization|. When steam condenses, it releases 2.26×10^6 J/kg of heat. When 100°C water drops its temperature by 1 C°, it releases only 4186 J/kg.

36. The melting point of lead is 328°C. So the temperature of the lead has to be increased to 328°C first. The total heat required is then

$$Q = cm\Delta T + mL_f = [130 \text{ J/(kg.C°)}](0.75 \text{ kg})(328°C - 20°C) + (0.75 \text{ kg})(0.25 \times 10^5 \text{ J/kg})$$

$$= \boxed{4.9 \times 10^4 \text{ J}}.$$

42. To completely melt the ice to water at 0°C it requires

$$Q_1 = cm\Delta T + mL_f = [2100 \text{ J/(kg·C°)}](0.60 \text{ kg})[0°C - (-10°C)] + (0.60 \text{ kg})(3.3 \times 10^5 \text{ J/kg}) = 2.11 \times 10^5 \text{ J}.$$

For 0.30 kg of water not to freeze into ice, it can only release

$$Q_2 = [4186 \text{ J/(kg)}](0.30 \text{ kg})(0°C - 50°C) = -6.28 \times 10^4 \text{ J} < Q_1.$$

So the temperature of the ice will increase to 0°C and then a portion of it will be melted.

Let the amount of ice melted be M. From calorimetry, $-Q_{lost} = Q_{gained}$,

we have $6.28 \times 10^4 \text{ J} = [2100 \text{ J/(kg·C°)}](0.60 \text{ kg})[0°C - (-10°C)] + M(3.3 \times 10^5 \text{ J/kg})$,

solving, $M = 0.15$ kg. Hence the total amount of water is $0.30 \text{ kg} + 0.15 \text{ kg} = \boxed{0.45 \text{ kg}}$.

45. If enough ice is added, the equilibrium temperature is 0°C.

Note 1 Liter of water has a mass of 1 kg.

The heat lost by the tea is $Q_{lost} = cm\Delta T = [4186 \text{ J/(kg·C°)}](0.75 \text{ kg})(0°C - 20°C) = 6.279 \times 10^4 \text{ J}.$

Let the mass of ice melted be M. $Q_{gained} = ML_f = (3.3 \times 10^5 M)$ J/kg.

From calorimetry, $-Q_{lost} = Q_{gained}$, we have $M = \dfrac{6.279 \times 10^4 \text{ J}}{3.3 \times 10^5 \text{ J/kg}} = 0.190$ kg.

So the total amount of liquid is $0.75 \text{ kg} + 0.19 \text{ kg} = \boxed{0.94 \text{ kg or } 0.94 \text{ L}}$.

49. The heat lost by the ceramic is

$$Q_{lost} = cm\Delta T = [840 \text{ J/(kg·C°)}](0.150 \text{ kg})(-196°C - 20°C) = 2.72 \times 10^4 \text{ J}.$$

Let the mass of nitrogen boiled be M. $Q_{gained} = ML_v = (2.0 \times 10^5 M)$ J/kg.

From calorimetry, $-Q_{lost} = Q_{gained}$, we have $M = \dfrac{2.72 \times 10^4 \text{ J}}{2.0 \times 10^5 \text{ J/kg}} = 0.136$ kg.

Note that 1 liter of liquid nitrogen has a mass of 0.80 kg (the density of liquid nitrogen is 0.80 kg/L).

From $\rho = \dfrac{m}{V}$, we have $V = \dfrac{m}{\rho} = \dfrac{0.136 \text{ kg}}{0.80 \text{ kg/L}} = \boxed{0.17 \text{ L}}$.

54. The bridge is exposed to the cold air above and below while the road is exposed only above. So more heat is removed from the bridge than the road. This results in a fast freeze of the water on the bridge.

59. From $\dfrac{\Delta Q}{\Delta t} = \dfrac{kA\Delta T}{d}$, we have $\dfrac{(\Delta Q/\Delta t)_t}{(\Delta Q/\Delta t)_o} = \dfrac{k_t}{k_o} = \dfrac{0.67 \text{ J/(m·s·C°)}}{0.15 \text{ J/(m·s·C°)}} = \boxed{4.5 \text{ times}}$.

62. The normal body temperature is 37°C.

$$\frac{\Delta Q}{\Delta t} = \frac{kA\Delta T}{d} = \frac{[0.20 \text{ J/(m·s·C°)}](1.5 \text{ m}^2)(37°C - 33°C)}{0.040 \text{ m}} = \boxed{30 \text{ J/s}}.$$

64. $\boxed{d_{Cu} = 1.6\, d_{Al}}$.

66. (a) The greater the R-value, the greater the insulation value.

(b) (1) $R = \dfrac{L}{k}$, ☞ $L = R\,k$.

So: $\dfrac{L_{\text{fiberboard}}}{L_{\text{foam plastic}}} = \dfrac{k_{\text{fiberboard}}}{k_{\text{foam plastic}}} = \dfrac{0.059 \text{ J/(m·s·C°)}}{0.042 \text{ J/(m·s·C°)}} = 1.40$.

Therefore $L_{\text{fiberboard}} = (1.40)(3.0 \text{ in}) = \boxed{4.2 \text{ in}}$.

(2) $\dfrac{L_{\text{brick}}}{L_{\text{foam plastic}}} = \dfrac{0.71 \text{ J/(m·s·C°)}}{0.042 \text{ J/(m·s·C°)}} = 16.9$. So $L_{\text{brick}} = 16.9(3.0 \text{ in}) = \boxed{51 \text{ in}}$.

71. $\dfrac{\Delta Q}{\Delta t} = \dfrac{949 \text{ H/s}}{2} = 474.5 \text{ J/s} =$

$$\frac{(3.5 \text{ m})(5.0 \text{ m})[20°C-(-10°C)]}{(0.020 \text{ m})/[0.059 \text{ J/(m·s·C°)}]+d/[0.042 \text{ J/(m·s·C°)}]+(0.150 \text{ m})/[1.3 \text{ J/(m·s·C°)}]+(0.070 \text{ m})/[0.71 \text{ J/(m·s·C°)}]}$$

Solving for $d = 0.023 \text{ m} = \boxed{2.3 \text{ cm}}$.

79. Let the initial temperature of the shot be T.

The heat lost by the shot is

$Q_{\text{copper}} = cm\Delta T = [390 \text{ J/(kg·C°)}](0.150 \text{ kg})(28°C - T) = (58.5 \text{ J/C°})(28°C - T)$,

the heat gained by water is

$Q_{\text{water}} = [4186 \text{ J/(kg·C°)}](0.200 \text{ kg})(28°C - 25°C) = 2512 \text{ J}$,

the heat gained by the cup is

$Q_{\text{cup}} = [920 \text{ J/(kg·C°)}](0.375 \text{ kg})(28°C - 25°C) = 1035 \text{ J}$.

From calorimetry, $-Q_{\text{lost}} = Q_{\text{gained}}$, we have $-(58.5 \text{ J/C°})(28°C - T) = 2512 \text{ J} + 1035 \text{ J}$.

Solving, $T = \boxed{88.7°C}$.

81. 25 km/h $= 6.95$ m/s.   The kinetic energy is $K = \frac{1}{2}mv^2 = \frac{1}{2}(65 \text{ kg})(6.95 \text{ m/s})^2 = 1.57 \times 10^3$ J.

So $Q = 0.40K = 627$ J.

From $Q = mL_f$, we have $m = \dfrac{Q}{L_f} = \dfrac{627 \text{ J}}{3.3 \times 10^5 \text{ J/kg}} = 1.9 \times 10^{-3}$ kg $= \boxed{1.9 \text{ g}}$.

The rest of the energy goes to heating the skates and losing to the environment.

VI.   Practice Quiz

1. The physiological process of body cooling by perspiration is based on

   (a) specific heat.   (b) latent heat of fusion.   (c) latent heat of vaporization.   (d) heat capacity.

2. A machine gear consists of 0.10 kg of iron and 0.16 kg of copper. How much heat is added to the gear if its temperature increases by 35 C°. [The specific heats for iron and copper are 460 and 390 J/(kg·C°).]

   (a) 9.1×10^2 J   (b) 3.8×10^3 J   (c) 4.0×10^3 J   (d) 4.4×10^3 J   (e) none of the above

3. How much heat is required to change 0.500 kg of ice at $0°$C to water at $50°$C?

   (a) 1.05×10^5 J   (b) 1.67×10^5 J   (c) 2.71×10^5 J   (d) 4.38×10^5 J   (e) 1.13×10^6 J

4. A 0.10 kg piece of cooper, initially at $95°$C, is dropped into 0.20 kg of water contained in a 0.28-kg aluminum calorimeter; the water and calorimeter are initially at $15°$C. What is the final temperature of the system?

   (a) 17.8 °C   (b) $18.3°$C   (c) $19.2°$C   (d) $23.7°$C   (e) $25.0°$C

5. If the absolute temperature of a radiator is doubled, by what factor does the radiating power change?

   (a) 2 (b) 4 (c) 8 (d) 16 (e) 32

6. The reason large lake temperatures do not vary drastically is that

   (a) water has a relatively high rate of conduction.   (b) water is a good radiator.

   (c) water is a poor heat conductor.   (d) water has a lower specific heat.

   (e) water has a relatively high specific heat.

7. You are designing a water heater so it can heat 20 kg of water from $20°$C to $100°$C in 20 minutes. What should be the power of the heating element? (Assume no heat loss.)

   (a) 2.8×10^2 J/s   (b) 1.2×10^3 J/s   (c) 5.6×10^3 J/s   (d) 3.3×10^5 J/s   (e) 6.7×10^6 J/s

8.   On a cold day, a piece of metal feels much colder to the touch than a piece of wood. This is due to the difference in which of the following physical properties?

(a) density   (b) specific heat   (c) latent heat   (d) thermal conductivity   (e) temperature

9.   Equal masses of water at 20°C and 80°C are mixed. What is the final temperature of the mixture?

(a) 65°C   (b) 50°C   (c) 40°C   (d) 30°C   (e) 25°C

10.   A person makes ice tea by adding ice to 1.8 kg of hot tea (water), initially at 80°C. How many kilograms of ice, initially at 0°C, are required to bring the mixture to 10°C?

(a) 0.78 kg   (b) 1.0 kg   (c) 1.2 kg   (d) 1.5 kg   (e) 1.8 kg

Answers to Practice Quiz:

1. c 2. b 3. c 4. a 5. d 6. e 7. c 8. d 9. b 10. d

CHAPTER 12

Thermodynamics

I. Chapter Objectives

Upon completion of this chapter, you should be able to:

1. define thermodynamic systems and states of systems, and explain how thermal processes affect such systems.

2. explain the relationship among internal energy, heat, and work as expressed by the first law, and analyze various thermal processes.

3. state and explain the second law of thermodynamics in several forms and explain the concept of entropy.

4. explain the concept of a heat engine and compute thermal efficiency, and explain the concept of a thermal pump and compute coefficient of performance.

5. explain how the Carnot cycle applies to heat engines, compute the ideal Carnot efficiency, and state the third law of thermodynamics.

II. Key Terms

Upon completion of this chapter, you should be able to define and/or explain the following key terms:

thermodynamics	second law of thermodynamics
system	entropy
thermally isolated system	isentropic process
heat reservoir	heat engine
equations of state	thermal cycle
process	thermal efficiency
irreversible process	thermal pump
reversible process	coefficient of performance (COP)
first law of thermodynamics	heat pump
isobaric process	Carnot cycle
isometric process	Carnot (ideal) efficiency
isothermal process	relative efficiency
adiabatic process	third law of thermodynamics

III.   Chapter Summary and Discussion

1.   Thermodynamic Systems, States, and Processes (Section 12.1)

Thermodynamics deals with the transfer or action (dynamics) of heat. A **system** is a definite quantity of matter enclosed by boundaries or surfaces, either real or imaginary. An *open system* or *closed system* refers to whether or not mass can be transferred into or out of a system.

There is no heat transfer in or out of a **thermally isolated system**. A **completely isolated system** has no energy exchange (that is, no interaction at all) with its surroundings.

A **heat reservoir** is a system with an unlimited heat capacity so that the addition or removal of heat causes no change in its temperature.

A thermodynamic **equation of state** is a mathematical relationship of the thermodynamic or *state variables*, such as pressure, volume and temperature. For an ideal gas, $pV = Nk_BT$ is the equation of state. A **process** is a *change* in the state or the thermodynamic variables of a system. A **reversible process** is one with a known path through equilibrium states, whereas an **irreversible process** is one for which the intermediate steps are nonequilibrium states. Irreversible does not mean that the system can't be taken back to the initial state; it only means the process path can't be retraced, because of the nonequilibrium conditions.

2.   The First Law of Thermodynamics (Section 12.2)

The **first law of thermodynamics** is a statement of the conservation of energy for thermodynamic systems. It relates the change in internal energy, heat, and work: $Q = \Delta U + W$, where Q is heat, ΔU is the change in internal energy, and W is work.

Note: Be aware of the strict sign convention for work and heat: $+Q$, means heat *added*; $-Q$, means heat *removed*; $+W$, means work done *by* system; $-W$, means work done *on* system.

The work done by an expanding gas at constant pressure is given by $W = p\Delta V = p(V_2 - V_1)$. For processes in which the pressure is not a constant, work can be calculated graphically by computing the area under the curve in a pressure versus volume graph.

"Iso-" processes are ones in which one of the thermodynamic variables is held constant. For example: **isobaric** (constant pressure), **isometric** (constant volume), **isothermal** (constant temperature or constant internal energy for an ideal gas), **adiabatic** ($Q = 0$ or no heat transfer). In solving problems involving these iso processes, make sure to identify the quantity which remains constant in the process.

Example 12.1   A system consists of 3.0 kg of water at 80°C. 25 J of work is done on the system by stirring with a paddle wheel, while 63 J of heat is removed. What is the change in internal energy of the system?

Solution: Be careful about the sign convention for heat and work.

Given: $W = -25$ J (done on system), $Q = -63$ J (heat removed).

Find: ΔU.

From the first law of thermodynamics, $Q = \Delta U + W$,

we have $\Delta U = Q - W = -63$ J $- (-25$ J$) = -63$ J $+ 25$ J $= -38$ J.

Since $\Delta U < 0$, the internal energy of the system decreases.

Example 12.2   During an adiabatic process, an ideal gas gained 5.0 J of internal energy.

(a) What is the heat involved in the process? (b) What is the work done in the process?

Solution: Given: adiabatic, $\Delta U = +5.0$ J (gained).

Find: (a) Q (b) W.

(a) In an adiabatic process, there is no heat exchange. So $Q = 0$.

(b) From the first law of thermodynamics, $Q = \Delta U + W$,

we have $W = Q - \Delta U = 0 - 5.0$ J $= -5.0$ J.

That is, 5.0 J of work is done on the system (since it is a gas, this means it is compressed).

Example 12.3   In an isometric process, the internal energy of a system decreases by 50 J.

(a) What is the work done? (b) What is the heat exchange?

Solution: Given: isometric process, $\Delta U = -50$ J (decrease).

Find: (a) W (b) Q.

(a) In an isometric process, the volume is a constant, so $\Delta V = 0$.

Therefore, $W = p\Delta V = 0$.

(b) $Q = \Delta U + W = -50$ J $+ 0 = -50$ J. That is, 50 J of heat is removed from the system.

---

3. The Second Law of Thermodynamics and Entropy (Section 12.3)

The **second law of thermodynamics** specifies the direction in which a process can naturally or spontaneously take place. Many equivalent statements are used to describe the second law according to the situation or application.

1. *Heat does not flow spontaneously from a colder body to a warmer body.* This does not mean heat cannot flow from a colder body to a warmer body, it just cannot take place spontaneously.

2. *In a thermal cycle, heat energy cannot be completely transformed into mechanical work.* This does not mean energy is not conserved. the concept is that 100% of heat energy cannot be converted to mechanical work in a cycle.

3. *It is impossible to construct an operational perpetual motion engine.*

4. *The total entropy of the universe increases in every natural process.*

Entropy is a measure of the disorder of a system and is related to heat and temperature. The change in entropy (ΔS) when an amount of heat (Q) is added to (or removed from) an object by a reversible process at a constant temperature (T) is given by $\Delta S = \dfrac{Q}{T}$, where T is the Kelvin temperature and the units for entropy are joules per kelvin (J/K). For a system, the total change in entropy is the addition of the changes in entropy of the objects in the system, $\Delta S = \sum\limits_{i} (\Delta S_i)$, where ΔS_i is the change in entropy of the ith object in the system.

Example 12.4   1.0 kg of water at 0°C freezes to ice at 0°C. What is the change in entropy in the process?

Solution: Given: $m = 1.0$ kg, $T = 0°C = 273$ K, $L_f = 3.33 \times 10^5$ J/kg.

Find: ΔS.

Since the heat removed in the process, we have $Q = -mL_f = -(1.0 \text{ kg})(3.33 \times 10^5 \text{ J/kg}) = -3.33 \times 10^5$ J.

So $\Delta S = \dfrac{Q}{T} = \dfrac{-3.33 \times 10^5 \text{ J}}{273 \text{ K}} = -1.2 \times 10^3$ J/K. The entropy decreases (more order in the final state).

Example 12.5   Show that it is a violation of the second law of thermodynamics to have some heat , say 10 J, to transfer spontaneously from a cold reservoir at 300 K to a hot reservoir at 350 K.

Solution: Given: $Q_{cold} = -10$ J, $Q_{hot} = 10$ J, $T_{cold} = 300$ K, $T_{hot} = 350$ K.

Find: ΔS (system).

According to the second law of thermodynamics, the total entropy of an isolated system increases for every natural process. That is, the change in entropy for all natural process must be greater than zero.

For the cold reservoir: $\Delta S_{cold} = \dfrac{Q_{cold}}{T_{cold}} = \dfrac{-10 \text{ J}}{300 \text{ K}} = -3.33 \times 10^{-2} \text{ J/K}$.

For the hot reservoir: $\Delta S_{hot} = \dfrac{Q_{hot}}{T_{hot}} = \dfrac{10 \text{ J}}{350 \text{ K}} = 2.86 \times 10^{-2} \text{ J/K}$.

The total change in entropy for the system (both reservoirs) is then

$\Delta S = \Delta S_{cold} + \Delta S_{hot} = -3.33 \times 10^{-2} \text{ J/K} + 2.86 \times 10^{-2} \text{ J/K} = -4.7 \times 10^{-3} \text{ J/K}$.

The total change in entropy for the process is negative, a decrease. This is a violation of the second law of thermodynamics, so the process described in the example *cannot* spontaneously take place in isolation.

---

4. Heat Engines and Thermal Pumps (Section 12.4)

A **heat engine** is any cyclic device that converts heat to work. It absorbs heat (Q_{hot} or Q_{in}) from a high-temperature reservoir, does net work (W_{net}), and exhausts heat (Q_{cold} or Q_{out}) to a low-temperature reservoir. After one thermal cycle, every thermodynamic variable returns to its original value including the internal energy U, so $\Delta U = 0$. Therefore, from the first law of thermodynamics, $W_{net} = Q_{hot} - Q_{cold}$. The **thermal efficiency** of a heat

engine is defined as the ratio of its work output and its heat input: $\varepsilon_{th} = \dfrac{W_{net}}{Q_{hot}} = \dfrac{Q_{hot} - Q_{cold}}{Q_{hot}} = 1 - \dfrac{Q_{cold}}{Q_{hot}}$.

A **thermal pump** is a device that transfers heat energy from a low temperature reservoir to a high temperature one. To do this, work must be done, according to the second law of thermodynamics, since it will not happen on its own. The **coefficient of performance** (COP) for a refrigerator or air conditioner is expressed as

$COP_{ref} = \dfrac{Q_{cold}}{W_{in}} = \dfrac{Q_{cold}}{Q_{hot} - Q_{cold}}$. The coefficient of performance for a heat pump in its heating mode is defined as

$COP_{hp} = \dfrac{Q_{hot}}{W_{in}} = \dfrac{Q_{hot}}{Q_{hot} - Q_{cold}}$. These two COP's are defined differently because for a refrigerator, performance is measured by how much heat is removed from the cold reservoir, however for a heat pump, performance is measured by how much heat is delivered to the hot reservoir. For either case, for a cycle, $Q_{hot} = W_{in} + Q_{cold}$, according to the first law of thermodynamics.

Example 12.6   A heat engine has an efficiency of 25% and extracts 120 J of heat from a hot reservoir per cycle.

(a) How much net work does it perform in each cycle?

(b) How much heat does it exhaust in each cycle?

Solution: Given: $\varepsilon_{th} = 25\% = 0.25,$ $Q_{hot} = 120$ J.

Find: (a) W_{net} (b) Q_{cold}.

(a) From $\varepsilon_{th} = \dfrac{W_{net}}{Q_{hot}}$, we have $W_{net} = \varepsilon_{th} Q_{hot} = (0.25)(120 \text{ J}) = 30$ J.

(b) Since $W_{net} = Q_{hot} - Q_{cold}$, we have $Q_{cold} = Q_{hot} - W_{net} = 120$ J $- 30$ J $= 90$ J.

Example 12.7   A refrigerator removes heat from the freezing compartment of a freezer at the rate of 20 kJ per cycle and ejects 24 kJ per cycle into a room.

(a) How much work input is required in each cycle?

(b) What is the coefficient of performance of this refrigerator?

Solution: Given: $Q_{cold} = 20$ kJ $= 20 \times 10^3$ J, $Q_{hot} = 24$ kJ $= 24 \times 10^3$ J.

Find: (a) W_{in} (b) COP_{ref}.

(a) $Q_{hot} = W_{in} + Q_{cold}$,   so $W_{in} = Q_{hot} - Q_{cold} = 24 \times 10^3$ J $- 20 \times 10^3$ J $= 4.0 \times 10^3$ J.

(b) $COP_{ref} = \dfrac{Q_{cold}}{W_{in}} = \dfrac{20 \times 10^3 \text{ J}}{4.0 \times 10^3 \text{ J}} = 5.0,$

or $COP_{ref} = \dfrac{Q_{cold}}{Q_{hot} - Q_{cold}} = \dfrac{20 \times 10^3 \text{ J}}{24 \times 10^3 \text{ J} - 20 \times 10^3 \text{ J}} = 5.0.$

---

5.   The Carnot Cycle and Ideal Heat Engines (Section 12.5)

The ideal **Carnot cycle** for a heat engine consists of two isothermal and two adiabatic processes. This is the most efficient engine, and the **Carnot efficiency** ε_C gives the upper limit of efficiency (unattainable in practice): $\varepsilon_C = 1 - \dfrac{T_{cold}}{T_{hot}} = \dfrac{T_{hot} - T_{cold}}{T_{hot}}$, where T_{cold} and T_{hot} are the absolute (Kelvin) temperatures of the low-temperature and high-temperature reservoirs, respectively. **Relative efficiency** is the ratio of the thermal efficiency to the Carnot efficiency. For example, if a gasoline engine has a thermal efficiency of 20% and a Carnot efficiency of 30%, then its relative efficiency is $\varepsilon_{rel} = \dfrac{\varepsilon_{th}}{\varepsilon_C} = \dfrac{0.20}{0.30} = 67\%.$

The third law of thermodynamics states that absolute zero has never been observed experimentally.

Example 12.8   What is the Carnot efficiency of a gasoline engine which operates between 450 K and room temperature, 300 K?

Solution: **Given:** $T_{hot} = 450$ K, $T_{cold} = 300$K.

                  Find: ε_C.

$$\varepsilon_C = 1 - \frac{T_{cold}}{T_{hot}} = 1 - \frac{300 \text{ K}}{450 \text{ K}} = 0.33 = 33\%.$$

Remember this is the unattainable, upper limit! No gasoline engine can have a higher efficiency if it is operating between these two temperatures!

IV. Mathematical Summary

First Law of Thermodynamics	$Q = \Delta U + W$ (12.1)	Relates heat, change in internal energy, and work. $+Q$, heat *added* to system $-Q$, heat *removed* from system $+W$, work done *by* system $-W$, work done *on* system
Work Done by a Gas (constant pressure)	$W = p\Delta V = p(V_2 - V_1)$ (12.2)	Calculates the work done by an expanding gas at constant pressure.
Change in Entropy (constant temperature)	$\Delta S = \dfrac{Q}{T}$ (12.7)	Defines the change in entropy at constant temperature.
Thermal Efficiency of a heat engine	$\varepsilon_{th} = \dfrac{W_{net}}{Q_{in}} = \dfrac{Q_{hot} - Q_{cold}}{Q_{hot}}$ $= 1 - \dfrac{Q_{cold}}{Q_{hot}}$ (12.9)	Defines and computes the thermal efficiency of a heat engine.
Coefficient of Performance of a refrigerator or air conditioner	$COP_{ref} = \dfrac{Q_{cold}}{W_{in}} = \dfrac{Q_{cold}}{Q_{hot} - Q_{cold}}$ (12.10)	Defines and calculates the coefficient of performance of a refrigerator or air conditioner.
Coefficients of Performance of a Heat Pump in Heating Mode	$COP_{hp} = \dfrac{Q_{hot}}{W_{in}} = \dfrac{Q_{hot}}{Q_{hot} - Q_{cold}}$ (12.11)	Defines and computes the coefficient of performance of a heat pump in heating mode. A heat pump in cooling mode works like a refrigerator or air conditioner.
Carnot Efficiency of an Ideal Heat Engine	$\varepsilon_C = 1 - \dfrac{T_{cold}}{T_{hot}} = \dfrac{T_{hot} - T_{cold}}{T_{hot}}$ (12.13)	Defines and calculates the Carnot efficiency of an ideal heat engine.

V. Solutions of Selected Exercises and Paired/Trio Exercises

5.

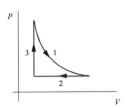

6. (a) isothermal   (b) isobaric   (c) isometric

   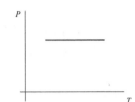   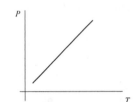

12. Since the container is rigid, $\Delta V = 0$, so $W = 0$. $Q = \Delta U + W = \Delta U = \boxed{2.0 \times 10^4 \text{ J}}$.

17. (a) $W = p\Delta V = (1.65 \times 10^4 \text{ Pa})(0.40 \text{ m}^3 - 0.20 \text{ m}^3) = \boxed{3.3 \times 10^3 \text{ J}}$.

(b) $\boxed{\text{Yes}}$, the internal energy of the system changes.

From the first law of thermodynamics, $Q = \Delta U + W$,

we have $\Delta U = Q - W = 1000 \text{ J} - 3.3 \times 10^3 \text{ J} = \boxed{-2.3 \times 10^3 \text{ J}}$.

22. Work equals the area under the curve.

From 1–2, $W = \boxed{0}$;   from 2–3, $W = (0.50 \times 10^5 \text{ Pa})(0.50 \text{ m}^3) = \boxed{2.5 \times 10^4 \text{ J}}$;

from 3–4, $W = \boxed{0}$;   from 4–5, $W = (1.00 \times 10^5 \text{ Pa})(0.25 \text{ m}^3) = \boxed{2.5 \times 10^4 \text{ J}}$.

24. $\boxed{-1.8 \times 10^5 \text{ J}}$.

29. There must be energy created for the change in entropy to be negative.

32. $\Delta S = \dfrac{Q}{T} = \dfrac{-mL_v}{T} = \dfrac{-(0.50 \text{ kg})(22.6 \times 10^5 \text{ J/kg})}{373 \text{ K}} = \boxed{-3.0 \times 10^3 \text{ J/K}}$.

38. (a) $\Delta S_h = \dfrac{Q_h}{T_h} = \dfrac{1000 \text{ J}}{373 \text{ K}} = \boxed{2.68 \text{ J/K}}$.

 (b) $\Delta S_c = \dfrac{-1000 \text{ J}}{273 \text{ K}} = \boxed{-3.66 \text{ J/K}}$.

 (c) $\Delta S = \Delta S_h + \Delta S_c = 2.68 \text{ J/K} - 3.66 \text{ J/K} = \boxed{-0.98 \text{ J/K}}$.

43. (a) $\Delta S = \dfrac{Q}{T} = \dfrac{mL_f}{T} = \dfrac{(0.0500 \text{ kg})(3.33 \times 10^5 \text{ J/kg})}{273 \text{ K}} = \boxed{61.0 \text{ J/K}}$.

 (b) From Calorimetry, the heat lost by water is also $(0.0500 \text{ kg})(3.33 \times 10^5 \text{ J/kg}) = 1.67 \times 10^4 \text{ J}$.

 The temperature change of the water is $\Delta T = \dfrac{Q}{cm} = \dfrac{-1.67 \times 10^4 \text{ J}}{[4186 \text{ J/(kg·C°)}](0.500 \text{ kg})} = -7.98 \text{ C°}$.

 So the final temperature of the water is 12.0°C.

 Therefore the average water temperature is $(20 + 12)/2 = 16°C$.

 Use the average temperature of the water in entropy calculation.

 $\Delta S = \dfrac{-1.67 \times 10^4 \text{ J}}{289 \text{ K}} = \boxed{-57.8 \text{ J/K}}$.

 (c) The total change in entropy is $\Delta S = 61.0 \text{ J/K} - 57.8 \text{ J/K} = \boxed{3.2 \text{ J/K}}$.

49. $\boxed{\text{No}}$, the warm air rises to the higher altitude and gravity and buoyancy do work. Since it is a natural process with work input, the entropy increases and the second law is not violated.

52. From the definition of thermal efficiency, $\varepsilon_{th} = \dfrac{W_{net}}{Q_{in}}$,

 we have $Q_{in} = \dfrac{W_{net}}{\varepsilon_{th}} = \dfrac{800 \text{ J}}{0.40} = 2000 \text{ J}$.

 So $Q_{out} = Q_{in} - W_{net} = 2000 \text{ J} = 2000 \text{ J} - 800 \text{ J} = \boxed{1.20 \times 10^3 \text{ J}}$.

54. $\boxed{35\%}$.

58. Since $Q_{in} = W_{net} + Q_{out}$, $P_{in} = P_{net} + P_{out} = 2.5 \text{ kW} + 7.5 \text{ kW} = \boxed{10 \text{ kW}}$.

63. (a) $Q_{\text{out}} = W_{\text{in}} + Q_{\text{in}} = 3.0 \times 10^4 \text{ J} + 2.1 \times 10^5 \text{ J} = \boxed{2.4 \times 10^5 \text{ J}}$.

(b) The heat lost by water is equal to $Q_{\text{in}} = -2.1 \times 10^5 \text{ J} = cm\Delta T$,

so $\Delta T = \dfrac{-2.1 \times 10^5 \text{ J}}{[4186 \text{ J/(kg·C°)}](5.0 \text{ kg})} = \boxed{-10 \text{ C°}}$.

68. (a) No, the product of p and V has to be a constant for T to be constant. $Q_{\text{in}} = +Q_4$.

(b) No. $Q_{\text{out}} = -Q_2$.

(c) Compression, negative work.   Expansion, positive work.   Net work

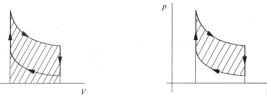

72. $\varepsilon_C = 1 - \dfrac{T_{\text{cold}}}{T_{\text{hot}}} = 1 - \dfrac{(273 + 20) \text{ K}}{(273 + 100) \text{ K}} = \boxed{21\%}$.

79. From $\varepsilon_c = 1 - \dfrac{T_{\text{cold}}}{T_{\text{hot}}}$, we have $T_{\text{hot}} = \dfrac{T_{\text{cold}}}{1 - \varepsilon_c}$. So $\dfrac{T'_{\text{hot}}}{T_{\text{hot}}} = \dfrac{1 - \varepsilon_c}{1 - \varepsilon'_c} = \dfrac{1 - 0.30}{1 - 0.40} = 1.17$.

Therefore $\Delta T_{\text{hot}} = \dfrac{T'_{\text{hot}} - T_{\text{hot}}}{T_{\text{hot}}} = 0.17 T_{\text{hot}} = 0.17(273 + 327) \text{ K} = \boxed{100 \text{ C°}}$.

82. The ideal efficiency is $\varepsilon_c = 1 - \dfrac{T_{\text{cold}}}{T_{\text{hot}}} = 1 - \dfrac{(273 + 100) \text{ K}}{(273 + 400) \text{ K}} = \boxed{44.6\%}$.

So the thermal efficiency is $\varepsilon_{\text{th}} = \varepsilon_{\text{rel}} \varepsilon_C = 0.45(44.6\%) = \boxed{20.1\%}$.

88. (a) $\text{COP}_C = \dfrac{Q_{\text{hot}}}{Q_{\text{hot}} - Q_{\text{cold}}} = \dfrac{T_{\text{hot}}}{T_{\text{hot}} - T_{\text{cold}}}$.

(b) The efficiency improves as the temperature difference between the two reservoirs decreases. The Carnot coefficient of performance, COC_C, of a refrigerator should be $\dfrac{T_{\text{cold}}}{T_{\text{hot}} - T_{\text{cold}}}$

95. (a) It is an $\boxed{\text{isobaric expansion}}$.

(b) From $Q = \Delta U + W$, we have $\Delta U = Q - W = Q - p\Delta V = Q - pA\Delta d$

$$= 420 \text{ J} - (1.01 \times 10^5 \text{ Pa})(\pi)(0.120 \text{ m})^2(0.0600 \text{ m}) = \boxed{146 \text{ J}}.$$

99. In an isothermal process, $\Delta U = 0$ for an ideal gas. So $Q = \Delta U + W = W = 30$ J.

$$\Delta S = \frac{Q}{T} = \frac{30 \text{ J}}{(273 + 27) \text{ K}} = \boxed{0.10 \text{ J/K}}.$$

102. $\Delta S = \dfrac{Q}{T} = \dfrac{mL_f}{T} = \dfrac{(0.75 \text{ kg})(3.3 \times 10^5 \text{ J/kg})}{273 \text{ K}} = \boxed{9.1 \times 10^2 \text{ J/K}}.$

VI. Practice Quiz

1. An ideal gas undergoes an *isothermal* process in doing 25 J of work. What is the change in internal energy?

   (a) 0   (b) 50 J   (c) 25 J   (d) −50 J   (e) −25 J

2. A heat engine receives 6000 J of heat from its combustion process and loses 4000 J through the exhaust and friction. What is its efficiency?

   (a) 33%   (b) 40%   (c) 60%   (d) 67%   (e) 73%

3. According to the second law of thermodynamics, for any process that may occur within an isolated system, which of the following choices applies?

   (a) entropy remains constant   (b) entropy increases   (c) entropy decreases

   (d) both (a) and (b)   (e) both (a) and (c)

4. The efficiency of a Carnot engine is 26.0%. What is the temperature of the cold reservoir if the temperature of the hot reservoir is 450 K?

   (a) 113 K   (b) 225 K   (c) 333 K   (d) 359 K   (e) 424 K

5. In an *adiabatic* process, there is

   (a) no pressure change.   (b) no volume change.   (c) no temperature change.   (d) no heat change.

   (e) no internal energy change.

6. When 0.50 kg of water at 100°C vaporizes, the change in entropy is

   (a) 3.0×10^4 J/K.   (b) 6.0×10^4 J/K.   (c) 1.1×10^5 J/K.   (d) 2.3×10^5 J/K.   (e) 1.1×10^6 J/K.

7. The work done on an ideal gas system during an isothermal process is −400 J. What is the change in internal energy?

   (a) zero   (b) −400 J   (c) −800 J   (d) 400 J   (e) 800 J

8. If a heat engine has an efficiency of 25.0% and its heat input 600 J per cycle from the high-temperature reservoir. What is the rate of heat output to the low-temperature reservoir per cycle?

    (a) 150 J   (b) 450 J   (c) 575 J   (d) 800 J   (e) 2400 J

9. During each cycle of operation, a refrigerator absorbs 230 J of heat from the freezer compartment and expels 356 J of heat to the room. How much work input is required in each cycle?

    (a) zero   (b) 460 J   (c) 712 J   (d) 586 J   (e) 126 J

10. What is the work done by the gas as it expands from pressure p_1 and volume V_1 to pressure p_2 and volume V_2 along a straight line path as shown?

    (a) zero   (b) $\frac{1}{2}(p_1 + p_2)(V_1 + V_2)$ (c) $\frac{1}{2}(p_1 - p_2)(V_1 + V_2)$

    (d) $\frac{1}{2}(p_1 + p_2)(V_1 - V_2)$ (e) $\frac{1}{2}(p_1 - p_2)(V_1 - V_2)$

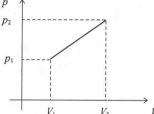

Answers to Practice Quiz:

1.a 2.a 3.d 4.c 5.d 6.a 7.a 8.e 9.b 10.d

CHAPTER 13

Vibrations and Waves

I. Chapter Objectives

Upon completion of this chapter, you should be able to:

1. describe simple harmonic motion and relate energy and speed in such motion.

2. understand the equation of motion for SHM and explain what is meant by phase and phase differences.

3. describe wave motion in terms of various parameters and identify different types of waves.

4. explain various wave properties and resulting phenomena.

5. describe the formation and characteristics of standing waves and explain the phenomenon of resonance.

II. Key Terms

Upon completion of this chapter, you should be able to define and/or explain the following key terms:

Hooke's law	interference
simple harmonic motion (SHM)	constructive interference
displacement	destructive interference
amplitude	total constructive interference
period	total destructive interference
frequency	reflection
hertz (Hz)	refraction
equation of motion	dispersion
phase constant	diffraction
damped harmonic motion	standing wave
wave	node
wave motion	antinode
wavelength	natural (resonant) frequencies
wave speed	fundamental frequency
transverse wave	harmonic series
longitudinal wave	resonance
principle of superposition	

III. Chapter Summary and Discussion

1. Simple Harmonic Motion (Section 13.1)

The motion of an oscillating object depends on the restoring forces that make it go back and forth. The simplest type of restoring force is a spring force described by Hooke's law, $F_s = -kx$, where k is the spring constant, and the negative sign indicates that the force is opposite to the displacement from the spring's relaxed position.

Motion under the influence of the type of force described by Hooke's law is called **simple harmonic motion (SHM)**, because this force is the simplest restoring force and because the motion can be described by harmonic functions (sines and cosines). SHM is described by the following parameters:

amplitude (A):   the magnitude of the maximum displacement of a mass from its equilibrium position,

period (T):   the time needed to complete one cycle of oscillation,

frequency (f):   the number of cycles per second.

The frequency and period are related by $f = \dfrac{1}{T}$. The SI unit of frequency is 1/s, or hertz (Hz). We sometimes descriptively say cycles/s (cycles per second, cps).

The total mechanical energy of an object in SHM is directly proportional to the square of its amplitude. For example, for a mass m oscillating on a spring of spring constant k, $E = \frac{1}{2}kA^2$. The maximum speed of the mass is directly proportional to the amplitude, $v_{max} = \sqrt{\dfrac{k}{m}}\, A$, because $E = \frac{1}{2}kA^2 = \frac{1}{2}mv_{max}^2$.

Example 13.1   A 0.50-kilogram object is attached to a spring of spring constant 20 N/m along a horizontal, frictionless surface. The object oscillates in simple harmonic motion and has a speed of 1.5 m/s at the equilibrium position.

(a) What is the total energy of the system?

(b) What is the amplitude?

(c) At what location are the values for the potential and kinetic energies the same?

Solution: Given: $m = 0.50$ kg, $v_{max} = 1.5$ m/s, $k = 20$ N/m.

Find: (a) E (b) A (c) x when $K = U$.

(a) At the equilibrium position, $x = 0$, so the potential energy is zero. Therefore, the total energy is equal to kinetic energy. $E = K = \frac{1}{2}mv_{max}^2 = \frac{1}{2}(0.50 \text{ kg})(1.5 \text{ m/s})^2 = 0.56$ J.

(b) At the amplitude $(x = \pm A)$, $v = 0$, so the total energy is all in the form of potential energy.

Thus at $x = \pm A$, we have $E = U = \frac{1}{2}kx^2 = \frac{1}{2}kA^2$,

therefore $A = \sqrt{\dfrac{2E}{k}} = \sqrt{\dfrac{2(0.56\ \text{J})}{20\ \text{N/m}}} = 0.24$ m.

(c) When $K = U$, $E = K + U = U + U = 2U = 2\frac{1}{2}kx^2 = kx^2$,

so $x = \sqrt{\dfrac{E}{k}} = \sqrt{\dfrac{0.56\ \text{J}}{20\ \text{N/m}}} = 0.17$ m.

Example 13.2   An object is attached to a spring of spring constant 60 N/m along a horizontal, frictionless surface. The spring is initially stretched by a force of 5.0 N on the object and let go. It takes the object 0.50 s first get back to its equilibrium position after its release.

(a) What is the amplitude?

(b) What is the period?

(c) What is the frequency?

Solution: Given: $k = 60$ N/m, $F = 5.0$ N, $t = 0.50$ s.

Find: (a) A (b) T (c) f.

(a) The initial stretch is also the maximum distance from the equilibrium position. (The spring force is the reaction force of the stretching force and thus equal in magnitude to it.) From Hooke's law,

$F_s = -kx$, we have $A = |x_{max}| = \dfrac{F_s}{k} = \dfrac{5.0\ \text{N}}{60\ \text{N/m}} = 0.083$ m.

(b) The motion from the amplitude to the equilibrium position is only 1/4 of a complete oscillation. So the period is $T = 4t = 4(0.50\ \text{s}) = 2.0$ s.

(c) $f = \dfrac{1}{T} = \dfrac{1}{2.0\ \text{s}} = 0.50$ Hz.

---

2.   Equation of Motion (Section 13.2)

The **equation of motion** of an object in SHM has the general form $y = A \sin(\omega t + \delta)$, where A is the amplitude, ω is the *angular* frequency of the motion $(\omega = 2\pi f = \dfrac{2\pi}{T})$, and δ is the **phase constant**. The amplitude and the phase constant depend on the initial conditions of the motion. The frequency f depends on the intrinsic properties of the system (for example, the stiffness and the inertia). From this equation of motion, we can express the velocity and acceleration of an object in SHM as

$v = \omega A \cos(\omega t + \delta)$ and $a = -\omega^2 A \sin(\omega t + \delta) = -\omega^2 y$.

For a mass-spring system, $T = 2\pi \sqrt{\dfrac{m}{k}}$, and $f = \dfrac{1}{2\pi} \sqrt{\dfrac{k}{m}}$, or $\omega = \sqrt{\dfrac{k}{m}}$, where m is the mass of the object and k is the spring constant.

For a simple pendulum, $T = 2\pi \sqrt{\dfrac{L}{g}}$, and $f = \dfrac{1}{2\pi} \sqrt{\dfrac{g}{L}}$, where L is the length of the pendulum and g is the acceleration due to gravity. (**Note**: the mass of the pendulum is not in the equation so the period is independent of mass.)

Without a driving force and with frictional losses, the amplitude and the energy of an oscillator will decrease with time, giving rise to **damped harmonic motion**.

Example 13.3 An object with a mass of 1.0 kg is attached to a spring with a spring constant of 10 N/m. The object is displaced by 3.0 cm from the equilibrium position and let go.

(a) What is the amplitude A?

(b) What is the period T?

(c) What is the frequency f?

Solution: Given: $m = 1.0$ kg, $k = 10$ N/m, initial $x = 3.0$ cm.

Find: (a) A (b) T (c) f.

(a) The initial stretch is the maximum displacement. So $A = 3.0$ cm.

(b) $T = 2\pi \sqrt{\dfrac{m}{k}} = 2\pi \sqrt{\dfrac{1.0 \text{ kg}}{10 \text{ N/m}}} = 2.0$ s.

(c) $f = \dfrac{1}{T} = \dfrac{1}{2.0 \text{ s}} = 0.50$ Hz.

Example 13.4 The pendulum of a grandfather clock is 1.0 m long.

(a) What is its period on the Earth?

(b) What would its period be on the Moon where the acceleration due to gravity is 1.7 m/s²?

Solution: Given: $L = 1.0$ m, $g_E = 9.8$ m/s², $g_M = 1.7$ m/s².

Find: (a) T_E (b) T_M.

(a) $T_E = 2\pi \sqrt{\dfrac{L}{g}} = 2\pi \sqrt{\dfrac{1.0 \text{ m}}{9.8 \text{ m/s}^2}} = 2.0$ s.

(b) $T_M = 2\pi \sqrt{\dfrac{1.0 \text{ m}}{1.7 \text{ m/s}^2}} = 4.8$ s.

Example 13.5   The position of an object in simple harmonic motion is described by $y = (0.25 \text{ m}) \sin\left(\dfrac{\pi}{2} t\right)$. Find

(a) the amplitude A, (b) the period T, (c) the maximum speed.

Solution: Given: $y = (0.25 \text{ m}) \sin\left(\dfrac{\pi}{2} t\right)$.

Find: (a) A (b) T (c) v_{max}.

(a) Comparing the equation of motion $y = (0.25 \text{ m}) \sin\left(\dfrac{\pi}{2} t\right)$ with the general form

$y = A \sin(\omega t + \delta)$, we see that $A = 0.25$ m.

(b) Again, comparing the equation of motion with the general form, $\omega t = 2\pi f t = \dfrac{\pi}{2} t$.

So $2\pi f = \pi/2$,   or $f = 1/4 \ 1/\text{s} = 0.25$ Hz.   Therefore $T = \dfrac{1}{f} = \dfrac{1}{0.25 \text{ Hz}} = 4.0$ s.

(c) From the equation of velocity as a function of time $v = A\omega \cos(\omega t + \delta)$, we know that the maximum speed is $v_{max} = \omega A = (2\pi f) A = 2\pi(1/4.0 \text{ Hz})(0.25 \text{ m}) = 0.39$ m/s.

3.   Wave Motion (Section 13.3)

Wave motion is the propagation of a disturbance (energy and momentum) through a material. The medium though which the wave propagates is *not* transferred in wave motion. It simply support the wave's passage. A periodic (as opposed to a pulse) wave can be characterized by the following quantities:

amplitude: the magnitude of the maximum displacement of the particles of the material from their equilibrium positions,

wavelength: the distance between two successive crests or troughs, (see note below),

frequency: the number of wavelengths that passes by a given point in a second,

wave speed: the speed of the wave motion (speed of a crest or trough) given by $v = \lambda f$.

Note: The wavelength is actually the distance between any two successive particles that are in phase (that is, at identical points on the wave form). Crests are used just for convenience.

In a wave, there are two motions, the motion of the particle of the medium (medium motion) and the motion of the travelling disturbance (wave motion). **Waves** are divided into two types based on the direction of the medium motion relative to the wave motion. For a **transverse wave** (sometimes called a *shear wave*) the particle motion is *perpendicular* to the direction of the wave velocity. A wave on a rope is an example of a transverse wave. For a **longitudinal wave** (or sometimes called a *compressional wave*) the particle motion is *parallel* to the direction of the wave velocity. Sound waves are examples of longitudinal waves.

Some waves are, however, combinations of transverse waves and longitudinal waves. Two good examples of this are water waves and seismic waves.

Example 13.6   The diagram shown is a "snapshot" of a wave on a rope at a given time. The frequency of the wave is 60 Hz.

(a) What is the amplitude?

(b) What is the wavelength?

(c) What is the wave speed?

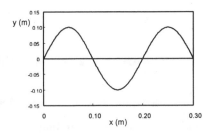

Solution: Given: $f = 60$ Hz   and   information in the diagram.

Find: (a) A (b) λ (c) v.

(a) Amplitude is the maximum distance from the equilibrium position.

From the diagram, $A = 0.10$ m.

(b) Wavelength is the distance between successive crests.

From the diagram, $\lambda = 0.20$ m.

(c) $v = \lambda f = (0.20 \text{ m})(60 \text{ Hz}) = 12$ m/s.

---

4.   Wave Phenomena (Section 13.4)

Interference occurs when waves meet or overlap. The **principle of superposition** states: *at any time, the combined waveform of two or more interfering waves is given by the sum of the displacements of the individual waves at each point in the medium.* If the amplitude of the combined wave is greater than that of any of the individual waves, we have what is called **constructive interference**. If the amplitude of the combined wave is smaller than that of any of the individual waves, it is called **destructive interference**. **Total constructive interference** occurs when two waves of the same frequency and amplitude are exactly in phase (the crest of one wave is aligned with the crest of the other) and **total destructive interference** takes place if two waves of the same frequency and amplitude are completely 180° out of phase (the crest of one wave is aligned with the trough of the other and vice versa).

Reflection occurs when a wave strikes an object or comes to a boundary of another medium and is at least partly bounced back. **Refraction** occurs when a wave crosses a boundary into another medium and the transmitted wave moves in a different direction. Generally, when a wave strikes the boundary, both reflection and refraction occur. **Dispersion** is exhibited when the wave speed depends on the wavelength or frequency. Waves of different wavelength or frequency travel at different speeds so if they start together they spread apart from one another. The rainbow we sometimes see is an excellent example of wave dispersion for light. **Diffraction** refers to the bending of waves around an edge of an object. A person in a room with an open door can hear sound from outside the room, which is partially due to diffraction (there may be some reflection).

5. Standing Waves and Resonance (Section 13.5)

Interfering waves of the same frequency and amplitude traveling in opposite directions in a rope or other conditions can produce a **standing wave**. In this case, some points on the rope are always stationary (destructive interference) and they are called **nodes**. The points of maximum amplitude, where constructive interference is greatest, are called **antinodes**.

The frequencies at which large-amplitude standing waves are produced are called **natural frequencies**, or **resonant frequencies**. The lowest frequency f_1 is called the **fundamental frequency**. In a stretched string fixed at both ends, all of the other natural frequencies are integral multiples of the fundamental frequency $f_n = nf_1$, where $n = 1, 2, 3, \ldots$, is an integer. The set of frequencies $f_1, f_2 = 2f_1, f_3 = 3f_1$, and so on, is called a **harmonic series**: f_1 is the *first harmonic*, f_2 the *second harmonic*, and so on.

Note: The names fundamental frequency and first harmonic frequency are interchangeably used.

The speed of a wave in a stretched string is given by $v = \sqrt{\dfrac{F_T}{\mu}}$, where F_T is the tension in the string, and μ the linear mass density (mass per unit length, $\mu = m/L$). Therefore, the frequencies of the harmonic series *in a stretched string* is given by $f_n = n\dfrac{v}{2L} = \dfrac{n}{2L}\sqrt{\dfrac{F_T}{\mu}}$ (for $n = 1, 2, 3, \ldots$).

Resonance is a phenomena in which the vibrational amplitude of a system is greatly enhanced. It happens when the driving frequency of an external source matches a natural frequency of the system.

Example 13.7   A 50-meter long string has a mass of 0.010 kg. A 2.0-meter segment of the string is fixed at both ends and when a tension of 20 N is applied to the string, three loops are produced. What is the frequency of the standing wave?

Solution: Three loops correspond to the 3rd harmonic.

Given: $L_o = 50$ m, $m = 0.010$ kg, $L = 2.0$ m, $F_T = 20$ N.

Find: f_3.

$$\mu = \frac{m}{L_o} = \frac{0.010 \text{ kg}}{50 \text{ m}} = 2.0 \times 10^{-4} \text{ kg/m}. \quad v = \sqrt{\frac{F_T}{\mu}} = \sqrt{\frac{20 \text{ N}}{2.0 \times 10^{-4} \text{ kg/m}}} = 3.2 \times 10^2 \text{ m/s}.$$

$$f_3 = 3f_1 = 3\frac{v}{\lambda_1} = 3\frac{v}{2L} = \frac{3(3.2 \times 10^2 \text{ m/s})}{2(2.0 \text{ m})} = 2.4 \times 10^2 \text{ Hz}.$$

IV.   Mathematical Summary

Hooke's Law	$F_s = -kx$ (13.1)	Relates spring force to displacement of spring from equilibrium.
Frequency and Period for SHM	$f = \frac{1}{T}$ (13.2)	Relates frequency and period for simpler harmonic motion.
Total Energy of a Spring-Mass in SHM	$E = \frac{1}{2}kA^2 = \frac{1}{2}mv^2 + \frac{1}{2}kx^2$ (13.4–5)	Gives the total energy of a spring-mass system in simple harmonic motion.
Velocity of Oscillating Mass on a Spring	$v = \pm\sqrt{\frac{k}{m}(A^2 - x^2)}$ (13.6)	Gives the velocity of an oscillating mass on a spring as a function of position.
Period of a Mass Oscillating on a Spring	$T = 2\pi\sqrt{\frac{m}{k}}$ (13.11)	Computes the period of a mass on a spring in simple harmonic motion.
Angular Frequency of a Mass Oscillating in a Spring	$\omega = 2\pi f = \sqrt{\frac{k}{m}}$ (13.13)	Calculates the angular frequency of a mass on a spring in simple harmonic motion.
Period of a Simple Pendulum	$T = 2\pi\sqrt{\frac{L}{g}}$ (13.14)	Computes the period of a simple pendulum at small-angle approximation.
Displacement of a Mass in SHM	$y = A\sin(\omega t + \delta)$ (13.15) (with $\delta = 0$) $y = A\sin\omega t$ $= A\sin 2\pi f t = A\sin\frac{2\pi t}{T}$	Defines the equation of motion for a simple harmonic oscillator.
Velocity of a Mass in SHM ($\delta = 0$)	$v = \omega A\cos\omega t$ (13.16)	Calculates the velocity of a simple harmonic oscillator as a function of time.

Acceleration of a Mass in SHM ($\delta = 0$)	$a = -\omega^2 A \sin \omega t$ $= -\omega^2 y \qquad (13.17)$	Calculates the acceleration of a simple harmonic oscillator as a function of time.
Wave Speed	$v = \dfrac{\lambda}{T} = \lambda f \qquad (13.18)$	Relates wave speed to wavelength and frequency.
Natural Frequencies in a Stretched String	$f_n = n\dfrac{v}{2L} = \dfrac{n}{2L}\sqrt{\dfrac{F_T}{\mu}}$ $(n = 1, 2, 3, \ldots) \qquad (13.21)$	Gives the natural frequencies in a stretched string as a harmonic series.

V. Solutions of Selected Exercises and Paired/Trio Exercises

3. (a) $E = \frac{1}{2} kA^2$, so $\boxed{\text{four times as large}}$.

   (b) $v_{\text{max}} = \sqrt{\dfrac{k}{m}}\, A$, so $\boxed{\text{twice as large}}$.

10. From $T = \dfrac{1}{f}$, we have $\Delta T = \dfrac{1}{f_2} - \dfrac{1}{f_1} = \dfrac{1}{0.50 \text{ s}} - \dfrac{1}{0.25 \text{ s}} = -2.0 \text{ s}$.

   That is a $\boxed{\text{decrease of 2.0 s}}$.

14. (a) $F = kx = (150 \text{ N/m})(0.150 \text{ m}) = \boxed{22.5 \text{ N}}$; $a = \dfrac{F}{m} = \dfrac{22.5 \text{ N}}{0.500 \text{ kg}} = \boxed{45.0 \text{ m/s}^2}$.

   (b) $F = (150 \text{ N/m})(0.050 \text{ m}) = \boxed{7.50 \text{ N}}$; $a = \dfrac{7.50 \text{ N}}{0.500 \text{ kg}} = \boxed{15.0 \text{ m/s}^2}$.

   (c) $F = \boxed{0}$; $a = \boxed{0}$.

18. $T = 2\pi\sqrt{\dfrac{m}{k}} \propto \sqrt{m}$, so the ratio is $\boxed{\sqrt{2}}$.

20. Since $f = \dfrac{1}{2\pi}\sqrt{\dfrac{k}{m}}$, $\dfrac{f_2}{f_1} = \sqrt{\dfrac{m_1}{m_2}} = \sqrt{\dfrac{0.25 \text{ kg}}{0.50 \text{ kg}}} = 0.707$.

   So $f_2 = (0.707)(1.0 \text{ Hz}) = \boxed{0.71 \text{ Hz}}$.

22. (a) $\boxed{0.90 \text{ J}}$.

   (b) $\boxed{\text{No}}$.

25. (a) From conservation of energy (choose the position of the object when the spring is compressed as $U_g = 0$), $E = \frac{1}{2}kA^2 = U = mgh$, we have $\frac{1}{2}(60.0 \text{ N/m})A^2 = (0.250 \text{ kg})(9.80 \text{ m/s}^2)(0.100 \text{ m} + A)$.

Reduce to a quadratic equation: $30.0A^2 - 2.45A - 0.245 = 0$.

Solving, $A = \boxed{0.14 \text{ m}}$ or -0.058 m, which is discarded.

(b) From energy conservation, the object will go to a height of $\boxed{10.0 \text{ cm}}$ (original position).

32. (a) $T = 2\pi\sqrt{\dfrac{m}{k}} = 2\pi\sqrt{\dfrac{0.50 \text{ kg}}{200 \text{ N/m}}} = \boxed{0.31 \text{ s}}$.

(b) $f = \dfrac{1}{T} = \dfrac{1}{0.31 \text{ s}} = \boxed{3.2 \text{ Hz}}$.

36. (a) Compare to $y = A \sin\omega t$. $A = \boxed{0.10 \text{ m}}$.

(b) $\omega = 2\pi f = 100$ rad/s, so $f = \dfrac{100}{2\pi} = \boxed{16 \text{ Hz}}$.

(c) $T = \dfrac{1}{f} = \dfrac{1}{16 \text{ Hz}} = \boxed{0.063 \text{ s}}$.

41. From $T = 2\pi\sqrt{\dfrac{L}{g}}$, we have $g = \dfrac{4\pi^2 L}{T^2} = \dfrac{4\pi^2(0.3690 \text{ m})}{(1.220 \text{ s})^2} = \boxed{9.787 \text{ m/s}^2}$.

44. From $T = 2\pi\sqrt{\dfrac{m}{k}}$, we have $\dfrac{T_1}{T_2} = \sqrt{\dfrac{k_2}{k_1}} = \sqrt{2}$. So $\boxed{\text{the first one by } \sqrt{2} \text{ times}}$.

46. $\boxed{2.5 \text{ N/m}}$.

50. (a) Since $T = 2\pi\sqrt{\dfrac{L}{g}}$ and the length is shorter, T is also shorter.

So the clock runs faster or $\boxed{\text{gains time}}$.

(b) $\Delta T = 2\pi\sqrt{\dfrac{0.7500 \text{ m}}{9.80 \text{ m/s}^2}} - 2\pi\sqrt{\dfrac{0.7480 \text{ m}}{9.80 \text{ m/s}^2}} = 2.32 \times 10^{-3}$ s.

So in 24 hours = 86 400 s (or 43 200 periods), the time difference is

$43\,200(2.32 \times 10^{-3} \text{ s}) = 100 \text{ s} = \boxed{1.7 \text{ min}}$.

(c) $\boxed{\text{Yes}}$. Because of linear thermal expansion, the length depends on the temperature.

54. (a) ⌐Transverse⌐.   (b) ⌐Longitudinal⌐.

60. $\lambda_{max} = \dfrac{v}{f} = \dfrac{345 \text{ m/s}}{20 \text{ Hz}} = \boxed{17 \text{ m}}$, $\lambda_{min} = \dfrac{345 \text{ m/s}}{20 \times 10^3 \text{ Hz}} = \boxed{0.017 \text{ m}}$.

65. (a) 90° latitude represents one quarter of the Earth's circonference. The straight line distance between the
     locations is

$$d = \sqrt{R^2 + R^2} = \sqrt{2}\, R = \sqrt{2}\, (6.4 \times 10^3 \text{ km}) = 9.05 \times 10^3 \text{ km}.$$

$$\Delta t = \dfrac{d}{v_S} - \dfrac{d}{v_P} = \dfrac{9.05 \times 10^3 \text{ km}}{6.0 \text{ km/s}} - \dfrac{9.05 \times 10^3 \text{ km}}{8.0 \text{ km/s}} = \boxed{3.8 \times 10^2 \text{ s}}.$$

(b) $r = R \cos 45°$.   So the depth under the surface is

$R - r = R(1 - \cos 45°) = (6.4 \times 10^3 \text{ km})(1 - \cos 45°) = 1.9 \times 10^3 \text{ km} > 30 \text{ km}.$

Therefore, ⌐yes⌐, the waves do cross the boundary of the mantle.

(c) $t = \dfrac{2(6.4 \times 10^3 \text{ km})}{8.0 \text{ km/s}} = \boxed{1.6 \times 10^3 \text{ s; S waves do not go through the liquid core}}$.

66. (a) $v = \sqrt{\dfrac{Y}{\rho}} = \sqrt{\dfrac{7.0 \times 10^{10} \text{ N/m}^2}{2.7 \times 10^3 \text{ kg/m}^3}} = 5091 \text{ m/s}.$ $\lambda = \dfrac{v}{f} = \dfrac{5091 \text{ m/s}}{40 \text{ Hz}} = \boxed{1.3 \times 10^2 \text{ m}}$.

(b) $v = \sqrt{\dfrac{11 \times 10^{10} \text{ N/m}^2}{8.9 \times 10^3 \text{ kg/m}^3}} = 3516 \text{ m/s}.$ $\lambda = \dfrac{3516 \text{ m/s}}{40 \text{ Hz}} = \boxed{88 \text{ m}}$.

76. ⌐5⌐ nodes.

78. From $f_3 = 3f_1$, we have $f_1 = \dfrac{f_3}{3} = \dfrac{450 \text{ Hz}}{3} = \boxed{150 \text{ Hz}}$.

80. (a) $f_1 = \dfrac{v}{2L} = \dfrac{12 \text{ m/s}}{2(4.0 \text{ m})} = 1.5 \text{ Hz}.$ So ⌐yes⌐, 15 Hz is the 10th harmonic.

(b) ⌐No⌐, 20 Hz is not a harmonic.

83. $v = \sqrt{\dfrac{F_T}{\mu}} = \sqrt{\dfrac{40 \text{ N}}{2.5 \times 10^{-2} \text{ kg/m}}} = 40 \text{ m/s}.$ $f_n = \dfrac{nv}{2L} = \dfrac{40 \text{ m/s}}{2(2.0 \text{ m})}\, n = 10n \text{ Hz}.$

So the frequencies of the first four harmonics are $\boxed{10 \text{ Hz, } 20 \text{ Hz, } 30 \text{ Hz, and } 40 \text{ Hz}}$.

86. The first harmonic, $\lambda_1 = 4L$. $f_1 = \dfrac{v}{\lambda} = \dfrac{v}{4L} = 1 \times \dfrac{v}{4L}$;

the next harmonic (3rd), $\lambda_3 = \dfrac{4L}{3}$, $f_3 = \dfrac{3v}{4L} = 3 \times \dfrac{v}{4L}$;

the next harmonic (5th), $\lambda_5 = \dfrac{4L}{5}$, $f_5 = \dfrac{5v}{4L} = 5 \times \dfrac{v}{4L}$.

Therefore $f_m = \dfrac{mv}{4L} = m\,\dfrac{3.5 \times 10^3 \text{ m/s}}{4(1.0 \text{ m})} = \boxed{m(8.8 \times 10^2 \text{ Hz}), \; m = 1, 3, 5, \ldots}$.

91. $x = A \cos\omega t = A \sin(\omega t + 90°)$.

So $v = \omega A \cos(\omega t + 90°) = -\omega A \sin\omega t$

and $a = -\omega^2 x = -\omega^2 A \cos\omega t$.

$v_{max} = \omega A = (50 \text{ rad/s})(0.10 \text{ m}) = \boxed{5.0 \text{ m/s}}$.

$a_{max} = \omega^2 A = (50 \text{ rad/s})^2(0.10 \text{ m}) = \boxed{2.5 \times 10^2 \text{ m/s}^2}$.

95. Since $v = \sqrt{\dfrac{F_T}{\mu}}$, doubling the tension will make the speed $\sqrt{2}$ times larger.

The wavelength is unchanged since it is determined by the length of the string and the mode of vibration. However, the frequency is increased by a factor of $\sqrt{2}$ due to the speed increase.

VI. Practice Quiz

1. A 2.0-kilogram object is attached to a spring of spring constant 1.8 N/m. If the object is displaced slightly from its equilibrium position and released. What is the period of vibration?
   (a) zero (b) 0.90 s (c) 0.95 s (d) 1.9 s (e) 6.0 s.

2. An object attached to the free end of a spring executes simple harmonic motion according to the equation $y = (0.50 \text{ m}) \sin(18\pi t)$ where y is in meters and t in seconds. What is the frequency of the vibration?
   (a) 0 (b) 3.0 Hz (c) 9.0 Hz (d) 18 Hz (e) 18π Hz

3. The wave speed on a string is 12 m/s. What is the wavelength of a wave of frequency 3.0 Hz traveling on the string?
   (a) 3.0 m (b) 4.0 m (c) 9.0 m (d) 15 m (e) 36 m

4. Resonance in a system, such as a string fixed at both ends or a suspension bridge, occurs when

(a) it is oscillating in simple harmonic motion.

(b) it is oscillating in damped simple harmonic motion.

(c) a natural frequency of the system is the same as the external driving frequency.

(d) a natural frequency of the system is smaller than the external driving frequency.

(e) a natural frequency of the system is larger than the external driving frequency.

5. In seismology, the S wave is a transverse wave. As an S wave travels through the Earth, the relative motion between the S wave and the particles of the Earth's interior is

(a) perpendicular.   (b) first perpendicular, then parallel.

(c) not at any particular angle to each other.   (d) parallel.   (e) antiparallel.

6. The mass of a simple pendulum is doubled. By what factor does the frequency change?

(a) It does not change.   (b) It doubles.   (c) It quadruples.

(d) It becomes half as large.   (e) It becomes 1/4 as large.

7. When an object of mass m oscillates on a spring, its frequency is f. A mass $2m$ is now placed on the same spring and it again oscillates in SHM. What is its frequency?

(a) $0.25f$ (b) $0.50f$ (c) $0.71f$ (d) $1.4f$ (e) $2f$

8. A piano string of linear mass density 0.0050 kg/m is under a tension of 1350 N. What is the wave speed?

(a) 130 m/s   (b) 260 m/s   (c) 520 m/s   (d) 1040 m/s   (e) 2080 m/s

9. If a guitar string has a third harmonic frequency of 1500 Hz, which of the following frequencies can set the string into resonant vibration?

(a) 125 Hz   (b) 250 Hz   (c) 500 Hz   (d) 625 Hz   (e) 750 Hz

10. An object on a spring oscillates in simple harmonic motion with a frequency of 1.00 Hz and an amplitude of 5.00 cm. If a timer is started when the mass is at the equilibrium position, what is its distance from its equilibrium position at $t = 2.25$ s?

(a) 7.50 cm   (b) 5.00 cm   (c) 2.50 cm   (d) 1.25 cm   (e) zero

Answers to Practice Quiz:

1. e 2. c 3. b 4. c 5. a 6. a 7. c 8. c 9. c 10. b

CHAPTER 14

Sound

I.   Chapter Objectives

Upon completion of this chapter, you should be able to:

1.   define sound and explain the sound frequency spectrum.

2.   tell how the speed of sound differs in different media and describe the temperature dependence of the speed of sound in air.

3.   define sound intensity and explain how it varies with distance from a point source, and calculate sound intensity levels on the decibel scale.

4.   explain sound reflection, refraction, and diffraction and distinguish between constructive and destructive interference.

5.   describe and explain the Doppler effect and give some examples of its occurrences and applications.

6.   explain some of the sound characteristics of musical instruments in physical terms.

II.   Key Terms

Upon completion of this chapter, you should be able to define and/or explain the following key terms:

sound waves	destructive interference
audible region	phase difference
sound frequency spectrum	path-length difference
infrasonic region	beats
ultrasonic region	beat frequency
intensity	pitch
loudness	Doppler effect
bel (B)	sonic boom
decibel (dB)	Mach number
sound intensity level (decibel level)	quality
constructive interference	

The definitions and/or explanations of the most important key terms can be found in the following section:

III. Chapter Summary and Discussion.

III. Chapter Summary and Discussion

1. Sound Waves (Section 14.1)

Sound waves in fluids are primarily longitudinal waves (or compressional waves). The high- and low-density pressure regions (analogous to crests and troughs of transverse wave) of a sound wave are called *condensations* and *rarefactions*, respectively. All sound waves are produced by vibrating sources such as the human vocal cords.

The **sound frequency spectrum** consists of three regions: frequencies lower than 20 Hz ($f < 20$ Hz) are in the **infrasonic region**, frequencies between about 20 Hz and 20 kHz (20 Hz $< f <$ 20 kHz) are in the **audible region**, and above 20 kHz ($f > 20$ kHz) is the **ultrasonic region**.

2. The Speed of Sound (Section 14.2)

The **speed of sound** in a medium depends on the elasticity or the intermolecular interactions of the medium and the mass or density of its particles. Generally, the speeds of sound in solids and liquids are given by

$$v = \sqrt{\frac{Y}{\rho}} \quad \text{and} \quad v = \sqrt{\frac{B}{\rho}},$$ respectively, where Y is the Young's modulus, B is the bulk modulus, and ρ is the density.

For *normal environment temperatures*, the speed of sound in air increases by about 0.6 m/s for each Celsius degree above 0°C. A good approximation is given by $v = (331 + 0.6T_C)$ m/s, where T_C is the air temperature in degrees Celsius and 331 m/s is the speed of sound in air at 0°C. A useful general value for the speed of sound in air is $\frac{1}{3}$ km/s (or $\frac{1}{5}$ mi/s).

Example 14.1 The speed of an ultrasonic sound of frequency 45 kHz in air is 342 m/s.

(a) What is the air temperature?

(b) What is the wavelength of the sound wave?

Solution: Given: $v = 342$ m/s, $f = 45$ kHz $= 45 \times 10^3$ Hz.

Find: (a) T_C (b) λ.

(a) From $v = (331 + 0.6T_C)$ m/s, we have $T_C = \dfrac{v - 331}{0.6} = \dfrac{342 - 331}{0.6} = 18°C$.

(b) From $v = \lambda f$, we have $\lambda = \dfrac{v}{f} = \dfrac{342 \text{ m/s}}{45 \times 10^3 \text{ Hz}} = 7.6 \times 10^{-3}$ m = 7.6 mm.

Example 14.2   Find the speed of sound in an aluminum rod.

Solution: Given: $Y_{Al} = 7.0 \times 10^{10}$ N/m^2 (Table 9.1), $\rho = 2.7 \times 10^3$ kg/m^3 (Table 9.2)

   Find: v.

$$v = \sqrt{\dfrac{Y}{\rho}} = \sqrt{\dfrac{7.0 \times 10^{10} \text{ N/m}^2}{2.7 \times 10^3 \text{ kg/m}^3}} = 5.1 \times 10^3 \text{ m/s}.$$

---

3.   Sound Intensity and Sound Intensity Level (Section 14.3)

Sound **intensity** is the rate of the sound energy transfer or the energy transported per unit time across a unit area: $\text{intensity} = \dfrac{\text{energy/time}}{\text{area}} = \dfrac{\text{power}}{\text{area}}$. The SI unit of intensity is W/m^2.

The intensity of a point source of power P at a distance R from the source is given by $I = \dfrac{P}{A} = \dfrac{P}{4\pi R^2}$.

Note that *the intensity is inversely proportional to the square of the distance from the point source.*

Sound intensity is perceived by the ear as **loudness**. On the average, the human ear can detect sound waves (at 1 kHz) with an intensity as low as $I_o = 10^{-12}$ W/m^2 (the **threshold of hearing**). At an intensity of $I_p = 1.0$ W/m^2 (**threshold of pain**) the sound is uncomfortably loud and may be painful to the ear. The ratio between the two intensities is $\dfrac{I_p}{I_o} = \dfrac{1.0 \text{ W/m}^2}{10^{-12} \text{ W/m}^2} = 10^{12}$, or 12 orders of magnitude. For a sound to be audible, it must have a frequency between 20 Hz and 20 kHz, *and* have an intensity greater than I_o.

The **sound intensity level**, or **decibel level** (β) of a sound of intensity I is defined as $\beta = 10 \log \dfrac{I}{I_o}$, where $I_o = 10^{-12}$ W/m^2 is the threshold of hearing. At the threshold of hearing, $\beta = 10 \log \dfrac{I_o}{I_o} = 10 \log 1 = 0$ dB and at the threshold of pain, $\beta = 10 \log \dfrac{1.0 \text{ W/m}^2}{10^{-12} \text{ W/m}^2} = 10 \log 10^{12} = 10(12) = 120$ dB. The 12 orders of magnitude of sound intensity is only a difference of 120 dB − 0 dB = 120 dB, on the decibel scale.

Note: Sound intensity and sound intensity level are two very different things. Intensity is a direct way to measure energy and is additive, that is, the sum of a 1.0 W/m² sound and a 2.0 W/m² sound will result in a sound of intensity 1.0 W/m² + 2.0 W/m² = 3.0 W/m². Intensity level is based on a logarithmic scale and therefore is not additive, that is, the sum of a 10 dB sound and a 20 dB sound will *not* make a sound of 10 dB + 20 dB ≠ 30 dB. Therefore, always use the definition of intensity level before answering any related questions.

Example 14.3   Your professor's lecturing voice has a power of about 0.50 mW. If this power is assumed to be uniformly distributed in all directions,

(a) what is the sound intensity at a distance of 5.00 m from the professor?

(b) what is the intensity level at a distance of 5.00 m from the professor?

(c) what would be the intensity level if the professor raises his voice so the intensity doubles to emphasize a concept?

Solution: Given: $R = 5.00$ m, $P = 0.50$ mW $= 0.50 \times 0^{-3}$ W.

Find: (a) I (b) β (c) β for twice the intensity.

(a) $I = \dfrac{P}{4\pi R^2} = \dfrac{0.50 \times 10^{-3} \text{ W}}{4\pi(5.00 \text{ m})^2} = 1.6 \times 10^{-6}$ W/m².

(b) $\beta = 10 \log \dfrac{I}{I_0} = 10 \log \dfrac{1.6 \times 10^{-6} \text{ W/m}^2}{10^{-12} \text{ W/m}^2} = 10 \log 1.6 \times 10^6 = 10(6.2) = 62$ dB.

(c) If the intensity doubles, the new intensity is $I = 2(1.6 \times 10^{-6}$ W/m²$) = 3.2 \times 10^{-6}$ W/m².

Therefore $\beta = 10 \log \dfrac{3.2 \times 10^{-6} \text{ W/m}^2}{10^{-12} \text{ W/m}^2} = 10 \log 3.2 \times 10^6 = 10(6.5) = 65$ dB.

Doubling the intensity only increases the intensity level by 65 dB − 62 dB = 3.0 dB! This is always true because $10 \log 2I = 10 \, (\log I + \log 2) = 10 \, (\log I + 0.30) = 10 \log I + 3.0$.

Example 14.4   The intensity level generated by 25 computer keyboards in a mail-order company is 70 dB. What is the intensity level generated by one keyboard?

Solution: Given: $\beta \, (25) = 70$ dB, $n = 25$.

Find: $\beta \, (1)$.

Intensity level is not additive (the intensity level of one keyboard is *not* 70/25 = 2.8 dB). We need to first find the intensity generated by 25 keyboards, then the intensity of one keyboard, and finally the intensity level of one keyboard.

From $\beta = 10 \log \dfrac{I}{I_0}$, we have $\dfrac{\beta}{10} = \log \dfrac{I}{I_0}$,   so $\dfrac{I}{I_0} = 10^{\beta/10}$ (because if $y = \log x$, then $x = 10^y$).

Therefore $I(25) = I_o\, 10^{\beta/10} = (10^{-12}\ \text{W/m}^2)\, 10^{70/10} = (10^{-12}\ \text{W/m}^2)\, 10^7 = 10^{-5}\ \text{W/m}^2$, and the intensity generated by one keyboard is $I(1) = \dfrac{I(25)}{25} = \dfrac{10^{-5}\ \text{W/m}^2}{25} = 4.0 \times 10^{-7}\ \text{W/m}^2$ (intensity is additive).

Hence $\beta(1) = 10 \log \dfrac{4.0 \times 10^{-7}\ \text{W/m}^2}{10^{-12}\ \text{W/m}^2} = 10 \log 4.0 \times 10^5 = 56$ dB. (Are you surprised?)

4. Sound Phenomena (Section 14.4)

Since sound is a wave, it has all the wave characteristics. It can be *reflected*, *refracted*, and *diffracted*: that is, bounce off objects or surfaces, change in direction due to a medium or density change, and bend around corners or objects, respectively.

Sound waves *interfere* when they meet. There can be **constructive interference** or **destructive interference** depending on the **path-length difference** (ΔL) of the waves. If the path-length difference between two waves of the same frequency is zero or an integer (whole) number of wavelength, $\Delta L = n\lambda$ with $n = 0, 1, 2, \ldots$, constructive interference (two crests or two troughs coincide) occurs. Conversely, if the path-length difference between two waves of the same frequency is an odd number of half wavelength, $\Delta L = m\lambda/2$ with $m = 1, 3, 5, \ldots$, destructive interference (crest and trough coincide) takes place. The **phase difference** $\Delta\theta$ is related to path-length difference by the simple relationship: $\Delta\theta = \dfrac{2\pi}{\lambda}(\Delta L)$. In terms of the phase difference, the conditions for interference are

$\Delta\theta = n(2\pi)$, with $n = 0, 1, 2, \ldots$ for constructive interference,

$\Delta\theta = m(\pi)$, with $m = 1, 3, 5, \ldots$ for destructive interference.

Another interesting interference effect occurs when two tones of nearly the same frequency ($f_1 \approx f_2$) sound simultaneously. The ear senses pulsations in loudness known as **beats**. The **beat frequency** is equal to the difference between the two frequencies, or $f_b = |f_1 - f_2|$.

Example 14.5   A person stands between two loud speakers driven by an identical source. Each speaker produces a tone with a frequency of 155 Hz on a day when the speed of sound is 341 m/s. The person is 1.65 m from one speaker and 4.95 m from the other speaker. What type of interference does the person sense?

Solution: Given: $f = 155$ Hz, $v = 341$ m/s, $L_1 = 1.65$ m, $L_2 = 4.96$ m.

Find: Interference type.

To determine what type interference the person senses, we need to find the path-length difference as a function of wavelength. From $v = \lambda f$, we have $\lambda = \dfrac{v}{f} = \dfrac{341 \text{ m/s}}{155 \text{ Hz}} = 220$ m.

The path-length difference is $\Delta L = L_2 - L_1 = 4.95$ m $- 1.65$ m $= 3.30$ m $= \frac{3}{2}(2.20$ m$) = \frac{3}{2}\lambda$. So the path-length difference is an odd number of half wavelength and therefore the interference is destructive.

Example 14.6   A music tuner uses a 256-hertz tuning fork to tune the frequency of sound from a musical instrument. If the tuner hears a beat frequency of 2.0 Hz, what is the frequency of the sound produced by the instrument?

Solution: Given: $f_b = 2.0$ Hz, $f_1 = 256$ Hz.

Find: f_2.

From $f_b = |f_1 - f_2|$, we have $f_2 = |f_1 \pm f_b|$ (Note the absolute sign!).

So $f_2 = |256$ Hz ± 2.0 Hz$| = 254$ Hz or 258 Hz.

There are two answers because we do not know which one (the tuning fork or the musical instrument) has a higher frequency.

---

5.   The Doppler Effect (Section 14.5)

If there is relative motion between a sound source and an observer, the observer will detect a frequency which is different from the frequency of the source. This phenomena is called the **Doppler effect**. Generally, if the source and the observer are moving toward each other, the observed frequency is higher than the source frequency; and if the source and the observer are moving away from each other, the observed frequency is lower than the source frequency.

For a stationary observer, the observed frequency f_o by the observer due to a moving source of frequency f_s is given by $f_o = \dfrac{v}{v \pm v_s} f_s = \dfrac{1}{1 \pm \dfrac{v_s}{v}} f_s$, where v_s is the speed of source, v is the speed of sound, the minus sign corresponds to source moving toward stationary observer, and the plus sign corresponds to source moving away from stationary observer.

For a stationary source of frequency f_s, the observed frequency f_o by an observer due to the moving observer is given by $f_o = \dfrac{v \pm v_o}{v} f_s = \left(1 \pm \dfrac{v_o}{v}\right) f_s$, where v_o is the speed of observer, v is the speed of sound, the plus sign corresponds to observer moving toward the source, and the minus sign corresponds to observer moving away from the source.

Note: For source moving, use − for *toward* and use + for *away from* observer.

For observer moving, use − for *away from* and use + for *toward* the source. (Just opposite! Why?)

Objects traveling at supersonic (speed greater than the speed of sound) produce large pressure ridges or shock waves. This gives rise to the **sonic boom** heard when a supersonic aircraft passes. The ratio of the speed of the source v_s to the speed of sound v is called the **Mach number**: $M = \dfrac{v_s}{v}$ and $M > 1$ for supersonic speeds.

Example 14.7   The frequency of a train horn is 500 Hz. Assume the speed of sound in air is 340 m/s. What is the frequency heard by an observer if

(a) the observer is moving away from the stationary train with a speed of 30.0 m/s?

(b) the train is approaching the stationary observer with a speed of 30.0 m/s?

Solution: Given: $f_s = 500$ Hz, $v = 340$ m/s,   (a) $v_o = 30.0$ m/s,   (b) $v_s = 30.0$ m/s.

Find: (a) f_o (b) f_o.

(a) Since the observer is moving away from the source, we use the minus sign in $f_o = \dfrac{v \pm v_o}{v} f_s$.

$$f_o = \frac{v - v_o}{v} f_s = \frac{340 \text{ m/s} - 30.0 \text{ m/s}}{340 \text{ m/s}} (500 \text{ Hz}) = 456 \text{ Hz}.$$

(b) Since the train is approaching (toward), we use the minus sign in $f_o = \dfrac{v}{v \pm v_s} f_s$.

$$f_o = \frac{v}{v - v_s} f_s = \frac{340 \text{ m/s}}{340 \text{ m/s} - 30.0 \text{ m/s}} (500 \text{ Hz}) = 548 \text{ Hz}.$$

Example 14.8   The Concord airplane flies from the United States to Europe with a Mach number of 1.05 where the air temperature is 5.0°C. What is the speed of the plane?

Solution: Given: $M = 1.05$, $T_C = 5.0$°C.

Find: v_s.

The speed of sound in air is $v = (331 + 0.6 T_C)$ m/s $= [331 + 0.6(5.0)]$ m/s $= 334$ m/s.

From $M = \dfrac{v_s}{v}$, we have $v_s = Mv = (1.05)(334 \text{ m/s}) = 351$ m/s.

---

6. Musical Instruments and Sound Characteristics (Section 14.6)

Stringed musical instruments produce notes by setting up transverse standing waves in strings with different fundamental frequencies (see Chapter 13). The harmonic series is given by $f_n = n\frac{v}{2L} = nf_1$ (for $n = 1, 2, 3, \ldots$), where L is the length of the string.

Organ pipes and wind instruments produce notes by forming longitudinal standing waves in air columns. For an open organ pipe (open at both ends), the harmonic series is given by $f_n = n\frac{v}{2L} = nf_1$ (for $n = 1, 2, 3, \ldots$), where L is the length of the pipe. For a closed organ pipe (one end closed), the harmonic series is given by $f_m = m\frac{v}{4L} = mf_1$ (for $m = 1, 3, 5, \ldots$). Note the missing even harmonics in a closed pipe.

The secondary auditory effects of **loudness**, **pitch**, and **quality** are related to the physical wave properties of *intensity*, *frequency*, and *waveform* (harmonics), respectively.

Example 14.9   A 3.00-meter long pipe is in a room where the temperature is 20°C.

(a) What is the frequency of the fundamental if the pipe is open?

(b) What is the frequency of the second harmonic if the pipe is open?

(c) What is the frequency of the fundamental if the pipe is closed?

(d) What is the frequency of the second harmonic if the pipe is closed?

Solution: Given: $L = 3.00$ m, $T_C = 20°C$.

Find: (a) f_1 (open)   (b) f_2 (open)   (c) f_1 (closed)   (d) f_2 (closed).

The speed of sound in air is $v = (331 + 0.6T_C)$ m/s $= [331 + 0.6(20)]$ m/s $= 343$ m/s.

(a) For an open pipe, $f_n = n\frac{v}{2L} = nf_1$ (for $n = 1, 2, 3, \ldots$).

So $f_1 = 1\frac{v}{2L} = \frac{343 \text{ m/s}}{2(3.00 \text{ m})} = 57.2$ Hz.

(b) $f_2 = 2f_1 = 2(57.2 \text{ Hz}) = 114$ Hz.

(c) For a closed pipe, $f_n = m\frac{v}{4L} = mf_1$ (for $m = 1, 3, 5, \ldots$).

So $f_1 = 1\frac{v}{4L} = \frac{343 \text{ m/s}}{4(3.00 \text{ m})} = 28.6$ Hz.

(d) Even harmonics cannot exist in a closed pipe. There is no second harmonic.

IV.   Mathematical Summary

Speed of Sound	$v = (331 + 0.6T_C)$ m/s (14.1)	Calculates the speed of sound in air (in m/s).
Intensity of a Point Source	$I = \dfrac{P}{4\pi r^2}$ and $\dfrac{I_2}{I_1} = \left(\dfrac{R_1}{R_2}\right)^2$ (14.2–3)	Calculates the intensity of a point source as a function of distance from it.
Intensity Level (in dB)	$\beta = 10 \log \dfrac{I}{I_o}$ (12.4) where $I_o = 10^{-12}$ W/m²	Calculates the intensity level from intensity.
Phase Difference (ΔL is path-length difference)	$\Delta\theta = \dfrac{2\pi}{\lambda}(\Delta L)$ (14.5)	Computes the phase difference from path-length difference ΔL.
Condition for Constructive Interference	$\Delta L = n\lambda$ $(n = 0, 1, 2, 3, \ldots)$ (14.6)	Defines the condition for constructive interference.
Condition for Destructive Interference	$\Delta L = m\lambda/2$ $(m = 1, 3, 5, \ldots)$ (14.7)	Defines the condition for destructive interference.
Beat Frequency	$f_b = |f_1 - f_2|$ (14.8)	Calculates the beat frequency from two frequencies.
Doppler Effect: Source Moving	$f_o = \dfrac{v}{v \pm v_s}f_s = \dfrac{1}{1 \pm \dfrac{v_s}{v}}f_s$ v_s = speed of source v = speed of sound (14.11)	Relates observed frequency and source frequency. − for source moving toward stationary observer + for source moving away from stationary observer
Doppler Effect: Observer Moving	$f_o = \dfrac{v \pm v_o}{v}f_s = \left(1 \pm \dfrac{v_o}{v}\right)f_s$ v_o = speed of observer v = speed of sound (14.14)	Relates observed frequency and source frequency. + for observer moving toward stationary source − for observer moving away from stationary source
Mach Number	$M = \dfrac{v_s}{v}$ (14.16)	Defines the Mach number of a moving source.
Natural Frequencies of Organ Pipe Open at Both Ends	$f_n = n\dfrac{v}{2L} = nf_1$ $(n = 1, 2, 3, \ldots)$ (14.17)	Gives the natural frequencies of an open (both ends open) organ pipe.
Natural Frequencies of Organ Pipe Closed on One End	$f_n = m\dfrac{v}{4L} = mf_1$ $(n = 1, 3, 5, \ldots)$ (14.18)	Gives the natural frequencies of a closed (one end open) organ pipe.

V.   Solutions of Selected Exercises and Paired/Trio Exercises

7.   They reach $\boxed{\text{at the same time}}$ because sound is not dispersive, i.e., speed not depending on frequency.

10.   (a) $v = (331 + 0.6T_C)$ m/s = [331 + 0.6(10)] m/s = $\boxed{337 \text{ m/s}}$.

   (b) $v = [331 + 0.6(20)]$ m/s = $\boxed{343 \text{ m/s}}$.

14. $f = \dfrac{v}{\lambda} = \dfrac{1500 \text{ m/s}}{3.0 \times 10^{-4} \text{ m}} = \boxed{5.0 \times 10^6 \text{ Hz}}$, where $v = 1500$ m/s is the speed of sound in water.

17.   (a) $\Delta t = \dfrac{d}{v_a} - \dfrac{d}{v_s} = \dfrac{300 \text{ m}}{343 \text{ m/s}} - \dfrac{300 \text{ m}}{4500 \text{ m/s}} = \boxed{0.81 \text{ s}}$.

   (b) 36 km/h = 10 m/s.   So $\Delta t = \dfrac{300 \text{ m}}{343 \text{ m/s} + 10 \text{ m/s}} - \dfrac{300 \text{ m}}{4500 \text{ m/s}} = \boxed{0.78 \text{ s}}$.

20. $T_C = \frac{5}{9}(T_F - 32) = \frac{5}{9}(72 - 32) = 22.22°C$.

   $v_s = (331 + 0.6T_C)$ m/s = [331 + 0.6(22.22)] m/s = 344.3 m/s.

   $\Delta t = \dfrac{d}{v_b} + \dfrac{d}{v_s} = d\left(\dfrac{1}{v_b} + \dfrac{1}{v_s} \right)$.

   So $d = \dfrac{\Delta t}{\dfrac{1}{v_b} + \dfrac{1}{v_s}} = \dfrac{1.00 \text{ s}}{\dfrac{1}{200 \text{ m/s}} + \dfrac{1}{344.3 \text{ m/s}}} = \boxed{127 \text{ m}}$.

28.   (a) $I = \dfrac{P}{4\pi r^2} = \dfrac{1.0 \text{ W}}{4\pi(3.0 \text{ m})^2} = \boxed{8.8 \times 10^{-3} \text{ W/m}^2}$.

   (b) $I = \dfrac{1.0 \text{ W}}{4\pi(6.0 \text{ m})^2} = \boxed{2.2 \times 10^{-3} \text{ W/m}^2}$.

30. $\boxed{\text{1.4 times}}$.

32.   (a) $\beta = 10 \log \dfrac{I}{I_o} = 10 \log \dfrac{10^{-2}}{10^{-12}} = \boxed{100 \text{ dB}}$.

   (b) $\beta = 10 \log \dfrac{10^{-6}}{10^{-12}} = \boxed{60 \text{ dB}}$.

   (c) $\beta = 10 \log \dfrac{10^{-15}}{10^{-12}} = \boxed{-30 \text{ dB}}$.

37. (a) $\beta = 10 \log \dfrac{10{,}000\,I}{I_0} = 10 \log 10{,}000 + 10 \log \dfrac{I}{I_0} = 40 + 23 = \boxed{63 \text{ dB}}$.

Here we used the property $\log xy = \log x + \log y$.

(b) $\beta = 10 \log \dfrac{10^6\,I}{I_0} = 10 \log 10^6 + 10 \log \dfrac{I}{I_0} = 60 + 23 = \boxed{83 \text{ dB}}$.

(c) $\beta = 10 \log \dfrac{10^9\,I}{I_0} = 10 \log 10^9 + 10 \log \dfrac{I}{I_0} = 90 + 23 = \boxed{113 \text{ dB}}$.

42. From $\beta = 10 \log \dfrac{I}{I_0}$, we have $\Delta\beta = \beta_2 - \beta_1 = 10 \log \dfrac{I_2}{I_0} - 10 \log \dfrac{I_1}{I_0} = 10 \log \dfrac{I_2}{I_1}$.

$(\log x - \log y = \log \dfrac{x}{y})$.

So $\dfrac{I_2}{I_1} = 10^{\Delta\beta/10} = 10^{-3}$. Also $\dfrac{I_2}{I_1} = \dfrac{R_1^{\,2}}{R_2^{\,2}}$,

therefore $R_2 = \sqrt{\dfrac{I_1}{I_2}}\, R_1 = \sqrt{10^3}\,(10.0 \text{ m}) = \boxed{316 \text{ m}}$.

47. For one bee, $I = I_0$. Assume it takes n bees.

Since $\beta = 10 \log \dfrac{nI}{I_0} = 10 \log n$, $n = 10^{\beta/10} = 10^{5.0} = \boxed{10^5 \text{ bees}}$.

55. At the first destructive point, $\Delta L = \tfrac{1}{2}\lambda = \tfrac{1}{2}\dfrac{v}{f} = \dfrac{343 \text{ m/s}}{2(1000 \text{ Hz})} = \boxed{0.172 \text{ m}}$.

58. (a) Since the heard frequency is higher than the siren frequency, the person is moving $\boxed{\text{toward}}$ the siren.

(b) From $f_0 = \dfrac{v + v_0}{v} f_s$,

we have $v_0 = \dfrac{f_0 - f_s}{f_s}\, v = \dfrac{520 \text{ Hz} - 500 \text{ Hz}}{500 \text{ Hz}}\,(343 \text{ m/s}) = \boxed{13.7 \text{ m/s}}$.

60. (a) $\boxed{431 \text{ Hz}}$.

(b) $\boxed{373 \text{ Hz}}$.

63. from $\sin\theta = \dfrac{v}{v_s}$, we have $\theta = \sin^{-1}\left(\dfrac{v}{v_s}\right) = \sin^{-1}(1) = \boxed{90°}$.

66. The minimum separation for destructive interference corresponds to $\Delta L = \dfrac{\lambda}{2}$.

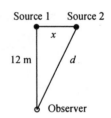

$$\Delta L = d - 12 \text{ m} = \sqrt{x^2 + (12.0 \text{ m})^2} - 12.0 \text{ m}$$

$$= \frac{\lambda}{2} = \frac{v}{2f} = \frac{340 \text{ m/s}}{2(1000 \text{ Hz})} = 0.170 \text{ m}.$$

So $x^2 + (12.0 \text{ m})^2 = (12.17 \text{ m})^2$. solving, $x = \boxed{2.03 \text{ m}}$.

74. (a) 378 Hz = 3(126 Hz) and 630 Hz = 5(126 Hz), so it is a $\boxed{\text{closed pipe}}$.

(b) From $f_n = \dfrac{mv}{4L}$, we have $L = \dfrac{v}{4f_1} = \dfrac{340 \text{ m/s}}{4(126 \text{ Hz})} = \boxed{0.675 \text{ m}}$; here we take $m = 1$.

79. If the observer were stationary, the frequency heard by the observer would have been

$(f_o)_1 = \dfrac{v}{v \mp v_s} f_s$, which would be the frequency of the "source" to the moving observer.

$(f_s)_2 = (f_o)_1$. So $(f_o)_2 = f_o = \dfrac{v \pm v_o}{v}(f_s)_2 = \dfrac{v \pm v_o}{v} \dfrac{v}{v \mp v_s} f_s = \dfrac{v \pm v_o}{v \mp v_s} f_s$.

83. $\lambda = \dfrac{v}{f} = \dfrac{[331 + 0.6(15)] \text{ m/s}}{440 \text{ Hz}} = 0.773 \text{ m}.$

$\Delta L = 8.90 \text{ m} - 6.97 \text{ m} = 1.93 \text{ m} = 2.5 \ (0.773) = 2.5\lambda.$

So they will interfere $\boxed{\text{destructively}}$.

89. From $\beta = 10 \log \dfrac{I}{I_o}$, we have $\dfrac{I}{I_o} = 10^{\beta/10}$.

So $I = 10^{9.0} \ (10^{-12} \text{ W/m}^2) = 10^{-3} \text{ W/m}^2.$

In 5.0 s, the energy is $E = IAt = (10^{-3} \text{ W/m}^2)(1.5 \text{ m}^2)(5.0 \text{ s}) = \boxed{7.5 \times 10^{-3} \text{ J}}$.

93. 90.0 km/h = 25 m/s.

Approaching: $f_o = \dfrac{v}{v - v_s} f_s = \dfrac{343 \text{ m/s}}{343 \text{ m/s} - 25 \text{ m/s}} (700 \text{ Hz}) = \boxed{755 \text{ Hz}}.$

Moving away: $f_o = \dfrac{v}{v + v_s} f_s = \dfrac{343 \text{ m/s}}{343 \text{ m/s} + 25 \text{ m/s}} (700 \text{ Hz}) = \boxed{652 \text{ Hz}}.$

VI.   Practice Quiz

1.   An echo is heard 2.0 s from a cliff on a day the temperature is 15°C. Approximately how far is the cliff from the observer?

   (a) 85 m   (b) 170 m   (c) 340 m   (d) 680 m   (e) 1360 m

2.   The third harmonic frequency in a pipe closed at one end is 330 Hz. What is the frequency of the fundamental?

   (a) 110 Hz   (b) 220 Hz   (c) 330 Hz   (d) 660 Hz   (e) 990 Hz

3.   The intensity of a point source at a distance d from the source is I. What is the intensity at half the distance from the source?

   (a) $I/4$ (b) $I/2$ (c) I (d) $2I$ (e) $4I$

4.   If the intensity level of a loud speaker is 40 dB. What is the intensity level of two identical speakers?

   (a) 20 dB   (b) 37 dB   (c) 40 dB   (d) 43 dB   (e) 80 dB

5.   Two tones have frequencies of 330 Hz and 332 Hz. What is the beat frequency?

   (a) 0 Hz   (b) 2 Hz   (c) 331 Hz   (d) 662 Hz   (e) none of the above

6.   The third harmonic frequency of a pipe open at both ends is 300 Hz. What is the length of the pipe? (Assume the speed of sound is 340 m/s.)

   (a) 3.40 m   (b) 1.70 m   (c) 1.13 m   (d) 0.567 m   (e) 0.378 m

7.   A sound source has a frequency of 500 Hz. If a listener moves at a speed of 30.0 m/s toward the source, what is the frequency heard by the listener? (The speed of sound is 340 m/s.)

   (a) 456 Hz   (b) 578 Hz   (c) 500 Hz   (d) 522 Hz   (e) 544 Hz

8.   Two loudspeakers are placed side by side and driven by the same frequency of 500 Hz. If the distance from a person to one of the speakers is 5.00 m and the person detects little or no sound, what is the distance from the person to the other speaker? (The speed of sound is 340 m/s.)

   (a) 7.72 m   (b) 8.06 m   (c) 8.40 m   (d) 9.08 m   (e) 5.0 m

9.   The sound intensity level 5.0 m from a point source is 70 dB. What is the sound intensity level at a distance of 10.0 m from the source?

   (a) 17.5 dB   (b) 35 dB   (c) 64 dB   (d) 67 m   (e) 140 dB

10.   What is the frequency heard by a stationary observer when a train approaches with a speed of 30.0 m/s? The frequency of the train horn is 600 Hz. (The speed of sound is 340 m/s.)

(a) 547 Hz   (b) 551 Hz   (c) 600 Hz   (d) 653 Hz   (e) 658 Hz

Answers to Practice Quiz:

1. c   2. a   3. e   4. d   5. b   6. b   7. e   8. b   9. c   10. e

CHAPTER 15

Electric Charge, Force, and Field

I.   Chapter Objectives

Upon completion of this chapter, you should be able to:

1.   distinguish between the two types of electric charge, state the force law that operates between charged objects, and understand and use the law of charge conservation.

2.   distinguish between conductors and insulators, explain the operation of the electroscope, and distinguish among charging by friction, conduction, induction, and polarization.

3.   understand Coulomb's law and use it to calculate the electric force between charged particles.

4.   understand the definition of the electric field, plot electric field lines and calculate electric fields for simple charge distributions.

5.   describe the electric field near the surface and in the interior of a conductor, determine where the highest concentration of excess charge accumulates on a charged conductor, and sketch the electric field line pattern outside a charged conductor.

*6.   state the physical basis of Gauss's law and use the law to make qualitative predictions.

II.   Key Terms

Upon completion of this chapter, you should be able to define and/or explain the following key terms:

electric charge	charging by contact *or* by conduction
law of charges *or* the charge-force law	charging by induction
coulomb	polarization
net charge	Coulomb's law
conservation of charge	electric field
conductors	electric lines of force
insulators	electric dipole
semiconductors	Gaussian surface
electrostatic charging	Gauss's law
charging by friction	

The definitions and/or explanations of the most important key terms can be found in the following section:

III. Chapter Summary and Discussion.

III. Chapter Summary and Discussion

1. Electric Charge (Section 15.1)

Electric charge is the property of an object that determines its electrical behavior: the electric force it can exert and the electric force it can experience. There are two types of charges, which are distinguished as positive (+) or negative (−). A positive charge is arbitrarily associated with the proton, and a negative charge with the electron. The magnitude of electron charge (fundamental unit of charge) is $|e| = 1.60 \times 10^{-19}$ C, where the SI unit of charge is the coulomb (C). Both electron and proton carry this charge but with opposite signs.

The directions of the electric forces on the charges of mutual interaction are given by the **law of charges** or the **charge-force law**: like charges repel and unlike charges attract.

An object with a **net charge** means that it has an *excess* of either positive or negative charges. The law of **conservation of charge** states that the net charge of an isolated system remains constant.

Example 15.1   A piece of glass has a net charge of −2.00 μC.

   (a) Are there more protons or electrons?

   (b) What happens if an identically charged piece of glass is placed near the first one?

   (c) How many fundamental units of excess charge does one piece of the glass contain?

Solution: Given: $q = -2.00\ \mu C = -2.00 \times 10^{-6}$ C.   Find: (c) n.

   (a) Since the glass has a negative charge, it must have more electrons than protons. Thus the piece of glass has gained some electrons somehow.

   (b) From the charge-force law, like charges repel. The two pieces of glass therefore repel each other.

   (c) From $q = ne$, we have $n = \dfrac{q}{|e|} = \dfrac{2.00 \times 10^{-6}\ \text{C}}{1.60 \times 10^{-19}\ \text{C/electron}} = 1.25 \times 10^{13}$ electrons.

So one piece of glass has 1.25×10^{13} excess electrons or most likely it has gained 1.25×10^{13} electrons, since protons are in the atomic nucleus and stay fixed in a solid.

---

2. Electrostatic Charging (Section 15.2)

A **conductor** is a material which has the ability to conduct or transmit electric charges while an **insulator** is a poor electric conductor.

Electrostatic charging can be done by the following four processes. In **charging by friction**, insulators are rubbed with different materials and the insulators and the materials acquire equal but opposite charges. In **charging by contact** or **by conduction**, a charged object makes a contact with a uncharged object and some of the charge on the charged object is transferred to the uncharged object. In **charging by induction**, a charged object is brought near (not touching) a uncharged object, the uncharged object is then "grounded," and the uncharged object acquires an opposite charge than the charged object. In **charging by polarization**, the positive and negative charges are simply separated or realigned within the object and the net charge of the object is still zero. Charges are merely separated so a portion of the object has excess positive charges and another has excess negative charges.

3. Electric Force (Section 15.3)

The magnitude of the electric force between two point charges q_1 and q_2 is given by **Coulomb's law**:

$F_e = \dfrac{kq_1q_2}{r^2}$, where $k \approx 9.00 \times 10^9$ N·m²/C² is called the Coulomb constant and r is the distance between the two

point charges. The direction of the electric force is determined from the law of charges. Like the universal law of

gravity ($F_g = \dfrac{Gm_1m_2}{r^2}$), Coulomb's law is also an *inverse square law* because the force is proportional to $\dfrac{1}{r^2}$.

Note: Since force is a vector, you have to use vector addition to calculate the net force on a charge if there is more than one force on it (from several other charges).

Example 15.2   Two charges are separated by a distance d and exert mutual attractive forces of F_1 on each other. What are the mutual attractive forces if the charges are separated by a distance of $3d$?

Solution: Given: $r_1 = d$, $r_2 = 3d$,   and F_1.   Find: F_2.

From Coulomb's law $F_e = \dfrac{kq_1q_2}{r^2}$, we have $\dfrac{F_2}{F_1} = \dfrac{r_1^2}{r_2^2}$ (since k, q_1, and q_2 remain constant).

So $F_2 = \dfrac{r_1^2}{r_2^2} F_1 = \dfrac{(d)^2}{(3d)^2} F_1 = \tfrac{1}{9} F_1$. The force decreases by a factor of 9 when the distance is increased to

3 times of its original value. This is an important feature of the inverse square law.

Example 15.3   A +4.0-coulomb charge is at the origin and a +9.0-coulomb charge is at $x = 4.0$ m. Where can a third charge q_3 be placed on the x axis so the net force on it is zero?

Solution: Given: $q_1 = 4.0$ C, $q_2 = 9.0$ C, $r_1 = d$, $r_2 = (4.0 \text{ m} - d)$, $F_1 = F_2$.

Find: d.

By the charge-force law, the third charge has to be placed in

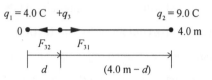

between the two charges. The third charge, however, can be either positive or negative. (Why?) Assume the third charge is placed at d from the +4.0 C charge, then it is $(4.0 \text{ m} - d)$ from the +9.0 C charge. For the force on q_3 to be zero, the forces by q_1 and q_2 on q_3 must equal each other in magnitude.

From Coulomb's law, we have $F_{31} = \dfrac{kq_1 q_3}{d^2}$ and $F_{32} = \dfrac{kq_2 q_3}{(4.0 \text{ m} - d)^2}$.

Equating F_{31} and F_{32} gives $\dfrac{k(4.0 \text{ C})q_3}{d^2} = \dfrac{k(9.0 \text{ C})q_3}{(4.0 \text{ m} - d)^2}$.

Simplifying and taking the square root on both sides $\dfrac{2.0}{d} = \dfrac{3.0}{(4.0 \text{ m} - d)}$.

Cross multiplying, we arrive at $2.0(4.0 \text{ m} - d) = 3.0d$, or $5.0d = 8.0$ m.

Solving, $d = \dfrac{8.0 \text{ m}}{5.0} = 1.6$ m from the +4.0 C charge.

Example 15.4 Consider three point charges located at the corners of a triangle as shown in the figure on the right. If $q_1 = 6.0$ nC, $q_2 = -1.0$ nC, and $q_3 = 5.0$ nC, what is the net force on q_3.

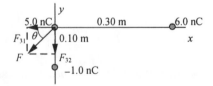

Solution: Given: $q_1 = 6.0$ nC $= 6.0 \times 10^{-9}$ C, $q_2 = -1.0$ nC $= -1.0 \times 10^{-9}$ C, $q_3 = 5.0$ nC $= 5.0 \times 10^{-9}$ C,

$r_1 = 0.30$ m, $r_2 = 0.10$ m.

Find: **F**.

By the charge-force law, the force on the 5.0 nC charge by the 6.0 nC charge is to the left and the force by the −3.0 nC charge is down.

From Coulomb's law,

we have $F_{31} = \dfrac{(9.00 \times 10^9 \text{ N·m}^2/\text{C}^2)(6.0 \times 10^{-9} \text{ C})(5.0 \times 10^{-9} \text{ C})}{(0.30 \text{ m})^2} = 3.0 \times 10^{-6}$ N,

and $F_{32} = \dfrac{(9.00 \times 10^9 \text{ N·m}^2/\text{C}^2)(1.0 \times 10^{-9} \text{ C})(5.0 \times 10^{-9} \text{ C})}{(0.10 \text{ m})^2} = 4.5 \times 10^{-6}$ N.

The negative sign of the negative charge does not have to be included here because the direction of the force is already determined by the charge-force law.

So $F = \sqrt{F_{31}^2 + F_{32}^2} = \sqrt{(3.0 \times 10^{-6}\,\text{N})^2 + (4.5 \times 10^{-6}\,\text{N})^2} = 5.4 \times 10^{-6}\,\text{N}$,

and $\theta = \tan^{-1}\dfrac{F_{32}}{F_{31}} = \tan^{-1}\dfrac{4.5 \times 10^{-6}\,\text{N}}{3.0 \times 10^{-6}\,\text{N}} = 56°$ below the $-x$ axis.

4. Electric Field (Section 15.4)

The **electric field** is a *vector field* that describes how nearby charges modify the space around them. It is defined as the electric force per unit positive charge: $\mathbf{F} = \dfrac{\mathbf{F}_{\text{on } q_0}}{q_0}$, where q_0 is a positive test charge. The direction of the electric field at any point is in the direction of the force experienced by the positive test charge. The SI unit of the electric field is N/C. The magnitude of the electric field due to a point charge is $E = \dfrac{kq}{r^2}$.

Note: Since electric field is a vector, you must use vector addition to calculate the net field at a point if there are more than one charge contributing to the fields.

Electric lines of force are the imaginary lines formed by connecting electric field vectors at many points. Their closeness and direction indicate the magnitude and direction of the electric field at any point. In general, the electric lines of force originate from positive charges (or infinity if there is no positive charge) and terminate at negative charges (or infinity if there is no negative charge). The number of lines leaving or entering a charge is proportional to the magnitude of that charge.

Note: Since the direction of electric lines of force indicates the direction of the electric field, no two electric lines of force can cross. (Why?)

The electric field between two closed spaced parallel plates of charge $\pm Q$ is a constant. Its magnitude is given by $E = \dfrac{4\pi kQ}{A}$, where Q is the magnitude of the charge on each plate and A is the surface area of the plate. The direction of the electric field points from the positive plate to the negative plate.

Example 15.5 A 2.0-coulomb charge is 10 m from a small test charge of 0.10 nC.

    (a) What is the electric field at the location of the test charge?

    (b) What is the direction of the electric field at the location of the test charge?

    (c) What is the magnitude of the force experienced by the test charge?

Solution: Given: $q = 2.0$ C, $q_o = 0.10$ nC $= 1.0 \times 10^{-10}$ C, $r = 10$ m.

Find: (a) E (b) direction of **E**   (c) F.

(a) $E = \dfrac{kq}{r^2} = \dfrac{(9.00 \times 10^9 \text{ N·m}^2/\text{C}^2)(2.0 \text{ C})}{(10 \text{ m})^2} = 1.8 \times 10^8$ N/C.

(b) Since the 2.0-coulomb charge is a positive charge, the direction of the electric field at the location of the test charge points radially outward from the +2.0-coulomb charge.

(c) From the definition of electric field $E = \dfrac{F}{q_o}$,

we have $F = q_o E = (1.0 \times 10^{-10} \text{ C})(1.8 \times 10^8 \text{ N/C}) = 0.045$ N.

Example 15.6   A charge of 5.0-μC is placed at the 0 cm mark of a meterstick and a charge of -4.0 μC is placed at the 50 cm mark.

(a) What is the electric field at the 30 cm mark?

(b) At what point along a line connecting the two charges is the electric field zero?

Solution: Given: $q_1 = 5.0$ μC $= 5.0 \times 10^{-6}$ C, $q_2 = -4.0$ μC $= 4.0 \times 10^{-6}$ C,

$r_1 = 0.30$ m, $r_2 = (0.50 \text{ m} - 0.30 \text{ m}) = 0.20$ m.

Find: (a) E (b) x (where $E = 0$).

(a) The electric fields due to the two point charges at the 30 cm mark are in the same direction (why?).

From $E = \dfrac{kq}{r^2}$, we have $E_1 = \dfrac{(9.00 \times 10^9 \text{ N·m}^2/\text{C}^2)(5.0 \times 10^{-6} \text{ C})}{(0.30 \text{ m})^2} = 5.0 \times 10^5$ N/C

and $E_2 = \dfrac{(9.00 \times 10^9 \text{ N·m}^2/\text{C}^2)(4.0 \times 10^{-6} \text{ C})}{(0.20 \text{ m})^2} = 9.0 \times 10^5$ N/C.

The negative sign of the negative charge does not have to be included here because the direction of the electric field is already determined.

$E = E_1 + E_2 = 5.0 \times 10^5$ N/C $+ 9.0 \times 10^5$ N/C $= 1.4 \times 10^6$ N/C. The direction is to the right.

(b) Since the direction of the fields between 0 and 50 cm marks are in the same direction, there is no point between them where the electric field is zero. We have to look outside the 0 and 50 cm marks along the meterstick. It is only possible to have zero electric field at a point to the right of the negative charge along the meterstick (why is it impossible to have zero electric field at a point to the left of the positive charge along the meterstick). Assume the point is a distance x from the 0 cm mark, and then it is $(x - 0.50$ m$)$ from q_2 at the 50 cm mark. The directions of the electric fields are opposite.

$E_1 = \dfrac{kq_1}{x^2}$ and $E_2 = \dfrac{kq_2}{(x - 0.50 \text{ m})^2}$.

To produce a zero net field, we get, $E_1 = E_2$, which gives $\dfrac{k(5.0 \times 10^{-6} \text{ C})}{x^2} = \dfrac{k(4.0 \times 10^{-6} \text{ C})}{(x - 0.50 \text{ m})^2}$.

Simplifying and taking the square root on both sides $\dfrac{2.24}{x} = \dfrac{2.0}{x - 0.50 \text{ m}}$.

Cross multiplying, we arrive at $(2.24)(x - 0.50 \text{ m}) = 2.0x$, or $0.24x = 1.12$ m.

Solving, $x = 4.7$ m from the 0-cm mark.

Example 15.7 Four charges occupy the corners of a square as shown in the figure on the right. Each charge has a magnitude of 3.0 μC and the side of the square is 0.50 m. Find the electric field at the geometrical center of the square.

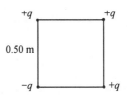

Solution: Given: $q = 3.0\ \mu$C, $L = 0.50$ m.

Find: **E**.

From symmetry, the fields due to the charges at the upper-left (1) and lower-right (4) corners cancel. The fields by the charges at the upper-right (2) and lower-left (3) are in the same direction (pointing toward the $-q$ charge). So the net field is simply the addition of these two fields.

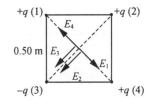

The length of the diagonal is $d = \sqrt{(0.50 \text{ m})^2 + (0.50 \text{ m})^2} = 0.707$ m.

So the distance from the charges to the center of the square is $\dfrac{d}{2} = \dfrac{0.707 \text{ m}}{2} = 0.354$ m.

From $E = \dfrac{kq}{r^2}$, we have $E_2 = E_3 = \dfrac{(9.00 \times 10^9 \text{ N·m}^2/\text{C}^2)(3.0 \times 10^{-6} \text{ C})}{(0.354 \text{ m})^2} = 2.15 \times 10^{15}$ N/C.

Therefore $E = E_2 + E_3 = 2(2.15 \times 10^{15} \text{ N/C}) = 4.3 \times 10^{15}$ N/C, toward the negative charge.

---

5. Conductors and Electric Fields (Section 15.5)

The electric fields associated with charged conductors in electrostatic equilibrium (isolated or insulated) have the following interesting properties:

(1) The electric field is zero everywhere inside a charged conductor.

(2) Any excess charge on an isolated conductor resides entirely on the surface of the conductor.

(3) The electric field at the surface of a charged conductor is perpendicular to the surface.

(4) Excess charge tends to accumulate at sharp points, or locations of the greatest curvature, on charged conductors, so the highest charge accumulations occur where the electric field from the conductor is the largest.

All the above properties can be explained with the properties of a conductor and the law of charges.

*6. Gauss's Law for Electric Fields: A Qualitative Approach (Section 15.6)

A qualitative statement of **Gauss's law** is that the net number of electric field lines passing through an imaginary closed surface is proportional to the amount of net charge enclosed within the surface. The surface is called a **Gaussian surface**. In general, by counting the number of lines entering or leaving a Gaussian surface, we can determine the sign and the *relative* magnitude of the charge enclosed within that surface. If there are a net number of lines leaving the surface (we label them "+"), the net charge enclosed is positive. If there are a net number of lines entering the surface (we label them "−"), the net charge enclosed is negative.

Example 15.8 A Gaussian surface encloses an object with a net charge of +2.0 C and there are 6 lines leaving the surface. Some charge is added to the object and now there are 18 lines *entering* the surface. How much charge was added?

Solution:

Since there are 6 lines when there is +2.0 C, therefore a charge of +1.0 C is equivalent to 3 lines. After charge is added, there are 18 lines entering.

So the net charge is now $-\dfrac{18 \text{ lines}}{3 \text{ lines/coulomb}} = -6.0$ C.

Therefore, the charge added was $\Delta Q = Q_f - Q_i = -6.0$ C $- 2.0$ C $= -8.0$ C.

IV. Mathematical Summary

Quantization of Electric Charge	$q = ne$ (15.1) $\|e\| = 1.60 \times 10^{-19}$ C	Relates charge and fundamental unit of charge.
Coulomb's Law	$F_e = \dfrac{kq_1 q_2}{r^2}$ (15.2) $k \approx 9.00 \times 10^9$ N·m²/C²	Computes the magnitude of the electric force between two point charges.

Electric Field (definition)	$E = \dfrac{F_{\text{on } q_0}}{q_0}$ (15.3)	Defines the electric field vector from the force vector on a charge q_0.
Electric Field due to a Point Charge q	$E = \dfrac{kq}{r^2}$ (15.4)	Calculates the magnitude of the electric field due to a point charge.
Electric Field Between Two Closely Spaced Parallel Plates	$E = \dfrac{4\pi kQ}{A}$ (15.5)	Calculates the magnitude of the electric field between two closely spaced parallel plates of charge $+-Q$.

V.   Solutions of Selected Exercises and Paired/Trio Exercises

8.   (a) $\boxed{-8.0 \times 10^{-10}\text{ C}}$ according to charge conservation.

   (b) $n = \dfrac{q}{|e|} = \dfrac{8.0 \times 10^{-10}\text{ C}}{1.6 \times 10^{-19}\text{ C/electron}} = \boxed{5.0 \times 10^{9}\text{ electrons}}$.

9.   (a) $\boxed{+4.8 \times 10^{-9}\text{ C}}$ according to charge conservation.

   (b) $n = \dfrac{q}{|e|} = \dfrac{4.8 \times 10^{-9}\text{ C}}{1.6 \times 10^{-19}\text{ C/electron}} = 3.0 \times 10^{10}\text{ electrons}$.

   So the mass is $(3.0 \times 10^{10}\text{ electron})(9.11 \times 10^{-31}\text{ kg/electron}) = \boxed{2.7 \times 10^{-20}\text{ kg}}$.

14.   If you bring a negatively charged object near the electroscope, the induction process will charge the electroscope with positive charges. You can prove the charges are positive by bringing the negatively charged object near the leaves and seeing if the leaves fall.

15.   The spheres can be charged through polarization by induction. For example, if you bring a positively charged object near one of the two spheres, the sphere near the charged object will have a net negative charge and the other sphere will have a net positive charge (polarization by induction). Then you separate the two spheres (while keep the positively charged object nearby) and the spheres will have opposite charges according to charge conservation.

20.   (a) Since $F = \dfrac{kq_1 q_2}{r^2} \propto \dfrac{1}{r^2}$, F is $\boxed{1/4\text{ as large}}$ if r is doubled.

   (b) F is $\boxed{9\text{ times as large}}$ if r is reduced to one-third.

24. From $F = \dfrac{kq_1q_2}{r^2}$, we have $\dfrac{F_2}{F_1} = \dfrac{r_1^2}{r_2^2}$. So $r_2 = \sqrt{\dfrac{F_1}{F_2}}\, r_1 = \sqrt{10}\ (30\ \text{cm}) = \boxed{95\ \text{cm}}$.

29. (a) If q_1 and q_2 are like charges, the third charge must be placed in between q_1 and q_2 for it to be in electrostatic equilibrium. Also q_1 and q_2 have the same magnitude; it must be at $\boxed{0.25\ \text{m}}$ from symmetry.

(b) If q_1 and q_2 are unlike charges, the third charge could be placed outside q_1 and q_3 ($x < 0$ and $x > 0.50$ m). However, since the magnitudes of q_1 and q_2 are equal, there will be $\boxed{\text{nowhere}}$ for it to happen according to Coulomb's law.

(c) Since $q_1 = +3.0\ \mu$C and $q_2 = -7.0\ \mu$C, a third charge of either type can be placed outside q_1 ($x < 0$) for it to be in electrostatic equilibrium. It cannot be placed outside q_2 ($x > 0.50$ m) since the force by q_2 will always be larger due to the closer distance from it. Assume the third charge is placed at d from q_1 (or $x = -d$).

From Coulomb's law, we have

$$\dfrac{kq_1q_3}{r_{13}^2} = \dfrac{k|q_2|q_3}{r_{23}^2}, \quad \text{or} \quad \dfrac{q_1}{d^2} = \dfrac{|q_2|}{(d+0.50\ \text{m})^2}.$$

Taking the square root on both sides gives $\dfrac{\sqrt{3.0}}{d} = \dfrac{\sqrt{7.0}}{d+0.50}$, so $1.73(d+0.50) = 2.65d$.

Solving, $d = 0.94$ m. Thus $\boxed{x = -0.94\ \text{m for either} \pm q_3}$.

32. $F_2 = F_3 = \dfrac{kq_1q_2}{r^2} = \dfrac{(9.0 \times 10^9\ \text{N·m}^2/\text{C}^2)(4.0 \times 10^{-6}\ \text{C})^2}{(0.20\ \text{m})^2} = 3.6$ N.

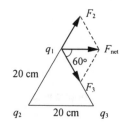

By symmetry, the net force on q_1 points in the $+x$ direction.

$F_{net} = F_x = F_2 \cos 60° + F_3 \cos 60°$

$\quad = 2(3.6\ \text{N}) \cos 60° = \boxed{3.6\ \text{N in the } +x \text{ direction}}$.

34. $\boxed{3.3 \times 10^{-8}\ \text{C}}$

39. It is determined by the relative density or spacing of the field lines. The closer the lines, the greater the magnitude.

44. $E = \dfrac{kq}{r^2} = \dfrac{(9.0 \times 10^9\ \text{N·m}^2/\text{C}^2)(2.0 \times 10^{-6}\ \text{C})}{(0.25 \times 10^{-2}\ \text{m})^2} = \boxed{2.9 \times 10^9\ \text{N/C}}$.

46. $\boxed{9.0 \times 10^5 \text{ N/C toward the } -5.0 \ \mu\text{C charge}}$.

52.   Due to symmetry, E_1 and E_2 cancel out.

So the net electric field is the electric field by q_3.

$$E = E_3 = \frac{kq_3}{r^2} = \frac{(9.0 \times 10^9 \text{ N·m}^2/\text{C}^2)(4.0 \times 10^{-6} \ \mu\text{C})}{[(0.20 \text{ m}) \cos 30°]^2}$$

$$= \boxed{1.2 \times 10^6 \text{ N/C toward the } -4.0 \ \mu\text{C charge}}.$$

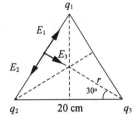

60. $\boxed{\text{Yes}}$, because the car (a metal frame) shields the electric field from reaching you.

62.   (a) since there is no electric field inside a conductor in electrostatic equilibrium.

65.   (a) It is $\boxed{\text{zero}}$ since there is no electric field inside a conductor in electrostatic equilibrium.

(b) Outside the sphere, the charge on the surface of the sphere can be considered as if it were concentrated at the center.

So $E = \frac{k(-Q)}{r^2} = -\boxed{\frac{kQ}{r^2}}$, the negative sign indicates it points toward the center.

(c) Again it is $\boxed{\text{zero}}$.

(d) All charges can be considered as if they were concentrated at the center.

So $E = \frac{k(-Q)}{r^2} + \frac{kQ}{r^2} + \frac{k(-Q)}{r^2} = -\boxed{\frac{kQ}{r^2}}$, the negative sign indicates it is toward the center.

72.   Since the number of lines is proportional to the net charge,

$$q_2 = \frac{75}{16}(+10.0 \ \mu\text{C}) = \boxed{+46.9 \ \mu\text{C}}.$$

79.   (a) We first need to find the acceleration from Newton's second law. $a = \frac{F}{m} = \frac{eE}{m}$

From the kinematic equation, $x = v_o t + \frac{1}{2}at^2 = \frac{1}{2}at^2$, we have

$$t = \sqrt{\frac{2x}{a}} = \sqrt{\frac{2x}{eE/m}} = \sqrt{\frac{2xm}{eE}} = \sqrt{\frac{2(1.0 \text{ m})(9.11 \times 10^{-31} \text{ kg})}{(1.6 \times 10^{-19} \text{ C})(450 \text{ N/C})}} = \boxed{1.6 \times 10^{-7} \text{ s}}.$$

(b) $x = \frac{1}{2}at^2 = \frac{1}{2}\frac{(-1.6 \times 10^{-19} \text{ C})(450 \text{ N/C})}{9.11 \times 10^{-31} \text{ kg}} \times (7.95 \times 10^{-8} \text{ s})^2 = -0.25 \text{ m}, y = 0.$

So it is at $\boxed{(-0.25 \text{ m}, 0)}$.

81.   Since $F = \dfrac{kq_1q_2}{r^2}$, $q_2 = \dfrac{Fr^2}{kq_1} = \dfrac{(1.8 \text{ N})(0.30 \text{ m})^2}{(9.0 \times 10^9 \text{ N·m}^2/\text{C}^2)(6.0 \times 10^{-6} \text{ C})} = 3.0 \times 10^{-6} \text{ C} = 3.0 \ \mu\text{C}.$

So the charge could be $\boxed{+3.0 \ \mu\text{C} \text{ on the } -y \text{ axis or } -3.0 \ \mu\text{C} \text{ on the } +y \text{ axis}}$ from the charge-force law.

86.

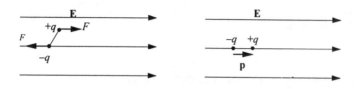

VI.   Practice Quiz

1.   Two point charges, initially 4.0 cm apart, experience a mutual force of magnitude 1.0 N. If they are moved to a new separation of 1.0 cm, what is the magnitude of the new electric force between them?

(a) 4.0 N   (b) 16 N   (c) 1/4 N   (d) 1/16 N   (e) none of the above

2.   A positively charged rod is brought near one end of an uncharged metal bar. The end of the metal bar farthest from the charged rod will be which of the following?

(a) positive   (b) negative   (c) neutral   (d) attracted   (e) repelled

3.   Two point charges each have a value of +3.0 C and are separated by a distance of 4.0 m. What is the magnitude of the electric field at a point midway between the two charges?

(a) zero   (b) 9.0×10^7 N/C   (c) 18.0×10^7 N/C   (d) 9.0×10^7 N/C   (e) 4.5×10^7 N/C

4.   Two point charges, separated by 1.5 cm, have charge values of +2.0 and −4.0 μC, respectively. What is the magnitude of the electric force midway between them?

(a) 160 N   (b) 320 N   (c) 3.6×10^{-8} N   (d) 8.0×10^{-12} N   (e) 3.1×10^{-3} N

5.   Which of the following is a vector?

(a) electric charge   (b) Gaussian surface   (c) electric field   (d) both (a) and (c)   (e) both (b) and (c)

6.   Three point charges are located at the following positions: $q_1 = 2.00 \ \mu$C at $x = 1.00$ m; $q_2 = 3.00 \ \mu$C at $x = 0$; $q_3 = -5.00 \ \mu$C at $x = -1.00$ m. What is the magnitude of the force on q_2?

(a) 5.40×10^{-2} N   (b) 0.135 N   (c) 8.10×10^{-2} N   (d) 0.189 N   (e) 9.0×10^{-2} N

7. A ball with a charge of $+4.0$ μC has a mass of 5.0×10^{-3} kg. What electric field directed upward will exactly balance the weight of the ball?

(a) 4.1×10^2 N/C (b) 8.2×10^2 N/C (c) 1.2×10^4 N/C (d) 5.1×10^6N/C (e) 2.0×10^{-7} N/C

8. If a conductor is in electrostatic equilibrium near an electric charge

(a) the total charge on the conductor must be zero.

(b) any charge on the conductor must be uniformly distributed.

(c) the force between the conductor and the charge must be zero.

(d) there may be some net excess charge on the surface of the conductor.

(e) the total electric field of the conductor must be zero.

9. Two point charges of $+3.0$ μC and -7.0 μC are placed at $x = 0$ and $x = 0.20$ m, respectively. What is the magnitude of the electric field at the point midway between them ($x = 0.10$ m)?

(a) 9.0×10^5 N/C (b) 1.8×10^6 N/C (c) 3.6×10^6 N/C (d) 4.5×10^6 N/C (e) 9.0×10^6 N/C

10. If the distance between two closely spaced parallel plates is halved, the electric field between them

(a) is also halved. (b) becomes only 1/4 as large. (c) remains the same (d) is doubled.

(e) becomes 4 times as large.

Answers to Practice Quiz:

1.b 2.a 3.a 4.b 5.c 6.d 7.d 8.c 9.e 10.c

CHAPTER 16

Electric Potential, Energy, and Capacitance

I. Chapter Objectives

Upon completion of this chapter, you should be able to:

1. understand the concept of electric potential difference ("voltage") and its relationship to electric potential energy and calculate electric potential differences and electric potential energy.

2. explain what is meant by an equipotential surface, sketch equipotential surfaces for simple charge configurations, and explain the relationship between equipotential surfaces and electric fields.

3. define capacitance and explain what it means physically, calculate the charge, voltage, electric field, and energy storage for parallel plate capacitors.

4. understand what a dielectric is and understand how it affects the physical properties of a capacitor.

5. find the equivalent capacitance of capacitors connected in series and in parallel, calculate the charges, voltages, and energy storage of individual capacitors in series and parallel configurations, and analyze capacitor networks that include both series and parallel arrangements.

II. Key Terms

Upon completion of this chapter, you should be able to define and/or explain the following key terms:

electric potential energy	capacitance
electric potential difference	farad
volt	dielectric
voltage	dielectric constant
electric potential	dielectric permittivity
equipotential surfaces	equivalent series capacitance
electron volt	equivalent parallel capacitance
capacitor	

The definitions and/or explanations of the most important key terms can be found in the following section: **III. Chapter Summary and Discussion.**

III. Chapter Summary and Discussion

1. Electric Potential Energy and Potential Difference (Section 16.1)

The **electric potential difference (voltage)** between two points is the work per unit positive charge done by an external force in moving charge between these two points, or the change in electric potential energy per unit positive charge: $\Delta V = V_B - V_A = \dfrac{W}{q_o} = \dfrac{\Delta U_e}{q_o}$, where q_o is the positive test charge. The SI unit of electric potential difference is joule/coulomb (J/C) or volt (V).

Note: Potential difference and potential energy difference are *different* quantities. Since potential difference is defined per unit charge, it does not depend on the amount of charge moved, whereas potential energy difference does.

In a uniform electric field E, the potential difference in moving a charge through a straight line distance d (in a direction opposite to the electric field, or from negative plate to positive plate in parallel plates) is $\Delta V = Ed$.

Example 16.1   An electron, initially at rest, is accelerated through an electric potential difference of 50.0 V.

(a) What is the kinetic energy of the electron?

(b) What is the speed of the electron?

Solution: Given: $\Delta V = 50.0$ V, $q = 1.60 \times 10^{-19}$ C, $m = 9.11 \times 10^{-31}$ kg.

Find: (a) K (b) v.

(a) When a negatively-charged electron is accelerated through a potential difference, work is done on the electron. The electron gains kinetic energy while losing potential energy. The electron will be moving from the lower potential location to the higher potential location but from the higher potential energy location to a lower potential energy location. Why?

From the conservation of energy, $\Delta K = K - K_o$, but $K_o = 0$, so $\Delta K = K = |\Delta U_e|$.

Since $\Delta V = \dfrac{\Delta U_e}{q}$, we can calculate the kinetic energy gained by the electron.

$K = \Delta U_e = q\Delta V = (1.60 \times 10^{-19}\ \text{C})(50.0\ \text{V}) = 8.00 \times 10^{-18}$ J.

(b) From $K = \frac{1}{2}mv^2$, we have $v = \sqrt{\dfrac{2K}{m}} = \sqrt{\dfrac{2(8.00 \times 10^{-18}\ \text{J})}{9.11 \times 10^{-31}\ \text{kg}}} = 4.19 \times 10^6$ m/s.

Example 16.2   A 12-volt battery maintains the electric potential difference between two parallel metal plates separated by 0.10 m. What is the electric field between the plates?

Solution: Given: $\Delta V = 12$ V, $d = 0.10$ m.

Find: E.

From $\Delta V = Ed$, we have $E = \dfrac{\Delta V}{d} = \dfrac{12\ \text{V}}{0.10\ \text{m}} = 1.2 \times 10^2$ V/m.

The electric potential difference due to a point charge is given by $\Delta V = \dfrac{kq}{r_B} - \dfrac{kq}{r_A}$, where r_A and r_B are the distances from the point charge to the two points A and B. If we define the potential at infinity as zero, then the **electric potential** at a distance r from a point charge is $V = \dfrac{kq}{r}$.

The electric potential changes according to the following general rules:

- *Electric potential increases for locations closer to positive charges or farther from negative charges.*
- *Electric potential decreases for locations farther from positive charges or nearer to negative charges.*

Note: Unlike electric field, electric potential is a scalar quantity. When adding potentials due to point charges, you only need to add them algebraically (including the + or − signs). Also, the potential is proportional to $\dfrac{1}{r}$ and the magnitude of the electric field is proportional to $\dfrac{1}{r^2}$.

For a configuration of multiple point charges, the total potential energy of the system U_t is

$U_t = U_{12} + U_{23} + U_{13} + \dots$, where $U_{ij} = \dfrac{kq_i q_j}{r_{ij}}$ is the potential energy due to any two of the point charges i and j.

Example 16.3   A charge of 5.0 nC is at (0,0) and a second charge of −2.0 nC is at (3.0 m, 0 m). If the potential is taken to be zero at infinity,

(a) what is the electric potential at point P (0, 4.0) m?

(b) what is the potential energy of a 1.0-nC charge at point P?

(c) what is the work required to bring a charge of 1.0 nC charge from infinity to point P?

(d) what is the total potential energy of the three charge system?

Solution: Given: $q_1 = 5.0$ nC $= 5.0 \times 10^{-9}$ C, $(x_1, y_1) = (0,0)$;

$q_2 = -2.0$ nC $= -2.0 \times 10^{-9}$ C, $(x_2, y_2) = (3.0,0)$ m;

$q_3 = 1.0$ nC $= 1.0 \times 10^{-9}$ C, $(x_P, y_P) = (0, 4.0)$ m.

Find: (a) V (b) U_e (c) W (d) U_{total}.

In the triangle shown, we calculate the distance from q_2 to q_3.

$r_{23} = \sqrt{(3.0\ \text{m})^2 + (4.0\ \text{m})^2} = 5.0$ m.

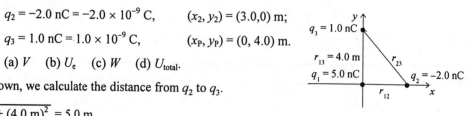

(a) From $V = \dfrac{kq}{r}$, the total potential, we have

$$V_t = V_1 + V_2 = \dfrac{(9.00 \times 10^9 \text{ N·m}^2/\text{C}^2)(5.0 \times 10^{-9} \text{ C})}{3.0 \text{ m}} + \dfrac{(9.00 \times 10^9 \text{ N·m}^2/\text{C}^2)(-2.0 \times 10^{-9} \text{ C})}{5.0 \text{ m}} = 11 \text{ V}.$$

(b) From $V = \dfrac{U_e}{q_0}$, and so we have $U_e = q_3 V = (1.0 \times 10^{-9} \text{ C})(11 \text{ V}) = 1.1 \times 10^{-8}$ J.

(c) From conservation of energy, the work required to bring the 1.0 nC charge from infinity is equal to its potential energy, $W = U_e = 1.1 \times 10^{-8}$ J.

(d) $U_{\text{total}} = U_{12} + U_{23} + U_{13} = \dfrac{kq_1 q_2}{r_{12}} + \dfrac{kq_2 q_3}{r_{23}} + \dfrac{kq_1 q_3}{r_{13}}$

$$= \dfrac{(9.00 \times 10^9 \text{ N·m}^2/\text{C}^2)(5.0 \times 10^{-9} \text{ C})(-2.0 \times 10^{-9} \text{ C})}{3.0 \text{ m}}$$

$$+ \dfrac{(9.00 \times 10^9 \text{ N·m}^2/\text{C}^2)(-2.0 \times 10^{-9} \text{ C})(1.0 \times 10^{-9} \text{ C})}{5.0 \text{ m}}$$

$$+ \dfrac{(9.00 \times 10^9 \text{ N·m}^2/\text{C}^2)(5.0 \times 10^{-9} \text{ C})(1.0 \times 10^{-9} \text{ C})}{4.0 \text{ m}} = -2.2 \times 10^{-8} \text{ J}.$$

2. Equipotential Surfaces and the Electric Field (Section 16.2)

Equipotential surfaces (equipotentials) are surfaces on which a charge experiences a constant electric potential V so it takes no work to move a charge along an equipotential. The electric field is *perpendicular* to such surfaces at all points. If the equipotentials are given, you can draw the electric field lines by simply connecting arrows perpendicular to the equipotentials. Or if the electric field lines are given, the equipotentials can be obtained by drawing surfaces perpendicular to the electric field lines.

The electric field is related to the electric potential difference as follows: $E = -\left(\dfrac{\Delta V}{\Delta x}\right)_{\text{max}}$, where the minus sign means that **E** is in the direction opposite that in which V increases most rapidly, or in the direction V decreases most rapidly. The units of electric field are volts per meter (V/m), which has the same dimensions as N/C introduced in Chapter 15.

An **electron volt** (eV) is the kinetic energy gained by an electron accelerated from rest though a potential difference of 1 V. One eV is equivalent to 1.60×10^{-19} J.

Example 16.4 Two parallel plates, separated by 0.10 m, are connected to a 6.0-volt battery. An electron is released from rest at the negative plate.

(a) What is the electric field in between the battery plates?

(b) What is the speed of the electron when the electron arrives at the positive plate?

Solution: Given: $\Delta V = 6.0$ V, $\Delta x = 0.10$ m.

Find: (a) E (b) v.

(a) $|E| = \left(\dfrac{\Delta V}{\Delta x}\right)_{max} = \dfrac{6.0 \text{ V}}{0.10 \text{ m}} = 60$ V/m.

(b) From energy conservation,

$$\Delta K = K - K_o = K - 0 = K = |\Delta U_e| = |q\Delta V| = (1.60 \times 10^{-19} \text{ C})(6.0 \text{ V}) = 9.6 \times 10^{-19} \text{ J}.$$

Since $K = \frac{1}{2}mv^2$, we have $v = \sqrt{\dfrac{2K}{m}} = \sqrt{\dfrac{2(9.6 \times 10^{-19} \text{ J})}{9.11 \times 10^{-31} \text{ kg}}} = 1.5 \times 10^6$ m/s.

---

3.   Capacitance (Section 16.3)

A **capacitor** consists of two conductors. It is used to store charge, and therefore electric energy in the form of an electric field. **Capacitance** (C) is a quantitative measure of how effective a capacitor is in storing charge and is defined as $C = \dfrac{Q}{V}$, where Q is the charge on each conductor (the two conductors have equal but opposite charge) and V is the potential difference between the two conductors. The SI unit of capacitance is farad (F).

Note: From now on the symbol V stands for potential *difference*. It means the same as ΔV used earlier.

A common capacitor is the parallel-plate capacitor. It consists of two parallel metal plates of area A and separated by a distance d. The capacitance of a parallel-plate capacitor is given by $C = \dfrac{\varepsilon_0 A}{d}$, where ε_0 is the *permittivity of free space* (vacuum) and is equal to 8.85×10^{-12} C^2/(N·m^2). ε_0 is a fundamental constant and is related to Coulomb's constant by $k = \dfrac{1}{4\pi\varepsilon_0} = 9.00 \times 10^9$ N·m^2/C^2.

The energy stored in a capacitor of capacitance C and potential difference V (therefore with a charge $Q = CV$) is given by $U = \frac{1}{2}QV = \dfrac{Q^2}{2C} = \frac{1}{2}CV^2$. The form $\frac{1}{2}CV^2$ is usually the most practical, since the capacitance and the applied voltage are often known or can be measured most easily.

Example 16.5   A parallel-plate capacitor consists of plates of area 1.5×10^{-4} m^2 and separated by 2.0 mm. The capacitor is connected to a 12-volt battery.

(a) What is the capacitance?

(b) What is the charge on the plates?

(c) How much energy is stored in the capacitor?

(d) What is the electric field between the plates?

Solution: Given: $A = 1.5 \times 10^{-4}$ m^2, $d = 2.0$ mm $= 2.0 \times 10^{-3}$ m, $V = 12$ V.

Find: (a) C (b) Q (c) U (d) E.

(a) $C = \dfrac{\varepsilon_0 A}{d} = \dfrac{[8.85 \times 10^{-12}\ \text{C}^2/(\text{N·m}^2)](1.5 \times 10^{-4}\ \text{m}^2)}{2.0 \times 10^{-3}\ \text{m}} = 6.6 \times 10^{-13}$ F $= 0.66$ pF.

(b) $Q = CV = (6.6 \times 10^{-13}$ F$)(12$ V$) = 7.9 \times 10^{-12}$ C.

(c) $U = \frac{1}{2}CV^2 = \frac{1}{2}(6.6 \times 10^{-13}$ F$)(12$ V$)^2 = 4.8 \times 10^{-11}$ J.

(d) $E = \dfrac{\Delta V}{d} = \dfrac{V}{d} = \dfrac{12\ \text{V}}{2.0 \times 10^{-3}\ \text{m}} = 6.0 \times 10^3$ V/m.

---

4. Dielectrics (Section 16.4)

A **dielectric** is any nonconducting material (insulator) capable of being partially polarized when placed in an electric field. When inserted between the plates of a capacitor, a dielectric raises the capacitance by a factor K, the **dielectric constant** (always greater than 1), which is defined as $K = \dfrac{C}{C_0}$, where C_0 is the capacitance without the dielectric (or vacuum is between the plates). As a result, the energy of the capacitor also increases if the potential difference is maintained constant across the capacitor, $U_C = \frac{1}{2}CV^2 = \frac{1}{2}(KC_0)V^2 = KU_0$. The product $K\varepsilon_0$ is called the **dielectric permittivity**, ε (that is, $\varepsilon = K\varepsilon_0$).

For an isolated capacitor (battery disconnected) with charge Q_0 and voltage V_0, the insertion of a dielectric decreases the voltage to V, while the charge remains constant. Therefore, capacitance increases to $C = \dfrac{Q_0}{V}$ $(V < V_0)$. For a capacitor connected to a battery of voltage V_0, the insertion of a dielectric increases the charge Q, while the voltage remains constant. Therefore, capacitance increases to $C = \dfrac{Q}{V_0}$ $(Q > Q_0)$.

Example 16.6 Repeat Example 16.5 if the capacitor is filled with paper. (Assume the capacitor remains connected to the battery.)

Solution: Additional given: $K = 3.5$ (from Table 16.1).

(a) $C = KC_0 = (3.5)(6.6 \times 10^{-13}$ F$) = 2.3 \times 10^{-12}$ F $= 2.3$ pF. Note C increases.

(b) $Q = CV = (2.3 \times 10^{-12}$ F$)(12$ V$) = 2.8 \times 10^{-11}$ C. Note Q increases.

(c) $U_C = \frac{1}{2}CV^2 = \frac{1}{2}(2.3 \times 10^{-12}$ F$)(12$ V$)^2 = 1.7 \times 10^{-10}$ J. Note U increases.

(d) $E = \dfrac{\Delta V}{d} = \dfrac{V}{d} = \dfrac{12\ \text{V}}{2.0 \times 10^{-3}\ \text{m}} = 6.0 \times 10^3$ V/m. E is the same. Why?

---

5. Capacitors in Series and Parallel (Section 16.5)

In a series combination, the **equivalent series capacitance** (C_s) is always less than that of the smallest capacitor in the combination. It is given by $\frac{1}{C_s} = \frac{1}{C_1} + \frac{1}{C_2} + \frac{1}{C_3} + \dots$. The charge on the equivalent capacitor is the same as that of each individual capacitor ($Q_s = Q_1 = Q_2 = Q_3 = \dots$) and the voltage across the equivalent capacitor is equal to the sum of the voltages on each capacitor ($V_s = V_1 + V_2 + V_3 + \dots$).

In a parallel combination, the **equivalent parallel capacitance** (C_p) is always larger than that of the largest capacitor in the combination. It is given by $C_p = C_1 + C_2 + C_3 + \dots$. The charge on the equivalent capacitor is equal to the sum of that of each capacitor ($Q_p = Q_1 + Q_2 + Q_3 + \dots$) and the voltage across the equivalent capacitor is the same as that of each individual capacitor ($V_p = V_1 = V_2 = V_3 = \dots$).

Note: If there are only two capacitors in a series combination, the equivalent series capacitance can be conveniently expressed as $C_s = \frac{1}{1/C_s} = \frac{1}{1/C_1 + 1/C_2} = \frac{C_1 C_2}{C_1 + C_2}$. If the two capacitors have the same capacitance $C_1 = C_2 = C$, then $C_s = \frac{CC}{C+C} = \frac{C}{2}$.

Example 16.7 For the capacitor network shown,

$C_1 = 6.00 \ \mu F, \quad C_2 = 8.00 \ \mu F, \quad C_3 = 14.0 \ \mu F.$

(a) What is the equivalent capacitance?

(b) What is the charge of each capacitor?

(a) What is the voltage across each capacitor?

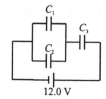

Solution: Given: $C_1 = 6.00 \ \mu F, \quad C_2 = 8.00 \ \mu F, \quad C_3 = 14.0 \ \mu F, \quad V = 12.0 \ V.$

Find: (a) C_{eq} (b) Q of each capacitor (c) V across each capacitor.

When working with capacitor combinations, always start with the combination you recognize. In this Example, C_1 and C_3 are *not* in series, neither are C_2 and C_3. However, C_1 and C_2 are in parallel. So we start from this combination.

$C_p = C_1 + C_2 = 6.00 \ \mu F + 8.00 \ \mu F = 14.0 \ \mu F$, and the circuit reduces to Figure 1. Now we can see that C_p and C_3 are in series.

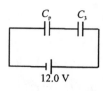

Figure 1

$$\frac{1}{C_s} = \frac{1}{C_p} + \frac{1}{C_3} = \frac{1}{14.0 \ \mu F} + \frac{1}{14.0 \ \mu F} = \frac{2}{14.0 \ \mu F} + \frac{1}{7.00 \ \mu F}.$$

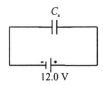

So $C_s = 7.00 \ \mu F$, or $C_s = \frac{C_p C_3}{C_p + C_3} = \frac{(14.0 \ \mu F)(14.0 \ \mu F)}{14.0 \ \mu F + 14.0 \ \mu F} = 7.00 \ \mu F.$

The circuit becomes Figure 2.

So the equivalent capacitance of the circuit is 7.00 μF.

Figure 2

(b) From Figure 2, the charge on the series capacitor (equivalent capacitor in this Example) is $Q_s = C_s V = (7.00 \ \mu F)(12.0 \ V) = 84.0 \ \mu C.$

Since series capacitors have the same charge, C_p and C_3 will have the same charge as C_s.

So $C_3 = 84.0 \ \mu C$ and $C_p = 84.0 \ \mu C$. $V_3 = \frac{Q_3}{C_3} = \frac{84.0 \ \mu C}{14.0 \ \mu F} = 6.00 \ V.$

Since the voltage on the series equivalent capacitor is the sum of that of each individual capacitor,

$V_s = V_p + V_3$, so $V_p = V_s - V_3 = 12.0 \ V - 6.00 \ V = 6.00 \ V.$

Parallel capacitors have the same voltage, so $V_1 = V_2 = V_p = 6.00 \ V.$

Therefore $Q_1 = C_1 V_1 = (6.00 \ \mu F)(6.00 \ V) = 36.0 \ \mu C$ and $Q_2 = C_2 V_2 = (8.00 \ \mu F)(6.00 \ V) = 48.0 \ \mu C.$

Hence, the answers are $Q_1 = 36.0 \ \mu C$, $Q_2 = 48.0 \ \mu C$, and $Q_3 = 84.0 \ \mu C.$

(c) $V_1 = V_2 = V_3 = 6.00 \ V.$

IV.   Mathematical Summary

Electric Potential Difference (voltage)	$\Delta V = \frac{\Delta U_e}{q_o} = \frac{W}{q_o}$ (16.1)	Defines the electric potential difference (voltage) in terms of electric potential energy difference.
Electric Potential difference between Parallel Plates	$\Delta V = Ed$ (16.2)	Calculate the voltage between two parallel plates in terms of the electric field in between and the distance between the plates.
Electric Potential Difference due to a Point charge	$V = \frac{kq}{r}$ (16.4) $(V = 0$ at $r = \infty)$	Computes the electric potential due to a point charge by assuming zero potential at infinity.
Electric Potential Energy of a Configuration of Point Charges	$U_t = U_{12} + U_{23} + U_{13} + \ldots$ (16.6)	Calculates the electric potential energy of a configuration of point charges.
Relationship between Potential and Electric Field	$E = -\left(\frac{\Delta V}{\Delta x}\right)_{max}$ (16.7)	Defines electric field in terms of electric potential gradient (difference).

Capacitance	$Q = CV$ or $C = \dfrac{Q}{V}$ (16.8)	Defines the capacitance of a capacitor.
Capacitance of a Parallel-Plate Capacitor (in air)	$C = \dfrac{\varepsilon_0 A}{d}$ (16.11) $\varepsilon_0 = 8.85 \times 10^{-12} \ \text{C}^2/(\text{N·m}^2)$	Computes the capacitance of a parallel-plate capacitor.
Energy in a Charged Capacitor	$U = \frac{1}{2} QV = \dfrac{Q^2}{2C} = \frac{1}{2} CV^2$ (16.12)	Calculates the energy stored in a capacitor.
Dielectric Effect on Capacitance	$C = KC_0$ (16.14)	Calculates the effect of dielectric on capacitance
Equivalent Capacitance for Capacitors in Series	$\dfrac{1}{C_s} = \dfrac{1}{C_1} + \dfrac{1}{C_2} + \dfrac{1}{C_3} + \ldots$ (16.16)	Computes the equivalent capacitance for capacitors connected in series.
Equivalent Capacitance for Capacitors in Parallel	$C_p = C_1 + C_2 + C_3 + \ldots$ (16.17)	Computes the equivalent capacitance for capacitors connected in parallel.

V.   Solutions of Selected Exercises and Paired/Trio Exercises

5.   Approaching a negative charge means moving towards a region of larger negative potential values, that is losing potential. Positive charges tend to move towards regions of lower potential, thus losing potential energy and gaining kinetic energy (speeding up)

10. $W_{\text{ext}} = q\Delta V = (-4.0 \times 10^6 \ \text{C})(-24 \ \text{V}) = \boxed{9.6 \times 10^{-5} \ \text{J}}$.

14.   The electron will accelerate downward because it is negatively charged.

From $a = \dfrac{F}{m} = \dfrac{qE}{m}$ and $v^2 = v_0^2 + 2ax = 2ax$,

we have $v = \sqrt{2ax} = \sqrt{\dfrac{2qEx}{m}} = \sqrt{\dfrac{2(1.6 \times 10^{-19} \ \text{C})(1000 \ \text{V/m})(0.0050 \ \text{m})}{9.11 \times 10^{-31} \ \text{kg}}}$

$= \boxed{1.3 \times 10^6 \ \text{m/s down}}$.

18.   (a) $\Delta V = \dfrac{kq}{r_B} - \dfrac{kq}{r_A} = \dfrac{(9.0 \times 10^9 \ \text{N·m}^2/\text{C}^2)(1.6 \times 10^{-19} \ \text{C})}{0.48 \times 10^{-9} \ \text{m}} - \dfrac{(9.0 \times 10^9 \ \text{N·m}^2/\text{C}^2)(1.6 \times 10^{-19} \ \text{C})}{0.21 \times 10^{-9} \ \text{m}}$

$= -3.9 \ \text{V}$.   So the difference is $\boxed{3.9 \ \text{V}}$.

(b) $\boxed{\text{The smaller orbit}}$ (closest to proton) at 0.21 nm is at a higher potential.

20. $\boxed{-2.2 \text{ J}}$.

24.   (a) $W = \Delta U_{\text{A-C}} = q\Delta V = qEd = (-1.60 \times 10^{-19} \text{ C})(15 \text{ V/m})(0.25 \text{ m}) = -\boxed{6.0 \times 10^{-19} \text{ J}}$.

        (b) $\Delta V_{\text{A-C}} = \dfrac{\Delta U_{\text{A-C}}}{q} = \dfrac{-6.0 \times 10^{-19} \text{ J}}{-1.60 \times 10^{-19} \text{ C}} = \boxed{3.8 \text{ V}}$.

        (c) Since $\Delta V_{\text{A-C}}$ is positive, $\boxed{\text{point C}}$ is at a higher potential.

29.   The distance from the charges to the center of the square is

        $r = \sqrt{(0.05 \text{ m})^2 + (0.05 \text{ m})^2} = 0.0707 \text{ m}$.

        $V = \Sigma \dfrac{kq}{r} = 2 \dfrac{(9.0 \times 10^9 \text{ N·m}^2/\text{C}^2)(-10 \times 10^{-6} \text{ C})}{0.0707 \text{ m}} + 2 \dfrac{(9.0 \times 10^9 \text{ N·m}^2/\text{C}^2)(5.0 \times 10^{-6} \text{ C})}{0.0707 \text{ m}}$

        $= \boxed{-1.3 \times 10^6 \text{ V}}$.

31.   (a) $\Delta K = K - K_{\text{o}} = K = \frac{1}{2} mv^2$.

        From work-energy theorem: $W = \Delta K = -\Delta U = -q\Delta V = e\Delta V = \frac{1}{2} mv^2$.

        we have $v = \sqrt{\dfrac{2e\Delta V}{m}} = \dfrac{2(1.6 \times 10^{-19} \text{ C})(10 \times 10^3 \text{ V})}{9.11 \times 10^{-31} \text{ kg}} = \boxed{5.9 \times 10^7 \text{ m/s}}$.

        (b) $t = \dfrac{\Delta x}{v} = \dfrac{0.35 \text{ m}}{5.93 \times 10^7 \text{ m/s}} = \boxed{5.9 \times 10^{-9} \text{ s}}$.

35.

        Higher gravitational potential energy

        Beach _____

               _____
                               |
               _____ | Ball accelerating
                               |
               _____ ▼
        Ocean _____

        Lower gravitational potential energy

43.   An electron volt is $e\Delta V = q\Delta V = \Delta U$. It is also the kinetic energy gained by an electron when it goes through a potential difference of 1 V. So it is a unit of energy. A GeV (10^9 eV) is larger than a MeV (10^6 eV) by 1000 times.

48. $d = \dfrac{7.0 \times 10^3 \text{ V}}{10 \times 10^3 \text{ V/m}} = 0.70 \text{ m} = \boxed{70 \text{ cm}}$.

50. $\boxed{\text{1.7 mm away from the positive plate toward the negative plate}}$.

54.   (a) $W = \Delta K = K - K_o = K = -(0.20)\Delta U = -(0.20)q\Delta V = (0.20)e\Delta V = (0.20)e(100 \times 10^6 \text{ V})$

   $= \boxed{2.00 \times 10^7 \text{ eV}}$.

   (b) $(2.00 \times 10^7 \text{ eV}) \times \dfrac{1.60 \times 10^{-19} \text{ J}}{1 \text{ eV}} = \boxed{3.20 \times 10^{-12} \text{ J}}$.

58.   (a) $\Delta V = \dfrac{\Delta U_e}{q} = \dfrac{\Delta U_e}{e} = \dfrac{3.5 \text{ eV}}{e} = \boxed{3.5 \text{ V}}$.   From $\frac{1}{2}mv^2 = K = \Delta U_e$,

   we have $v = \sqrt{\dfrac{2\Delta U_e}{m}} = \sqrt{\dfrac{2(3.5 \text{ eV})(1.6 \times 10^{-19} \text{ J/eV})}{1.67 \times 10^{-27} \text{ kg}}} = \boxed{2.6 \times 10^4 \text{ m/s}}$.

   (b) $\Delta V = \boxed{4.1 \text{ kV}}$. $v = \sqrt{\dfrac{2(4.1 \times 10^3 \text{ eV})(1.6 \times 10^{-19} \text{ J/eV})}{1.67 \times 10^{-27} \text{ kg}}} = \boxed{8.9 \times 10^5 \text{ m/s}}$.

   (c) $8.0 \times 10^{-16} \text{ J} = (8.0 \times 10^{-16} \text{ J}) \times \dfrac{1 \text{ eV}}{1.60 \times 10^{-19} \text{ J}} = 5.0 \times 10^3 \text{ eV} = 5.0 \text{ keV}$.

   So $\Delta V = \boxed{5.0 \text{ kV}}$. $v = \sqrt{\dfrac{2(5.0 \times 10^3 \text{ eV})(1.6 \times 10^{-19} \text{ J/eV})}{1.67 \times 10^{-27} \text{ kg}}} = \boxed{9.8 \times 10^5 \text{ m/s}}$.

61.   If it is moved parallel to the plates, $\theta = 90°$, so $\Delta V = 0$.

   Therefore it is still $\boxed{+2.8 \text{ V}}$.

66. $C = \dfrac{\varepsilon_o A}{d} = \dfrac{(8.85 \times 10^{-12} \text{ F/m})(0.50 \text{ m}^2)}{2.0 \times 10^{-3} \text{ m}} = \boxed{2.2 \times 10^{-9} \text{ F}}$.

71.   The energy supplied is $E = U_C = Pt = (0.50 \text{ W})(5.0 \text{ s}) = 2.5 \text{ J}$.

   Also $U_C = \frac{1}{2}CV^2$,   so $V = \sqrt{\dfrac{2U_C}{C}} = \sqrt{\dfrac{2(2.5 \text{ J})}{1.0 \text{ F}}} = \boxed{2.2 \text{ V}}$.

76. $K = \dfrac{C}{C_o} = \dfrac{150 \text{ pF}}{50 \text{ pF}} = \boxed{3.0}$.

85.   The charge is the same for both capacitors.

   From $U_t = \frac{1}{2}C_s V^2$, we have $C_s = \dfrac{2U_t}{V^2} = \dfrac{2(173 \ \mu J)}{(12 \text{ V})^2} = 2.40 \ \mu\text{F}$.

   Also $\dfrac{1}{C_s} = \dfrac{1}{C_1} + \dfrac{1}{C_2}$, so $C_2 = \dfrac{C_1 C_s}{C_1 - C_s} = \dfrac{(4.0 \ \mu C)(2.40 \ \mu F)}{4.0 \ \mu F - 2.40 \ \mu F} = \boxed{6.0 \ \mu F}$.

88.   Three in series: $\dfrac{1}{C_s} = \dfrac{1}{C_1} + \dfrac{1}{C_2} + \dfrac{1}{C_2} = \dfrac{3}{1.0 \ \mu F},$ so $C_s = \boxed{0.33 \ \mu F}.$

   Two in series: $\dfrac{1}{C_s} = \dfrac{1}{C_1} + \dfrac{1}{C_2} = \dfrac{2}{1.0 \ \mu F},$ so $C_s = \boxed{0.50 \ \mu F}.$

   Two parallel, then series: $C_p = C_1 + C_2 = 2.0 \ \mu F.$

   $\dfrac{1}{C_s} = \dfrac{1}{C_1} + \dfrac{1}{C_2},$ so $C_s = \dfrac{C_1 C_2}{C_1 + C_2} = C_s = \dfrac{(2.0 \ \mu F)(1.0 \ \mu F)}{2.0 \ \mu F + 1.0 \ \mu F} = \boxed{0.67 \ \mu F}.$

   Just one: $C = \boxed{1.0 \ \mu F}.$

   Two in series, then parallel: $C_p = 0.50 \ \mu F + 1.0 \ \mu F = \boxed{1.5 \ \mu F}.$

   Two in parallel: $C_p = 1.0 \ \mu F + 1.0 \ \mu F = \boxed{2.0 \ \mu F}.$

   Three in parallel: $C_p = 1.0 \ \mu F + 1.0 \ \mu F + 1.0 \ \mu F = \boxed{3.0 \ \mu F}.$

   There are a total of $\boxed{7}$ different values.

94.   Electric field lines are pointing toward the wire.

Looking at the end of the wire

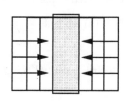
Side view

97.   (a) The electric force must balance the gravitational force.

   Since $|F| = qE = q \dfrac{\Delta V}{\Delta x} = mg,$

   $\Delta V = \dfrac{mg\Delta x}{q} = \dfrac{(9.11 \times 10^{-31} \ \text{kg})(9.80 \ \text{m/s}^2)(0.015 \ \text{m})}{1.6 \times 10^{-19} \ \text{C}} = \boxed{8.4 \times 10^{-13} \ \text{V}}.$

   (b) $\Delta V = \dfrac{(1.67 \times 10^{-27} \ \text{kg})(9.80 \ \text{m/s}^2)(0.015 \ \text{m})}{1.6 \times 10^{-19} \ \text{C}} = \boxed{1.5 \times 10^{-9} \ \text{V}}.$

   (c) $\boxed{\text{In (a), top is +; in (b), top is} -}.$

VI. Practice Quiz

1. For an electron moving in the direction of the electric field,

    (a) its potential energy increases and its potential increases.

    (b) its potential energy decreases and its potential increases.

    (c) its potential energy increases and its potential decreases.

    (d) its potential energy decreases and its potential decreases.

    (e) both the potential energy and potential remain constants.

2. The electric field between two parallel plates, separated by 0.50 cm, is 6.0×10^3 V/m. What voltage across the plates is required?

    (a) 3.0 V   (b) 30 V   (c) 300 V   (d) 1.2×10^4 V   (e) 1.2×10^6 V

3. A proton moves 0.10 m along the direction of an electric field line of magnitude 3.0 V/m. What is the magnitude of the change in potential energy of the proton?

    (a) 4.8×10^{-20} J   (b) 3.2×10^{-20} J   (c) 1.6×10^{-20} J   (d) 8.0×10^{-21} J   (e) zero

4. If three capacitors of 1.0, 1.5, and 2.0 μF each are connected in parallel, what is the equivalent capacitance?

    (a) 4.5 μF   (b) 4.0 μF   (c) 2.17 μF   (d) 0.46 μF   (e) 0.67 μF

5. If a capacitor is charged so that it stores 0.020 J of electric potential energy when connected to a 120-volt line, what is its capacitance?

    (a) 1.4 pF   (b) 3.8 pF   (c) 1.4 μF   (d) 2.8 μF   (e) 0.33 mF

6. A parallel plate capacitor has a capacitance C. The area of the plates is doubled and the distance between the plates is halved. What is the new capacitance?

    (a) $C/4$ (b) $C/2$ (c) C (d) $2C$ (e) $4C$

7. Two charges of 3.00 μC each are located on the ends of a meterstick. Find the electrical potential at the center of the meterstick.

    (a) zero   (b) 2.70×10^4 V   (c) 5.40×10^4 V   (d) 1.08×10^5 V   (e) 2.16×10^5 V

8. A uniform electric field, with a magnitude of 500 V/m, is directed parallel to the negative x axis. If the potential at $x = 5.0$ m is 2500 V, what is the potential at $x = 2.0$ m?

    (a) 500 V   (b) 1000 V   (c) 2000 V   (d) 4000 V   (e) 8000 V

---

9.   What is the equivalent capacitance of the combination shown?

(a) 10 μF

(b) 25 μF

(c) 29 μF

(d) 40 μF

(e) 68 μF

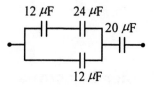

10.   If the combination in the previous problem is connected to a 12-volt battery, what is the charge on the capacitor of 20 μF?

(a) 6.0 μC   (b) 12 μC   (c) 18 μC   (d) 24 μC   (e) 36 μC

Answers to Practice Quiz:

1. c 2. b 3. a 4. a 5. d 6. e 7. d 8. b 9. a 10. b

CHAPTER 17

Electric Current and Resistance

I.   Chapter Objectives

Upon completion of this chapter, you should be able to:

1.   summarize the basic features of a battery and explain how a battery produces a direct current in a circuit.

2.   define electric current, distinguish between electron flow and conventional current, and explain the concept of drift velocity and electric energy transmission.

3.   define electrical resistance and explain what is meant by an ohmic resistor, summarize the factors that determine resistance, and calculate the effect of these factors in simple situations.

4.   define electric power, calculate the power delivery of simple electric circuits, and explain joule heating and its significance.

II.   Key Terms

Upon completion of this chapter, you should be able to define and/or explain the following key terms:

battery	resistance
cathode	Ohm's law
anode	ohm (Ω)
electromotive force (emf)	resistivity
terminal voltage	conductivity
direct current (dc)	temperature coefficient of resistivity
complete circuit	superconductivity
conventional current	electric power
electric current	joule heat ($I^2 R$) loss
ampere (A)	kilowatt-hour (kWh)
drift velocity	

The definitions and/or explanations of the most important key terms can be found in the following section: **III. Chapter Summary and Discussion.**

III.   Chapter Summary and Discussion

1.   Batteries and Direct Current (Section 17.1)

A **battery** is a device that converts chemical potential energy into electrical energy. The potential difference across the two terminals, **anode** (+) and **cathode** (−), of a battery (or any dc power supply) when not connected to an external circuit is called the **electromotive force** (emf). The **terminal voltage** (operating voltage) across a battery or power supply is the voltage when it is connected to an external circuit. Terminal voltage is always less than the emf because of internal resistance of the battery.

Electric current is the net rate at which charge flows past a given point. The direction of **conventional current** is that in which positive charge would move. In most materials (e.g., metals), the actual current is carried by electrons moving in the opposite direction to the conventional current due to the fact that electrons have negative charge. In a battery circuit, the electrons can only flow in one direction, from negative terminal to positive terminal, the current is called **direct current** (dc).

2.   Current and Drift Velocity (Section 17.2)

A battery or some other voltage source connected to a continuous conducting path forms a **complete circuit**. In general, we say **electric current** represents a flow of charge. Quantitatively, the electric current (I) in a wire is defined as the time rate of flow of net charge. If a net charge q passes a given point in a wire in time t at a constant rate, the current is given by $I = \frac{q}{t}$. The SI unit of current is coulomb/s (C/s) or ampere (A).

Example 17.1   If 3.0×10^{15} electrons flow through a section of a wire of diameter 2.0 mm in 4.0 s, what is the electric current in the wire?

Solution: Given: $n = 3.0 \times 10^{15}$ electrons, $t = 4.0$ s, $d = 2.0$ mm.

Find: I.

The charge in 3.0×10^{15} electrons is $q = ne = (3.0 \times 10^{15}$ electrons$)(1.60 \times 10^{-19}$ C/electron$) = 4.8 \times 10^{-4}$ C.

So $I = \frac{q}{t} = \frac{4.8 \times 10^{-4} \text{ C}}{4.0 \text{ s}} = 1.2 \times 10^{-4}$ A $= 0.12$ mA.

The diameter of the wire is not needed in this Example since the current is simply the rate of charge flow.

The electron flow in a metal wire is characterized by an average velocity called the **drift velocity**, which is much smaller than the random velocities of the electrons themselves. The drift velocity is usually very slow (approximately 0.10 cm/s), but the electric field (which is what pushes the charges in the wire) travels down the wire at a speed close to the speed of light (on the order of 10^8 m/s). Hence the current starts "instantly" in all parts of the circuits.

3.   Resistance and Ohm's Law (Section 17.3)

The electrical **resistance** of a circuit element is defined as the potential difference (voltage) across the element divided by the resulting current, $R = \dfrac{V}{I}$. The SI unit of resistance is volt/ampere (V/A), or ohm (Ω). A resistor that has constant resistance is (at a given temperature) said to obey **Ohm's law**, or to be ohmic. Ohm's law is usually written as $V = IR$.

Example 17.2   A resistor with a resistance of 20 Ω is connected to a 12-volt battery. What is the current through the resistor?

Solution: Given: $R = 20\ \Omega$, $V = 12$ V.

Find: I.

From Ohm's law, $I = \dfrac{V}{R} = \dfrac{12\ \text{V}}{20\ \Omega} = 0.60$ A.

The major factors affecting the resistance of a conductor of uniform cross-section are:

(1) the type of material or the intrinsic resistive properties,

(2) its length (L),

(3) its cross-sectional area (A),

(4) its temperature (T).

The resistive properties of a particular material are characterized by its **resistivity** (ρ), where for a conductor of uniform cross-section, $\rho = \dfrac{RA}{L}$. The SI unit of resistivity is $\Omega \cdot$m. This resistivity is somewhat temperature dependent, and over a small range of temperature change (ΔT), the resistivity changes according to $\rho = \rho_0 (1 + \alpha \Delta T)$, where ρ_0 is a reference resistivity at T_0 (usually 0°C) and α is the **temperature coefficient of resistivity**. The reciprocal of resistivity is called the **conductivity**, $\sigma = \dfrac{1}{\rho}$. The unit of conductivity is $(\Omega \cdot \text{m})^{-1}$.

For a conductor of uniform cross section, we may write the resistance as $R = \dfrac{\rho L}{A}$ and the variation of resistance with temperature as $R = R_0(1 + \alpha \Delta T)$. Most metallic materials have positive α's and their resistances increase with temperature.

Example 17.3   Calculate the current in a piece of 10.0-meter long 22-gauge (the radius is 0.321 mm) nichrome wire if it is connected to a source of 12.0 V. Assume the temperature is 20°C.

Solution: Given: $L = 10.0$ m, $r = 0.321$ mm $= 0.321 \times 10^{-3}$ m, $\rho = 100 \times 10^{-8}$ $\Omega \cdot$m (Table 17.1),

$V = 12.0$ V.

Find: I.

To find the current I, we first need to find the resistance R of the wire.

$$R = \frac{\rho L}{A} = \frac{\rho L}{\pi r^2} \frac{(100 \times 10^{-8}\ \Omega \cdot \text{m})(10.0\ \text{m})}{\pi(0.321 \times 10^{-3}\ \text{m})^2} = 30.9\ \Omega.$$

From Ohm's law, $I = \dfrac{V}{R} = \dfrac{12.0\ \text{V}}{30.9\ \Omega} = 0.388$ A.

Example 17.4   A carbon resistor has a resistance of 16 Ω at a temperature 20°C. What is the resistance if it is heated up to a temperature of 100°C?

Solution: Given: $R_0 = 16\ \Omega$, $T_0 = 20°C$, $T = 100°C$, $\alpha = -5.0 \times 10^{-4}$ C$°^{-1}$

Find: R.

$R = R_0(1 + \alpha \Delta T) = (16\ \Omega)[1 + (-5.0 \times 10^{-4}\ \text{C}°^{-1})(100°C - 20°C)] = 15\ \Omega.$

The resistance of carbon decreases with increasing temperature. When carbon is used with materials with positive temperature coefficient, we can make resistors which do not vary with temperature (why?).

---

4.   Electric Power (Section 17.4)

Electric power is the time rate at which work is done or electric energy is transferred. The power delivered to a circuit element depends on its resistance and the voltage across it. The SI unit of power is watt (W). Power can be expressed as $P = IV = \dfrac{V^2}{R} = I^2 R.$

The thermal energy expended in a current-carrying resistor is sometimes referred to as **joule heat**, or $I^2 R$ ("I squared R") **losses.**

Example 17.5   What is the operating resistance of a 100-watt household light bulb? The operating line voltage of household electricity is 120 V.

Solution: Given: $P = 100$ W, $V = 120$ V.

               Find: R.

From $P = \dfrac{V^2}{R}$, we have $R = \dfrac{V^2}{P} = \dfrac{(120 \text{ V})^2}{100 \text{ W}} = 144 \ \Omega$.

Example 17.6   A computer, including its monitor, is rated at 300 W. Assuming the power company charges 10 cents for each kilowatt-hour of electricity used and the computer is on 8.0 hours per day, estimate the annual cost to operate the computer.

Solution: Given: $P = 300$ W, $t = 8.0$ h every day,   cost = \$0.10 per kWh.

               Find: total annual cost.

In one year, the computer is on for $t = (8.0 \text{ h/d})(365 \text{ d}) = 2.93 \times 10^3$ h.

The total electrical energy expended is

$E = Pt = (300 \text{ W})(2.93 \times 10^3 \text{ h}) = (0.30 \text{ kW})(2.93 \times 10^3 \text{ h}) = 8.80 \times 10^2$ kWh.

So the total cost is (\$0.10/kWh)(8.80×10^2 kWh) = \$88.

IV.   Mathematical Summary

Electric Current	$I = \dfrac{q}{t}$ (17.1)	Defines electric current in terms of charge flow.
Electrical Resistance (definition)	$R = \dfrac{V}{I}$ (17.2)	Defines electrical resistance.
Ohm's Law	$V = IR$ (R = constant) (17.2)	Relates voltage, current, and resistance.
Resistivity	$\rho = \dfrac{RA}{L}$ (17.3)	Defines the resistivity of a material.
Conductivity	$\sigma = \dfrac{1}{\rho}$ (17.4)	Defines the conductivity of a material in terms of its resistivity.
Temperature Variation of Resistivity (α constant)	$\rho = \rho_0 (1 + \alpha \, \Delta T)$ or $\Delta \rho = \rho_0 \, \alpha \, \Delta T$ (17.5, 17.6) (where $\Delta \rho = \rho - \rho_0$)	Expresses the resistivity of a material as a function of temperature.

Temperature-dependence of Resistance	$R = R(1 + \alpha \Delta T)$ or $\Delta R = R_o \alpha \Delta T$ (17.7) (where $\Delta R = R - R_o$)	Expresses the resistance of a material as a function of temperature.
Electric Power	$P = IV = \dfrac{V^2}{R} = I^2 R$ (17.8)	Computes the electric power delivery to a resistor.

V. Solutions of Selected Exercises and Paired/Trio Exercises

5.　As the internal resistance increases, the voltage across it also increases. This decreases the terminal voltage.

8.　(a) $\mathscr{E} = 6(1.5 \text{ V}) = \boxed{9.0 \text{ V}}$.

　(b) $\boxed{1.5 \text{ V}}$.

14.　(a) $q = It = (0.50 \times 10^{-3} \text{ A})(600 \text{ s}) = \boxed{0.30 \text{ C}}$.

　(b) $E = \Delta U_e = q\Delta V = (0.30 \text{ C})(3.0 \text{ V}) = \boxed{0.90 \text{ J}}$.

19.　(a) In each second.

$$q = It = (9.5 \times 10^{-3} \text{ A})(1.0 \text{ s}) = 9.5 \times 10^{-3} \text{ C} = (9.5 \times 10^{-3} \text{ C}) \times \frac{1 \, e}{1.6 \times 10^{-19} \text{ C/proton}}$$

$$= \boxed{5.9 \times 10^{16} \text{ protons}}.$$

　(b) $P = \dfrac{E}{t} = \dfrac{(5.9 \times 10^{16} \text{ protons})(20 \times 10^6 \text{ eV/proton})(1.6 \times 10^{-19} \text{ J/eV})}{1.0 \text{ s}} = \boxed{1.9 \times 10^5 \text{ J/s}}$.

25.　(a) Since $R = \dfrac{\rho L}{A}$, $\dfrac{R_2}{R_1} = \dfrac{L_2}{L_1}\dfrac{A_1}{A_2} = (2)\left(\tfrac{1}{2}\right) = 1$.

　thus since $I = \dfrac{V}{R}$, the current remains the $\boxed{\text{same}}$.

　(b) Since $A = \dfrac{\pi d^2}{4}$, half the diameter means $\tfrac{1}{4}$ the area, so $\dfrac{A_2}{A_1} = \tfrac{1}{4}$.

　Therefore $\dfrac{R_2}{R_1} = (1)(4) = 4$.　Thus $\dfrac{I_2}{I_1} = \dfrac{R_1}{R_2} = \tfrac{1}{4}$, i.e., $\boxed{\text{1/4 the current}}$.

28. $I = \dfrac{V}{R} = \dfrac{12 \text{ V}}{15 \ \Omega} = \boxed{0.80 \text{ A}}.$

30. $\boxed{0.42 \text{ A}}$.

38. $\rho = \rho_0 (1 + \alpha \Delta T) = \rho_0 [1 + (6.80 \times 10^{-3} \text{ C}^{\circ -1})(80 \text{ C}^{\circ})] = 1.544 \rho_0.$

So the percentage variation is $\dfrac{\rho - \rho_0}{\rho_0} = \dfrac{\rho}{\rho_0} - 1 = 0.544.$

That is, a $\boxed{54.4\%}$ increase.

43.   The volume (amount of material) of the wire remains constant, or $V = A_1 L_1 = A_2 L_2$.

So $\dfrac{A_1}{A_2} = \dfrac{L_2}{L_1}.$

Since $R = \dfrac{\rho L}{A}$, $\dfrac{R_2}{R_1} = \dfrac{L_2}{L_1} \dfrac{A_1}{A_2} = \left(\dfrac{L_2}{L_1}\right)^2 = \left(\dfrac{1.25}{1}\right)^2 = 1.6.$

No, the resistance of the wire is different before and after. $\boxed{\text{It increases by 1.6 times}}$.

47.   From $I = \dfrac{V}{R} = \dfrac{V}{R_0(1 + \alpha \Delta T)} = \dfrac{1}{(1 + \alpha \Delta T)} I_0$, we have $\dfrac{I}{I_0} = \dfrac{1}{1 + \alpha \Delta T}.$

So $I = \dfrac{1}{1 + \alpha \Delta T} I_0 = \dfrac{1}{1 + (-7.0 \times 10^{-2} \text{ C}^{\circ -1})(5 \text{ C}^{\circ})} (0.50 \text{ A}) = \boxed{0.77 \text{ A}}.$

54. $P = \dfrac{V^2}{R} = \dfrac{(110 \text{ V})^2}{10 \ \Omega} = \boxed{1.2 \times 10^3 \text{ W}}.$

56. $\boxed{144 \ \Omega}$.

64. $E = Pt = (0.200 \text{ kW})(10 \text{ h/d})(365 \text{ d/y}) = 730 \text{ kWh/y}.$

So the annual cost is $(730 \text{ kWh})(15 \ ¢/\text{kWh}) = \boxed{\$110}.$

67.   (a) $I = \dfrac{V}{R} = \dfrac{15 \text{ V}}{100 \ \Omega} = \boxed{0.15 \text{ A}}.$

(b) From $R = \dfrac{\rho L}{A}$, we have $\rho = \dfrac{RA}{L} = \dfrac{(100 \ \Omega)(\pi)(1.5 \times 10^{-3} \text{ m})^2}{5.0 \text{ m}} = \boxed{1.4 \times 10^{-4} \ \Omega\cdot\text{m}}.$

(c) $P = IV = (0.15 \text{ A})(15 \text{ V}) = \boxed{2.3 \text{ W}}.$

70.　(a) $I = \dfrac{P}{V} = \dfrac{5.5 \times 10^3 \text{ W}}{240 \text{ V}} = 23$ A > 20 A.　So it should have a $\boxed{30\text{-amp}}$ circuit breaker.

(b) The heat (energy) required is

$Q = cm\Delta T = [4190 \text{ J/(kg·C°)}](55 \text{ gal})(3.785 \text{ kg/gal})(60 \text{ C°}) = 5.23 \times 10^7$ J.

The energy input is $E = \dfrac{5.23 \times 10^7 \text{ J}}{0.85} = 6.16 \times 10^7$ J.

$t = \dfrac{E}{P} = \dfrac{6.16 \times 10^7 \text{ J}}{5.5 \times 10^3 \text{ W}} = 1.12 \times 10^4 \text{ s} = \boxed{3.1 \text{ h}}$.

75.　$E = \Sigma Pt = (5.0 \text{ kW})(0.30)(24 \text{ h/d})(30 \text{ d}) + (0.8 \text{ kW})(0.50 \text{ h}) + (0.625 \text{ kW})(1/4 \text{ h/d})(30 \text{ d})$

$\qquad + (0.5 \text{ kW})(0.15)(24 \text{ h/d})(30 \text{ d}) + (10.5 \text{ kW})(10 \text{ h}) + (0.1 \text{ kW})(120 \text{ h}) = 1256$ kWh.

So it costs　$(1256 \text{ kWh})(\$0.12 \text{ /kWh}) = \boxed{\$151}$.

81.　Since $P = \dfrac{V^2}{R}$, $R_0 = \dfrac{V^2}{P_0} = \dfrac{(120 \text{ V})^2}{1600 \text{ W}} = 9.00 \ \Omega$.

Also since $R = \dfrac{\rho L}{A}$, R is $0.90R_0$.

Therefore $P = \dfrac{(120 \text{ V})^2}{0.90(9.00 \ \Omega)} = 1.78 \times 10^3 \text{ W} = \boxed{1.78 \text{ kW}}$.

89.　(a) $R = \dfrac{V}{I} = \dfrac{40 \text{ V}}{0.10 \text{ A}} = \boxed{4.0 \times 10^2 \ \Omega}$.

(b) $P = IV = (0.10 \text{ A})(40 \text{ V}) = \boxed{4.0 \text{ W}}$.

(c) $E = Pt = (4.0 \text{ W})(2.0)(60 \text{ s}) = \boxed{4.8 \times 10^2 \text{ J}}$.

91.　Since $P = IV$, $q = It = \dfrac{P}{V}\, t = \dfrac{(60 \text{ W})(3600 \text{ s})}{120 \text{ V}} = 1800$ C.

$n = \dfrac{q}{|e|} = \dfrac{1800 \text{ C}}{1.6 \times 10^{-19} \text{ C/electron}} = \boxed{1.1 \times 10^{22} \text{ electrons}}$.

93.　Since $P = IV$, a higher voltage means lower current if P is kept constant.　The resistance of the transmission lines is a fixed value.　With lower current, the power loss on the lines is reduced since

$P_{\text{loss}} = I^2 R = \dfrac{P^2}{V^2}\, R \propto \dfrac{1}{V^2}$.　So if V is raised by 10 times, P_{loss} will be reduced by a factor of 1/100.

VI. Practice Quiz

1. What is the current through a 5.0-ohm resistor if the voltage across it is 10 V?

   (a) zero   (b) 0.50 A   (c) 2.0 A   (d) 5.0 A   (e) 50 A

2. A household light bulb of 100 W is designed to operate at a voltage of 120 V, what is the current through the bulb?

   (a) zero (b) 0.83 A   (c) 1.2 A   (d) 12 A   (e) 144 A

3. A wire carries a steady current of 1.0 A over a period of 20 s. What total charge passes through the wire in this time interval?

   (a) 200 C   (b) 20 C   (c) 2.0 C   (d) 0.20 C   (e) 0.05 C

4. A nichrome wire has a radius of 0.50 mm and a resistivity of 100×10^{-8} Ω·m. If the wire carries a current 0.50 A, what is the voltage across 1.0 m of the wire?

   (a) 0.003 V   (b) 0.32 V   (c) 0.64 V   (d) 1.6 V   (e) 1.9 V

5. The length of a wire is doubled and the radius is doubled. By what factor does the resistance change?

   (a) 4 times as large   (b) twice as large   (c) unchanged   (d) half as large   (e) quarter as large

6. During a large power demand, the line voltage is reduced by 5.0%. What is the power output of a light bulb rated at 100 watts when the voltage is normal?

   (a) 2.5 W   (b) 5.0 W   (c) 10 W   (d) 90 W   (e) 95 W

7. A 12-volt battery is connected to a 50-ohm automobile dome light. How many electrons flow through the connecting wire in one minute?

   (a) 0.24   (b) 1.5×10^{18} (c) 3.8×10^{-20} (d) 3.1×10^{18} (e) 6.3×10^{18}

8. A 1500-watt heater is connected to a 120-volt source for 2.0 h. How much heat energy is produced?

   (a) 1.1×10^{7} J   (b) 1.8×10^{5} J   (c) 9.0×10^{4} J   (d) 3.0×10^{3} J   (e) 1.5×10^{3} J

9. A webserver rated at 400 W (computer plus monitor) is left on 24 hours per day. If electricity costs $0.10 per kWh, how much does it cost to run the server annually?

   (a) $147   (b) $289   (c) $350   (d) $877   (e) $3500

10.   A platinum wire is used to determine the melting point of indium. The resistance of the platinum wire is 2.000 Ω at 20°C and increases to 3.072 Ω just as the indium starts to melt. What is the melting temperature of indium?

(a) 117°C   (b) 137°C   (c) 157°C   (d) 351°C   (e) 731°C

Answers to Practice Quiz:

1. c 2. c 3. b 4. c 5. d 6. d 7. b 8. a 9. c 10. c

CHAPTER 18

Basic Electric Circuits

I.   Chapter Objectives

Upon completion of this chapter, you should be able to:

1.   determine the equivalent resistance of resistors in series, parallel, and series-parallel combinations, and use equivalent resistances to solve simple circuits.

2.   understand the physical principles that underlie Kirchhoff's circuit rules and apply these rules in the analysis of actual circuits.

3.   understand the charging and discharging of a capacitor through a resistor and calculate the current and voltage at specific times during these processes.

4.   understand how galvanometers are used as ammeters and voltmeters, how multirange versions of these devices are constructed, and how they are connected to measure current and voltage in real circuits.

5.   understand how household circuits are wired and the underlying principles that govern electric safety devices.

II.   Key Terms

Upon completion of this chapter, you should be able to define and/or explain the following key terms:

equivalent series resistance	time constant
equivalent parallel resistance	ammeter
Kirchhoff's rules	voltmeter
junction (node)	galvanometer
branch	fuse
Kirchhoff's first rule (junction theorem)	circuit breaker
Kirchhoff's second rule (loop theorem)	grounded plug
RC circuit	polarized plug

The definitions and/or explanations of the most important key terms can be found in the following section: **III. Chapter Summary and Discussion.**

III.   Chapter Summary and Discussion

1.   Resistors in Series, Parallel, and Series-Parallel Combinations (Section 18.1)

When resistors are connected in *series* in a circuit, the **equivalent series resistance** (R_s) is given by $R_s = R_1 + R_2 + R_3 + \ldots = \Sigma_i R_i$, or the equivalent series resistance is the algebraic sum of the individual resistances. The equivalent series resistance is larger than that of the largest resistor in the series combination.

In a series circuit, the current is the same through all resistors, $I = I_1 = I_2 = I_3 = \ldots$, and the total voltage is the sum of the voltages of the individual resistors, that is, $V = V_1 + V_2 + V_3 + \ldots = \Sigma_i V_i = \Sigma_i I R_i$.

When resistors are connected in *parallel* in a circuit, the **equivalent parallel resistance** (R_p) is given by $\frac{1}{R_p} = \frac{1}{R_1} + \frac{1}{R_2} + \frac{1}{R_3} + \ldots = \Sigma_i \frac{1}{R_i}$, or the reciprocal of the equivalent parallel resistance is equal to the sum of the reciprocals of individual resistances. The equivalent parallel resistance is less than that of the smallest resistor in the parallel combination.

In a parallel circuit, the voltage is the same across all resistors, $V = V_1 = V_2 = V_3 = \ldots$, and the total current is the sum of the currents of the individual resistors, that is, $I = I_1 + I_2 + I_3 + \ldots = \Sigma_i I_i$. If there are only two resistors in parallel, $\frac{1}{R_p} = \frac{1}{R_1} + \frac{1}{R_2} = \frac{R_1 + R_2}{R_1 R_2}$, so $R_p = \frac{R_1 R_2}{R_1 + R_2}$. If $R_1 = R_2 = R$, then $R_p = \frac{R R}{R + R} = \frac{R}{2}$.

In Chapter 16, the equivalent series capacitance is given by $\frac{1}{C_s} = \frac{1}{C_1} + \frac{1}{C_2} + \frac{1}{C_3} + \ldots = \Sigma_i \frac{1}{C_i}$; the equivalent parallel capacitance is equal to $C_p = C_1 + C_2 + C_3 + \ldots = \Sigma_i C_i$.

Example 18.1   In the circuit shown, find:

(a) the equivalent resistance between points A and B,

(b) the current through each resistor.

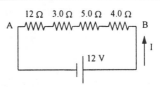

Solution: Given: $R_1 = 12\ \Omega$, $R_2 = 3.0\ \Omega$, $R_3 = 5.0\ \Omega$, $R_4 = 4.0\ \Omega$, $V = 12$ V.

Find: (a) R_s (b) I.

(a) All four resistors are in series combination, so

$R_s = R_1 + R_2 + R_3 + R_4 = 12\ \Omega + 3.0\ \Omega + 5.0\ \Omega + 4.0\ \Omega = 24\ \Omega$.

(b) The current through all resistors in series is the same. $I = \dfrac{V}{R} = \dfrac{V}{R_s} = \dfrac{12\ \text{V}}{24\ \Omega} = 0.50$ A.

Example 18.2   In the circuit shown, find:

(a) the equivalent resistance between points A and B,

(b) the current through the battery.

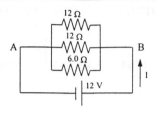

Solution: Given: $R_1 = 12\ \Omega$, $R_2 = 12\ \Omega$, $R_3 = 6.0\ \Omega$, $V = 12$ V.

Find: (a) R_p (b) I.

(a) The three resistors are in a parallel combination.

$$\frac{1}{R_p} = \frac{1}{R_1} + \frac{1}{R_2} + \frac{1}{R_3} = \frac{1}{12\ \Omega} + \frac{1}{12\ \Omega} + \frac{1}{6.0\ \Omega} = \frac{1}{12\ \Omega} \frac{1}{12\ \Omega} + \frac{2}{12\ \Omega} = \frac{4}{12\ \Omega} = \frac{1}{3.0\ \Omega}.$$

So $R_p = 3.0\ \Omega$.

Or using $R_p = \dfrac{R\,R}{R + R} = \dfrac{R}{2}$ for the two 12-ohm resistors in parallel, $R_{p1} = \dfrac{12\ \Omega}{2} = 6.0\ \Omega$.

Now the two 6.0 Ω resistances are in parallel and the equivalent resistance is $R_p = \dfrac{6.0\ \Omega}{2} = 3.0\ \Omega$.

(b) From Ohm's law, $I = \dfrac{V}{R} = \dfrac{V}{R_p} = \dfrac{12\ \text{V}}{3.0\ \Omega} \doteq 4.0$ A.

Resistors in a circuit may be connected in a variety of series-parallel combinations. The general procedure for analyzing circuits with different series-parallel combinations of resistors is to find the voltage across and the current through the various resistors as follows:

(1)   Start from the resistor combination farthest from the voltage source, find the equivalent series and parallel resistances.

(2)   Reduce the circuit until there is a single loop with one total equivalent resistance.

(3)   Find the total current delivered to the reduced circuit using $I = V/R$, where R is the total equivalent resistance.

(4)   Expand the reduced circuit in reverse order to **(1)** to find the currents and voltages for the resistors in each step.

Example 18.3   Find the equivalent resistance between points

(a) A and B,

(b) A and C,

(c) B and C.

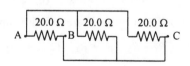

Solution: Given: $R_1 = R_2 = R_3 = R = 20.0\ \Omega$.

Find: (a) R_{AB} (b) R_{AC} (c) R_{BC}.

(a) All three resistors are in parallel between points A and B, that is, the ends of all three resistors are across points A and B. We redraw the circuit.

$$\frac{1}{R_{AB}} = \frac{1}{R_1} + \frac{1}{R_2} + \frac{1}{R_3} = \frac{3}{R},$$

or $R_{AB} = \dfrac{R}{3} = \dfrac{20.0\ \Omega}{3} = 6.67\ \Omega$.

(b) Since point C and point B are connected (they are the same point), $R_{AB} = R_{AC} = 6.67\ \Omega$.

(c) Again since points B and C are connected, there is no resistance between them (short circuit).

So $R_{BC} = 0$.

Example 18.4   In the circuit shown, find:

(a) the equivalent resistance between points A and B,

(b) the current through the battery,

(c) the current though the 4.0-ohm resistor.

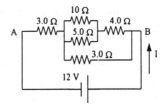

Solution:

Here we have a variety of series-parallel combinations. We follow the general procedures outlined in the text and above.

(a) The 10 Ω and the 5.0 Ω are in parallel.

$$R_{p1} = \frac{(10\ \Omega)(5.0\ \Omega)}{10\ \Omega + 5.0\ \Omega} = 3.33\ \Omega.$$

The circuit reduces to Figure 1.

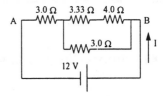

Figure 1

Now the 3.33 Ω and the 4.0 Ω are in series.

$R_{s1} = 3.33\ \Omega + 4.0\ \Omega = 7.33\ \Omega$.

The circuit reduces to Figure 2.

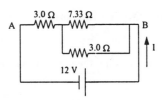

Figure 2

The 7.33 Ω and the 3.0 Ω are in parallel.

$$R_{p2} = \frac{(7.33\ \Omega)(3.0\ \Omega)}{7.33\ \Omega + 3.0\ \Omega} = 2.13\ \Omega.$$

The circuit reduces to Figure 3.

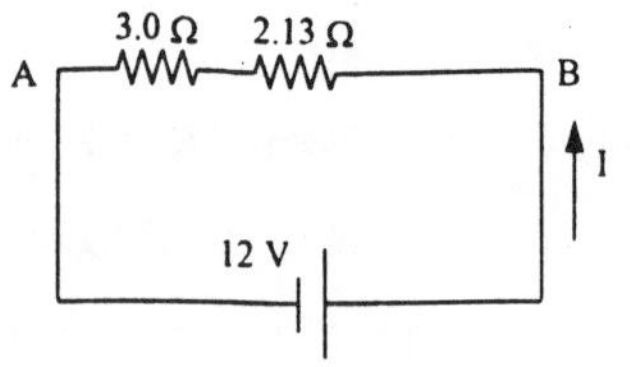

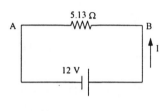

Figure 3

Finally, the 2.13 Ω and the 3.0 Ω are in series.

$R = R_{s2} = 2.13\ \Omega + 3.0\ \Omega = 5.13\ \Omega = 5.1\ \Omega$ (2 significant figures).

The circuit reduces to Figure 4.

Figure 4

(b) From Ohm's law, $I = \dfrac{V}{R} = \dfrac{12\ V}{5.13\ \Omega} = 2.33\ A = 2.3\ A$ (2 significant figures).

(c) To find the current through the 4.0-ohm resistor, we need to expand the combinations.

In Figure 3, the current through the 2.13 Ω and 3.0 Ω is the same as the total current, 2.33 A.

The voltage across the 2.13 Ω is then $V_{2.13} = (2.13\ \Omega)(2.33\ A) = 4.96\ V$.

In Figure 2, the voltage across the 7.33 Ω and 3.0 Ω is the same as that across the 2.13 Ω, 4.96 V.

So the current through the 7.33 Ω is $I_{7.33} = \dfrac{4.96\ V}{7.33\ \Omega} = 0.677\ A$.

In Figure 1, the current through the 3.33 Ω and the 4.0 Ω is the same as the current through the 7.33 Ω.

Therefore $I_{4.0} = 0.68\ A$ (2 significant figures).

In this Example, an extra digit is carried in intermediate results.

2. Multiloop Circuits and Kirchhoff's Rules (Section 18.2)

In a *multiloop circuit*, a **junction** or **node** is a point at which three or more connecting wires are joined together. A circuit path between two junctions is called a **branch** and may contain one or more circuit elements.

Kirchhoff's first rule, or **junction theorem**, states that the algebraic sum of the currents at any junction is zero, $\Sigma_i\ I_i = 0$. This means that the sum of the currents going into a junction (taken as positive) is equal to the sum of the currents leaving the junction (taken as negative).

Note: You have to *assume* current directions at a particular junction because you generally cannot tell whether a particular current is directed into or out of a junction simply by looking at a multiloop circuit diagram. These assumptions are totally arbitrary. If these assumptions are wrong, you will find out later from the negative sign in your mathematical results. A negative current will indicate the wrong choice of direction.

Kirchhoff's second rule, or **loop theorem**, states that the algebraic sum of the potential differences (voltages) across all elements of any *closed* loop is zero, $\Sigma_i V_i = 0$. This means that the sum of the voltages rises equal to the sum of the voltage drops across the voltage sources and resistors around a closed loop, which must be true if energy is conserved. Voltage drops may be positive or negative, and we will use the following convention.

Sign convention for voltages across circuit elements in traversing a loop:

$+V$: when a battery is traversed from its negative terminal to positive terminal (voltage rises).

$-V$: when a battery is traversed from its positive terminal to negative terminal (voltage drops).

$-IR$: when a resistor is traversed in the direction of the assigned branch current, (voltage drop).

$+IR$: when a resistor is traversed in the direction opposite to the assigned branch current, (voltage rise).

The general steps in applying Kirchhoff's rules are as follows:

(1) Assign a current and current direction for each branch in the circuit. This is done most conveniently at junctions.

(2) Indicate the loops and the arbitrarily chosen directions in which they are to be traversed. Every branch *must* be in at least one loop.

(3) Apply Kirchhoff's first rule and write equations for the currents, one for each junction that gives different equation. (In general, this gives a set of equations that includes all branch currents.)

(4) Traverse the number of loops necessary to include all branches. In traversing a loop, apply Kirchhoff's second rule and write equations using the adopted sign convention.

(5) Solve the simultaneous equations for the currents.

These steps may seem complicated, but are really are straightforward, as the following Examples show. However, if there are more than two simultaneous equations involved, the mathematical solution could be lengthy. With the better calculators (for example, the Texas Instruments series TI - 83 and later), you can solve the simultaneous equations by simply punching in the numbers. Ask your instructor if you are not sure how to use this feature of your calculator.

Example 18.5   Find the currents in the branches of the circuit shown.

Solution:

This is a simple circuit and can be solved with resistor combinations and Ohm's law. However, we use Kirchoff's rules here.

First, we assign junctions and currents.

From the junction rule:

for junction c, $I_1 = I_2 + I_3$.   Eq. (1)

From the loop rule:

for loop abceda, $+6.0 \text{ V} - I_1 (4.0 \ \Omega) - I_2 (5.0 \ \Omega) = 0$,

or $4.0I_1 + 5.0I_2 = 6.0$;   Eq. (2)

for loop abcfda, $+6.0 \text{ V} - I_1 (4.0 \ \Omega) - I_3 (9.0 \ \Omega) = 0$,

or $4.0I_1 + 9.0I_3 = 6.0$.   Eq. (3)

From Eq. (1), $I_3 = I_1 - I_2$. Substituting this into Eq. (3) to eliminate I_3 gives

$13I_1 - 9.0I_2 = 6.0$.   Eq. (4)

From Eq. (2), $I_1 = 1.5 - 1.25I_2$. Substituting this into Eq. (4) to eliminate I_1 yields $25.25I_2 = 13.5$.

Therefore, $I_2 = 0.535 \text{ A}$, $I_1 = 1.5 - 1.25I_2 = 1.5 \text{ A} - (1.25)(0.535 \text{ A}) = 0.832 \text{ A}$,

and $I_3 = I_1 - I_2 = 0.832 \text{ A} - 0.535 \text{ A} = 0.297 \text{ A}$.

Hence the currents are $I_1 = 0.53$ A, $I_2 = 0.83$ A, and $I_3 = 0.30$ A (two significant figures).

Example 18.6   Find the currents in the branches of the circuit shown.

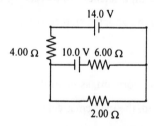

Solution:

You cannot solve this circuit with resistor combinations and Ohm's law (why?). You must use Kirchhoff's rules.

First, we assign junctions and currents.

From the junction rule:

for junction c, $I_1 + I_2 = I_3$.   Eq. (1)

From the junction rule:

for loop bcfeb, $+10.0 \text{ V} - (6.00 \ \Omega)I_1 + 14.0 \text{ V} + (4.00 \ \Omega)I_2 = 0$,

   or $(6.00)I_1 - (4.00)I_2 = 24.0 \text{ V}$;   Eq. (2)

for loop bcdab, $+10.0 \text{ V} - (6.00 \ \Omega)I_1 - (2.00 \ \Omega)I_3 = 0$,

or $(6.00)I_1 + (2.00)I_3 = 10.0 \text{ V}$.   Eq. (3)

Substituting Eq. (1) into Eq. (3) to eliminate I_3 gives $4.00I_1 + 2.00I_2 = 5.00$.   Eq. (4)

From Eq. (2), $I_2 = 1.50I_1 - 6.00$. Substituting this into Eq. (4) to eliminate I_2 yields $7.00I_1 = 17.0$.

Therefore, $I_1 = 2.43 \text{ A}$, $I_2 = 1.50I_1 - 6.00 \text{ A} = 1.50(2.43 \text{ A}) - 6.00 \text{ A} = -2.36 \text{ A}$,

and $I_3 = I_1 + I_2 = 2.43 \text{ A} + (-2.36 \text{ A}) = 0.0700 \text{ A}$.

I_2 is negative, meaning our assumption (downward) is wrong. It should be upward.

Hence the currents are $I_1 = 2.43 \text{ A}$, $I_2 = 2.36 \text{ A}$,   and $I_3 = 0.0700 \text{ A}$ (3 significant figures).

3.   RC Circuits (Section 18.3)

An **RC circuit** consists of a resistor and a capacitor connected in series. The **time constant** ($\tau = RC$) for an RC circuit is a "characteristic time" by which we measure the capacitor's charge and discharge. In one time constant, $t = \tau$, an uncharged capacitor charges to 63% of its maximum value when charging and a fully charged capacitor discharges 63% of its charge when discharging (or its charge decreases to 37% of its original value). In charging, both the voltage and the charge on the capacitor increase; in discharging, the voltage and the charge on the capacitor decrease. The current in the circuit always decreases with time regardless of charging or discharging. (Why?) The voltages and currents as functions of time in a capacitor are given by

$V_C = V_0\left(1 - e^{-t/RC}\right)$ voltage across capacitor when charging

$V_C = V_0\, e^{-t/RC}$ voltage across capacitor when discharging

$I = I_0\, e^{-t/RC}$ current in the circuit when charging or discharging,

where V_0 and I_0 are the initial voltage across the capacitor and the current in the circuit, respectively, and $\tau = RC$ is the time constant of the circuit.

Example 18.7   In the circuit shown, $C = 5.0 \ \mu\text{F}$, $R = 4.0 \ \text{M}\Omega$, and $V = 12 \text{ V}$. The switch S is open when the capacitor is initially uncharged.

(a) Immediately after the switch is closed, find

(1) the voltage across the resistor,

(2) the current through the resistor,

(b) What is the current in the circuit after 10 s?

(c) After the switch has been closed for a long period of time, find

(1) the current through the resistor,

(2) the charge and energy stored in the capacitor.

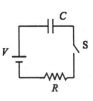

Solution:　　　Given:　$R = 4.0 \text{ M}\Omega = 4.0 \times 10^6 \ \Omega, \quad C = 5.0 \ \mu\text{F} = 5.0 \times 10^{-6} \text{ F}, \quad V = 12 \text{ V}.$

(a)　(1) Immediately after the switch is closed, the capacitor is not charged, so the voltage across the capacitor is zero ($V_C = 0$). Therefore, the voltage across the resistor is the battery voltage.

$V_R = V = 12 \text{ V}.$

(2) $I_o = \dfrac{V}{R} = \dfrac{12 \text{ V}}{4.0 \times 10^6 \ \Omega} = 3.0 \times 10^{-6} \text{ A}.$

(b) The time constant $\tau = RC = (4.0 \times 10^6 \ \Omega)(5.0 \times 10^{-6} \text{ F}) = 20 \text{ s}, \quad$ so $\quad \dfrac{t}{RC} = \dfrac{10 \text{ s}}{20 \text{ s}} = 0.50.$

So　$I = I_o \ e^{-t/RC} = (5.0 \times 10^{-6} \text{ A}) \ e^{-0.50} = 3.0 \times 10^{-6} \text{ A}. \quad (e^{-0.50} \approx 0.607)$

(c)　(1) After a long period of time, the capacitor is fully charged and the voltage across it is

$V_C = V = 12 \text{ V}$, so the voltage across the resistor is zero. Therefore the current through the resistor is zero.

(2) The charge on the capacitor is $Q = CV_C = (5.0 \times 10^{-6} \text{ F})(12 \text{ V}) = 6.0 \times 10^{-5} \text{ C}.$

The energy stored in the capacitor is $U = \frac{1}{2}CV^2 = \frac{1}{2}(5.0 \times 10^{-6} \text{ F})(12 \text{ V})^2 = 3.6 \times 10^{-4} \text{ J}.$

4.　Ammeters and Voltmeters (Section 18.4)

A **galvanometer** is a sensitive meter with an internal resistance r and whose needle deflection is proportional to the current through it. Only allow a very small maximum current I_g can exist in a galvanometer without destroying it.

An **ammeter** is a low resistance device to measure current; it consists of a shunt resistor in parallel with a galvanometer. An ammeter *must* be connected in series with the resistor whose current you want to measure. Since the ammeter has a low resistance, it does not appreciably affect the current measurement. A shunt resistor (R_s) of proper value needs to be selected in constructing an ammeter. The equation which allows us to calculate R_s is

$I_g = \dfrac{IR_s}{r + R_s}$, where I_g is the maximum current through the galvanometer (or the current at full needle deflection), r the internal resistance of the galvanometer, R_s the shunt resistance, and I the current through the ammeter.

A **voltmeter** is a high resistance device used to measure voltage. It consists of a galvanometer and a multiplier resistor in series with it. A voltmeter *must* be connected in parallel with a circuit element whose voltage we wish to measure. Since the voltmeter has a high resistance, it does not appreciably affect the voltage across an element. A multiplier resistor of proper value must be selected in constructing a voltmeter. The equation which allows us to calculate R_m is $I_g = \dfrac{V}{r + R_m}$, where I_g is the maximum current through the galvanometer, r the internal resistance of the galvanometer, R_m the multiplier resistance, and V the voltage of the voltmeter.

Example 18.8   A galvanometer has an internal resistance 25 Ω and can safely carry 150 μA of current.

(a) What shunt resistor is needed to make an ammeter with a range of 0 to 10 A?

(b) What multiplier resistor is needed to make a voltmeter with a range of 0 to 100 V?

Solution: Given: $r = 25\ \Omega$, $I_g = 150\ \mu A = 150 \times 10^{-6}$ A, $I = 10$ A, $V = 100$ V.

Find: (a) R_s (b) R_m.

(a) From $I_g = \dfrac{IR_s}{r + R_s}$, $R_s = \dfrac{I_g r}{I - I_g} = \dfrac{(150 \times 10^{-6}\ \text{A})(25\ \Omega)}{10\ \text{A} - 150 \times 10^{-6}\ \text{A}} = 3.8 \times 10^{-4}\ \Omega = 0.38\ \text{m}\Omega.$

(b) From $I_g = \dfrac{V}{r + R_m}$, $R_m = \dfrac{V - I_g r}{I_g} = \dfrac{100\ \text{V} - (150 \times 10^{-6}\ \text{A})(25\ \Omega)}{150 \times 10^{-6}\ \text{A}} = 6.7 \times 10^5\ \Omega = 670\ \text{k}\Omega.$

5.   Household Circuits and Electrical Safety (Section 18.5)

Fuses and **circuit breakers** are safety devices that open (or "break") a circuit when the current exceeds a preset value. If the current in a circuit exceeds a preset (safe) value, this is usually an indication of some problem (short circuit) in the circuit.

A **grounded plug** (three-prong plug) uses a dedicated grounding wire to ground objects that may become conductors and thus dangerous. A **polarized plug** identifies the ground side of the line for use as a grounding safety feature.

The first precaution to take for personal safety is to avoid coming into contact with an electrical conductor that might cause a voltage across a human body or part of it, thus causing a current through the body which could be dangerous.

IV.   Mathematical Summary

Equivalent Series Resistance	$R_s = R_1 + R_2 + R_3 + \ldots$ $= \Sigma_I\, R_i$ (18.2)	Computes the equivalent series resistance of resistors in series combination.
Equivalent Parallel Resistance	$\dfrac{1}{R_p} = \dfrac{1}{R_1} + \dfrac{1}{R_2} + \dfrac{1}{R_3} + \ldots$ $= \Sigma_i\, \dfrac{1}{R_i}$ (18.3)	Computes the equivalent parallel resistance of resistors in parallel combination.

Kirchhoff's Rules	(1) $\Sigma_I I_i = 0$ (18.4)	Applies the conservation of charge and energy in
(1) junction theorem	(2) $\Sigma_I V_i = 0$ (18.5)	electric circuits.
(2) loop theorem		
Charging Voltage across a capacitor in an RC Circuit	$V_C = V_0\left(1 - e^{-t/RC}\right)$ (18.6)	Gives the voltage on a charging capacitor as a function of time for an RC circuit.
Time Constant for an RC Circuit	$\tau = RC$ (18.8)	Defines the time constant in a resistor-capacitor series circuit.
Discharging Voltage for an RC Circuit	$V_C = V_0\, e^{-t/RC}$ (18.9)	Gives the voltage on a discharging capacitor as a function of time for an RC circuit.
Galvanometer Current (used as ammeter)	$I_g = \dfrac{IR_s}{r + R_s}$ (18.10)	Expresses the galvanometer current in terms of the shunt resistance in an ammeter.
Galvanometer Current (used as voltmeter)	$I_g = \dfrac{V}{r + R_m}$ (18.11)	Expresses the galvanometer current in terms of the multiplier resistance in a voltmeter.

V.   Solutions of Selected Exercises and Paired/Trio Exercises

7.   (a) Series gives maximum resistance:

$R_s = R_1 + R_2 + R_3 = 10\ \Omega + 20\ \Omega + 30\ \Omega = \boxed{60\ \Omega}$.

(b) Parallel gives minimum resistance:

$$\frac{1}{R_p} = \frac{1}{R_1} + \frac{1}{R_2} + \frac{1}{R_3} = \frac{1}{10\ \Omega} + \frac{1}{20\ \Omega} + \frac{1}{30\ \Omega} = \frac{11}{60\ \Omega},$$

so $R_p = \boxed{5.5\ \Omega}$.

8.   Series combination: $R_s = R_1 + R_2 = R + R = 2R$.

Parallel combination: $\dfrac{1}{R_p} = \dfrac{1}{R_s} + \dfrac{1}{R_3}$,

so $R_p = \dfrac{R_s R_3}{R_s + R_3} = 10\ \Omega = \dfrac{(2R)(20\ \Omega)}{2R + 20\ \Omega}$.

Simplifying, $20R + 200 = 40R$.   Solving, $R = \boxed{10\ \Omega}$.

10.   Parallel: $\dfrac{1}{R_p} = \dfrac{1}{R_1} + \dfrac{1}{R_2} + \dfrac{1}{R_3} = 3\,\dfrac{1}{4.0\ \Omega} = \dfrac{3}{4.0\ \Omega}$,   so $R_p = \boxed{1.3\ \Omega\text{ for parallel}}$.

Series: $R_s = R_1 + R_2 + R_3 = 3(4.0\ \Omega) = \boxed{12\ \Omega\text{ for series}}$.

parallel   series   series-parallel   parallel-series

Two in series-parallel: $\dfrac{1}{R_p} = \dfrac{1}{R_s} + \dfrac{1}{R} = \dfrac{1}{R+R} + \dfrac{1}{R} = \dfrac{3}{2(4.0\ \Omega)}$

so $R_p = \boxed{2.7\ \Omega\text{ for series-parallel}}$.

Two in parallel-series: $R_s = R + R_p = R + \dfrac{R\,R}{R+R} = \dfrac{3R}{2} = \dfrac{3(4.0\ \Omega)}{2} = \boxed{6.0\ \Omega\text{ for parallel-series}}$.

17.   (a) $I = \dfrac{V}{R_s} = \dfrac{12\ \text{V}}{2.0\ \Omega + 4.0\ \Omega + 6.0\ \Omega} = \boxed{1.0\ \text{A}}$.

(b) Series has the same current, $\boxed{1.0\ \text{A}}$.

(c) Since $P = I^2 R$, $P_{2\,\Omega} = (1.0\ \text{A})^2\,(2.0\ \Omega) = \boxed{2.0\ \text{W}}$; $P_{4\,\Omega} = (1.0\ \text{A})^2\,(4.0\ \Omega) = \boxed{4.0\ \text{W}}$;

and $P_{6\,\Omega} = (1.0\ \text{A})^2\,(6.0\ \Omega) = \boxed{6.0\ \text{W}}$.

(d) $P_{\text{total}} = (1.0\ \text{A})^2\,(12\ \Omega) = 12\ \text{W}$, $P_{\text{sum}} = 2.0\ \text{W} + 4.0\ \text{W} + 6.0\ \text{W} = 12\ \text{W}$.

So $\boxed{P_{\text{sum}} = P_{\text{total}} = 12\ \text{W}}$.

20. R_1 and R_2 are in series: $R_s = 2.0\ \Omega + 2.0\ \Omega = 4.0\ \Omega$.

R_s, R_3, and R_4 are in parallel: $\dfrac{1}{R_p} = \dfrac{1}{4.0\ \Omega} + \dfrac{1}{2.0\ \Omega} + \dfrac{1}{2.0\ \Omega} = \dfrac{5}{4.0\ \Omega}$,

so $R_p = \dfrac{4.0\ \Omega}{5} = \boxed{0.80\ \Omega}$.

21. R_2 and R_3 are in series: $R_s = 6.0\ \Omega + 4.0\ \Omega = 10\ \Omega$.

R_s, R_1, and R_4 are in parallel: $\dfrac{1}{R_p} = \dfrac{1}{10\ \Omega} + \dfrac{1}{6.0\ \Omega} + \dfrac{1}{10\ \Omega} = \dfrac{22}{60\ \Omega}$,   so $R_p = \boxed{2.7\ \Omega}$.

22. $\boxed{7.5\ \Omega}$.

27.   (a) $I_1 = \dfrac{V}{R_1} = \dfrac{20\text{ V}}{20\ \Omega} = \boxed{1.0\text{ A}}$. R_2 and R_3 are in series: $R_s = 20\ \Omega + 20\ \Omega = 40\ \Omega$.

    So $I_2 = I_3 = \dfrac{20\text{ V}}{40\ \Omega} = \boxed{0.50\text{ A}}$.

    (b) $V_1 = \boxed{20\text{ V}}$, $V_2 = V_3 = I_2\,R_2 = (0.50\text{ A})(20\ \Omega) = \boxed{10\text{ V}}$.

    (c) The total power is $P = \dfrac{V^2}{R_1} + \dfrac{V^2}{R_s} = \dfrac{(20\text{ V})^2}{20\ \Omega} + \dfrac{(20\text{ V})^2}{40\ \Omega} = \boxed{30\text{ W}}$.

32.   (a) $I_1 = \dfrac{V}{R_1} = \dfrac{6.0\text{ V}}{6.0\ \Omega} = \boxed{1.0\text{ A}}$ R_2 and R_3 are in series. $R_s = 4.0\ \Omega + 6.0\ \Omega = 10\ \Omega$.

    So $I_2 = I_3 = \dfrac{6.0\text{ V}}{10\ \Omega} = \boxed{0.60\text{ A}}$, $I_4 = \dfrac{6.0\text{ V}}{10\ \Omega} = \boxed{0.60\text{ A}}$.

    (b) $P_1 = I_1^2\,R_1 = (1.0\text{ A})^2(6.0\ \Omega) = \boxed{6.0\text{ W}}$, $P_2 = (0.60\text{ A})^2(4.0\ \Omega) = \boxed{1.4\text{ W}}$,

    $P_3 = (0.60\text{ A})^2(6.0\ \Omega) = \boxed{2.2\text{ W}}$, $P_4 = (0.60\text{ A})^2(10\ \Omega) = \boxed{3.6\text{ W}}$.

    (c) $P_{\text{sum}} = 6.0\text{ W} + 1.44\text{ W} + 2.16\text{ W} + 3.6\text{ W} = 13\text{ W}$.   From Exercise 18.21,

    $P_{\text{total}} = \dfrac{(6.0\text{ V})^2}{2.7\ \Omega} = 13\text{ W}$.   So $\boxed{P_{\text{sum}} = P_{\text{tot}} = 13\text{ W}}$.

34. R_1 and R_2 are in series. $R_{s1} = 10\ \Omega + 5.0\ \Omega = 15\ \Omega$.

    So $I_1 = I_2 = \dfrac{V}{R_{s1}} = \dfrac{10\text{ V}}{15\ \Omega} = \boxed{0.67\text{ A}}$ $I_3 = \dfrac{10\text{ V}}{10\ \Omega} = \boxed{1.0\text{ A}}$.

    R_4 and R_5 are in series. $R_{s2} = 5.0\ \Omega + 20\ \Omega = 25\ \Omega$.

    So $I_4 = I_5 = \dfrac{10\text{ V}}{25\ \Omega} = \boxed{0.40\text{ A}}$.

    (b) $V_1 = I_1\,R_1 = (0.667\text{ A})(10\ \Omega) = \boxed{6.7\text{ V}}$, $V_2 = (0.667\text{ A})(5.0\ \Omega) = \boxed{3.3\text{ V}}$, $V_3 = \boxed{10\text{ V}}$,

    $V_4 = (0.40\text{ A})(5.0\ \Omega) = \boxed{2.0\text{ V}}$, $V_5 = (0.40\text{ A})(20\ \Omega) = \boxed{8.0\text{ V}}$.

36. $\boxed{8.1\ \Omega}$.

44.   Around the inner loop:

    $10\text{ V} - I_2(2.0\ \Omega) - I_3(5.0\ \Omega) = 0$.   Eq. (1)

    Around the outer loop:

    $10\text{ V} - I_1(10\ \Omega) - I_3(5.0\ \Omega) = 0$.   Eq. (2)

    From junction theorem:

    $I_3 = I_1 + I_2$.   Eq. (3)

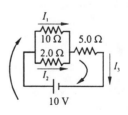

Substituting Eq. (3) into Eq. (1) and Eq. (2) gives

$(5.0\ \Omega)I_1 + (7.0\ \Omega)I_2 = 10\ \text{V}$ Eq. (3)

$(7.0\ \Omega)I_1 + (5.0\ \Omega)I_2 = 10\ \text{V}$ Eq. (4)

Solving, $I_1 = \boxed{0.25\ \text{A}}$, $I_2 = \boxed{1.25\ \text{A}}$,   and $I_3 = \boxed{1.5\ \text{A}}$.

46.   (a) Around the loop, $10\ \text{V} - 4\ \text{V} - I(12\ \Omega) = 0$,

so $I = \boxed{0.50\ \text{A}}$ and $P = I^2 R = (0.50\ \text{A})^2(12\ \Omega) = \boxed{3.0\ \text{W}}$.

(b) $P_{10} = I\mathcal{E} = (0.50\ \text{A})(10\ \text{V}) = \boxed{5.0\ \text{W output}}$,

$P_4 = (-4\ \text{V})(0.50\ \text{A}) = -2.0\ \text{W} = \boxed{2.0\ \text{W input}}$.

$P_{\text{net}} = 5.0\ \text{W} - 2.0\ \text{W} = \boxed{3.0\ \text{W to the resistor}}$

48.   (a) For the R_1 and R_2 connecting junction: $I = I_1 + I_2$.   Eq. (1)

Around the loop through R_1 in counterclockwise direction,

$12\ \text{V} - I(2.0\ \Omega) - I(8.0\ \Omega) + 6.0\ \text{V} - I(2.0\ \Omega) - I_1(4.0\ \Omega) = 0$,

or $-12I - 4I_1 + 18 = 0$.   Eq. (2)

Around the loop through R_2 in counterclockwise direction,

$12\ \text{V} - I(2.0\ \Omega) - I(8.0\ \Omega) + 6.0\ \text{V} - I(2.0\ \Omega) - I_2(6.0\ \Omega) = 0$,

or $-12I - 6I_2 + 18 = 0$.   Eq. (3)

Substituting Eq. (1) into Eq. (2) and Eq. (3) gives $-16I_1 - 12I_2 + 18 = 0$.   Eq. (4)

$-12I_1 - 18I_2 + 18 = 0$.   Eq. (5)

Solving, $\boxed{I_1 = 0.75\ \text{A left}}$, $\boxed{I_2 = 0.50\ \text{A left}}$, $I = I_3 = I_4 = I_5$,

so $\boxed{I_3 = 1.25\ \text{A up}}$, $\boxed{I_4 = 1.25\ \text{A right}}$,   and $\boxed{I_5 = 1.25\ \text{A down}}$.

(b) $P = I^2 R = (1.25\ \text{A})^2(8.0\ \Omega) = \boxed{13\ \text{W}}$.

21. R_2 and R_3 are in series: $R_s = 6.0\ \Omega + 4.0\ \Omega = 10\ \Omega$.

R_s, R_1, and R_4 are in parallel: $\dfrac{1}{R_p} = \dfrac{1}{10\ \Omega} + \dfrac{1}{6.0\ \Omega} + \dfrac{1}{10\ \Omega} = \dfrac{22}{60\ \Omega}$,

so $R_p = \boxed{2.7\ \Omega}$.

50. $\boxed{I_1 = 3.23\ \text{A down}}$, $\boxed{I_2 = 1.54\ \text{A down}}$, $\boxed{I_3 = 1.02\ \text{A down}}$, and $\boxed{I_4 = I_5 = I_6 = 1.93\ \text{A left}}$.

55. (a) Just after the switch is closed, the capacitor is not charged.

So $V_C = \boxed{0}$ and $V_R = V - V_C = \boxed{V_o}$.

(b) After one time constant, $V_C = \boxed{0.63V_o}$ and $V_R = V - V_C = \boxed{0.37V_o}$.

(c) After many time constants, the capacitor is fully charged. So $V_C = \boxed{V_o}$ and $V_R = \boxed{0}$.

60. (a) The potential difference on the capacitor is zero immediately after the switch is closed because

$V_C = V_o\left(1 - e^{-t/\tau}\right) = V_o\left(1 - e^0\right) = 0$. So the potential difference across the resistor is $\boxed{24\text{ V}}$.

(b) $\boxed{0}$.

(c) $I = \dfrac{\varepsilon}{R} = \dfrac{24\text{ V}}{6.0\ \Omega} = \boxed{4.0\text{ A}}$.

68. From $I_g = \dfrac{IR_s}{r + R_s}$, we have $R_s = \dfrac{I_g r}{I - I_g} = \dfrac{(2000 \times 10^{-6}\text{ A})(100\ \Omega)}{30\text{ A} - 2000 \times 10^{-6}\text{ A}} = 0.00667\ \Omega = \boxed{6.7\text{ m}\Omega}$.

74. (a) The current reading I is the total current through R and R_v.

So the voltage reading is $V = IR_p = I\dfrac{R_v R}{R_v + R}$.

Therefore $R = \dfrac{V}{I - (V/R_v)}$.

(b) If R_v is ∞, $R = \dfrac{V}{I}$, i.e., the measurement is "perfect."

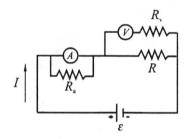

78. The fuse and the switch are on the ground side of the circuit. It were touched by a person, current could possibly flow through the person's body because an open switch or blown fuse would leave the motor at a high voltage.

80. A conductor has very low resistance. The resistance of the wire between the feet is very very small, so the voltage between the feet is also small and so is the current through the bird.

86. From the junction theorem, $I_1 = I_2 + I_3$.

Around the upper loop in clockwise direction, $6.0\text{ V} - I_1(2.0\ \Omega) - I_2(4.0\ \Omega) + 6.0\text{ V} = 0$,

or $2I_1 + 4I_2 = 12$.   Eq. (1)

Around the whole circuit in clockwise direction, $6.0\text{ V} - I_1(2.0\ \Omega) - I_3(8.0\ \Omega) + 6.0\text{ V} = 0$,

or $10I_1 - 8I_2 = 12$.   Eq. (2)

Solving, $\boxed{I_1 = 2.6\text{ A right}}$, $\boxed{I_2 = 1.7\text{ A left}}$, and $\boxed{I_3 = 0.86\text{ A left}}$.

90. The three R's are in series, $R_{s1} = 3R$. This R_{s1} and R are in parallel, $R_{p1} = \dfrac{(3R)R}{3R + R} = \dfrac{3}{4}R$.

This R_{p1} and two R are in series, $R_{s2} = 2R + \dfrac{3}{4}R = \dfrac{11}{4}R$.

This R_{s2} and R are in parallel, $R_{p2} = \dfrac{\frac{11}{4}R\,R}{\frac{11}{4}R + R} = \dfrac{11}{15}R$.

Finally this R_{p2} and two R are in series, $R_{s3} = 2R + \dfrac{11}{15}R = \boxed{\dfrac{41}{15}R = 2.73R}$.

97. (a) $\varepsilon = 3(1.5 \text{ V}) = 4.5 \text{ V}$. $I = \dfrac{\varepsilon}{R + r} = \dfrac{4.5 \text{ V}}{10 \ \Omega + 3(0.02 \ \Omega)} = 0.447 \text{ A}$.

So $V = IR = (0.447 \text{ A})(10 \ \Omega) = \boxed{4.47 \text{ V}}$.

(b) As determined in (a), $I = \boxed{0.447 \text{ A}}$.

VI. Practice Quiz

1. Three resistors of 4.0 Ω, 6.0 Ω, and 10.0 Ω are connected in series. What is their equivalent resistance?

    (a) 20 Ω   (b) 7.3 Ω   (c) 6.0 Ω   (d) 4.0 Ω   (e) 1.9 Ω

2. Three resistors having values of 4.0 Ω, 6.0 Ω, and 10.0 Ω are connected in parallel. If the circuit is connected in series to a battery of 12.0 V and a resistor of 2.0 Ω, what is the current through the 10 Ω resistor?

    (a) 0.59 A   (b) 1.0 A   (c) 2.7 A   (d) 11.2 A   (e) 16.0 A

    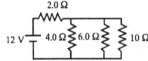

3. The following three appliances are connected to a 120-volt house circuit: (1) computer and printer, 350 W, (2) coffee pot, 650 W, and (3) microwave, 900 W. If all were operated at the same time what total current would they draw?

    (a) 0.063 A   (b) 2.9 A   (c) 5.4 A   (d) 7.5 A   (e) 16 A

4. Find the current in the 15-ohm resistor.

    (a) 0.10 A   (b) 0.13 A   (c) 0.20 A   (d) 0.26 A   (e) 0.30 A

    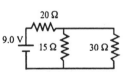

5. What is the maximum number of 75-watt lightbulbs you can connect in parallel in a 120-volt home circuit without tripping the 15-amp circuit breaker?

    (a) 15   (b) 18   (c) 21   (d) 24   (e) 27

6. What is the result from the Kirchhoff's junction rule for this figure?

(a) $I_2 = I_1 + I_3$

(b) $I_1 = I_2 + I_3$

(c) $I_3 = I_1 + I_2$,

(d) $I_1 + I_2 + I_3 = 0$

(e) none of the above

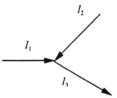

7. What is the current through the 2-ohm resistor in the circuit shown?

(a) 2 A

(b) 3 A

(c) 4 A

(d) 5 A

(e) 6 A

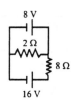

8. What is the equivalent resistance between points A and B of the resistors in the figure?

(a) 0.443R (b) 0.75R (c) R (d) 2.5R (e) 4R

9. In the following simple circuit, G is a galvanometer and R is a resistor. What is this arrangement likely to be used to measure in a circuit?

(a) voltage   (b) current   (c) resistance   (d) power   (e) energy

10. A voltage source of 10 V is connected to a series RC circuit with $R = 1.0$ MΩ and $C = 4.0$ μF. Initially the capacitor is fully charged. Find the time required for the current in the circuit to decrease to 10% of its original value.

(a) 0.42 s   (b) 2.3 s   (c) 4.0 s   (d) 9.2 s   (e) 18 s.

Answers to Practice Quiz:

1.a 2.a 3.e 4.c 5.d 6.c 7.c 8.b 9.b 10.d

CHAPTER 19

Magnetism

I.   Chapter Objectives

Upon completion of this chapter, you should be able to:

1.   state the force rule between magnetic poles and explain how the magnetic field direction is determined with a compass.

2.   define the magnetic field strength in terms of the force exerted on a moving charged particle and determine the magnetic force exerted by a magnetic field on such a particle.

3.   understand the origin of the magnetic field and calculate its strength for simple cases, and use the right-hand force rule to determine the direction of the magnetic field from the direction of the current that produces it.

4.   explain how ferromagnetic materials enhance external magnetic fields, how "permanent" magnets are produced, and how "permanent" magnetism can be destroyed.

5.   calculate the magnetic force on a current-carrying wire and the torque on a current-carrying loop, and explain the concept of a magnetic moment for such a loop.

6.   explain the operation of various instruments whose functions depend on electromagnetic interactions.

*7.   state some of the general characteristics of the Earth's magnetic field, explain the theory about its possible source, and discuss some of the ways in which it affects the Earth's local environment.

II.   Key Terms

Upon completion of this chapter, you should be able to define and/or explain the following key terms:

pole-force law or law of poles	magnetic permeability
magnetic field	Curie temperature
electromagnetism	right-hand force rule for a current-carrying wire
tesla (T)	magnetic moment
right-hand force rule	dc motor
right-hand source rule	cathode ray tube (CRT)
ferromagnetic materials	mass spectrometer
magnetic domains	

III.   Chapter Summary and Discussion

1.   Magnets, Magnetic Poles, and Magnetic Field Direction (Section 19.1)

Electricity and magnetism (electromagnetism) are manifestations of a single fundamental force or interaction, the electromagnetic force.

Magnets have two different poles or "centers" of force, which are designated as north and south poles. By the *law of poles*, like magnetic poles repel each other, and unlike magnetic poles attract each other. A magnet can create a **magnetic field**, similar to the electric field created by an electric charge.

The direction of a magnetic field (**B**) at any location is in the direction that the north pole of a compass at that location would point. Hence, the magnetic field lines outside of a bar magnet are directed away from a magnetic north pole and toward a magnetic south pole.

2.   Magnetic Field Strength and Magnetic Force (Section 19.2)

A **magnetic field** can exert forces only on *moving* charges. The *magnitude* of a magnetic field can be defined in terms of the magnetic force (F) on a moving charge (q) by, $B = \dfrac{F}{qv \sin \theta}$, where v is the speed of the charge and θ is the angle between the velocity vector and the magnetic field vector. The SI unit of magnetic field is the tesla (T). When **v** and **B** are perpendicular, the magnitude is at its maximum since when $\theta = 90°$, $F = qvB \sin \theta = qvB \sin 90° = qvB$ (maximum force).

The *direction* of the magnetic force on a charged particle is determined by the **force right-hand rule**: when the fingers of the right hand are pointed in the direction of **v** and then curled toward the vector **B**, the extended thumb points in the direction of **F** on a positive charge. (For a negative charge, the force is opposite this direction.).

Note: The magnetic force is always perpendicular to both the velocity vector and the magnetic field vector, or equivalently the magnetic force is perpendicular to a plane formed by the velocity and magnetic field vectors.

Example 19.1   An electron moves with a speed of 4.0×10^6 m/s along the $+x$-axis. It enters a region where there is a uniform magnetic field of 2.5 T, directed at an angle of 60° to the x axis and lying in the xy plane. Calculate the initial force and acceleration of the electron.

Solution: Given: $q = 1.60 \times 10^{-19}$ C, $m = 9.11 \times 10^{-31}$ kg, $v = 4.0 \times 10^{6}$ m/s, $B = 2.5$ T, $\theta = 60°$.

Find: **F** and **a**.

According to the right-hand rule, the direction of the magnetic force in the

diagram is into the page ($-z$) for the electron because it has negative charge.

$F = qvB\sin\theta = (1.60 \times 10^{-19}$ C$)(4.0 \times 10^{6}$ m/s$)(2.5$ T$)\sin 60° = 1.4 \times 10^{-12}$ N.

By Newton's second law, $a = \dfrac{F}{m} = \dfrac{1.4 \times 10^{-12} \text{ N}}{9.11 \times 10^{-31} \text{ kg}} = 1.5 \times 10^{18}$ m/s^2.

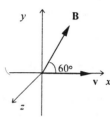

The directions of both the magnetic force and acceleration are initially along the $-z$ axis.

Example 19.2 A proton has a speed of 4.5×10^{6} m/s in a direction perpendicular to a uniform magnetic field, and the proton moves in a circle of radius 0.20 m. What is the magnitude of the magnetic field?

Solution: Given: $q = 1.60 \times 10^{-19}$ C, $m = 1.67 \times 10^{-27}$ kg, $v = 4.5 \times 10^{6}$ m/s, $r = 0.20$ m.

Find: B.

Since the magnetic force is always perpendicular to the velocity of the proton (right-hand rule), it is the

centripetal force that causes the particle to move in a circular path. That is, the centripetal force is provided

by the magnetic force. By Newton's second law: $F_c = ma_c = \dfrac{mv^2}{r} = qvB$,

so $B = \dfrac{mv}{qr} = \dfrac{(1.67 \times 10^{-27} \text{ kg})(4.5 \times 10^{6} \text{ m/s})}{(1.60 \times 10^{-19} \text{ C})(0.20 \text{ m})} = 0.23$ T.

---

3. Electromagnetism—The Source of Magnetic Fields (Section 19.3)

Generally, magnetic fields are produced by electric currents (moving charges). The magnitude of the

magnetic field near *a long, straight, current-carrying wire* is given by $B = \dfrac{\mu_0 I}{2\pi d}$, where $\mu_0 = 4\pi \times 10^{-7}$ T·m/A is a

constant called the *magnetic permeability of free space*, I the current in the wire, and d the perpendicular distance

from the wire. The direction of the magnetic field is determined by the **right-hand source rule**: if a current-

carrying wire is grasped with the right hand with the extended thumb pointing in the direction of the current (I), the

curled fingers indicate the circular sense of the magnetic field.

The magnitude of the magnetic field at the *center of a circular current-carrying wire loop* is given by

$B = \dfrac{\mu_0 I}{2r}$, where r is the radius of the loop. The direction of the magnetic field is determined by the right-hand

source rule and is perpendicular to the plane of the loop at its center.

The magnitude of the magnetic field near the center of *a current-carrying solenoid* is given by $B = \frac{\mu_0 NI}{L}$, where N is the number of turns (loops) in the solenoid and L is the length of the solenoid. The direction of the magnetic field is determined by the right-hand source rule as applied to one of the loops of the solenoid. The quantity $n = \frac{N}{L}$ is called the *linear turn density* (number of turns per unit length). The magnitude of the magnetic field can also be expressed in terms of n as $B = \mu_0 nI$. The magnetic field near the center of a solenoid is independent of the radius of the solenoid and is approximately constant if you stay away from the ends.

Note: Since magnetic field is a vector, you must use vector addition to find the net field if there are contributions from two or more sources.

Example 19.3   What current is required for a long straight wire to produce a magnetic field of magnitude equal to the strength of the Earth's magnetic field of about 5.0×10^{-5} T at a location 2.5 cm from the wire?

Solution: Given: $B = 5.0 \times 10^{-5}$ T, $d = 2.5$ cm $= 0.25$ m.

Find: I.

From $B = \frac{\mu_0 I}{2\pi d}$,   we have $I = \frac{2\pi Bd}{\mu_0} = \frac{2\pi(5.0 \times 10^{-5}\ \text{T})(0.025\ \text{m})}{4\pi \times 10^{-7}\ \text{T·m/A}} = 6.3$ A.

Example 19.4   Calculate the magnitude of the magnetic field at the center of a solenoid which has 100 turns, is 0.10 m long and carries a current of 2.0 A.

Solution: Given: $N = 100$, $L = 0.10$ m, $I = 2.0$ A.

Find: B.

$B = \frac{\mu_0 NI}{L} = \frac{(4\pi \times 10^{-7}\ \text{T·m/A})(100)(2.0\ \text{A})}{0.10\ \text{m}} = 2.5 \times 10^{-3}$ T $= 2.5$ mT.

Example 19.5   Two long parallel wires carry currents of 20 A and 5.0 A in opposite directions. The wires are separated by 0.20 m.

(a) What is the magnetic field midway between the two wires?

(b) At what point between the wires are the magnetic fields from the two wires the same?

Solution:

(a) According to the right-hand source rule, the magnetic fields between the wires due to the two wires are in the same direction (both are into the page). At midway between the two wires, the distance from either wire is 0.10 m.

$$B_1 = \frac{\mu_0 I_1}{2\pi d_1} = \frac{(4\pi \times 10^{-7} \text{ T·m/A})(20 \text{ A})}{2\pi(0.10 \text{ m})} = 4.0 \times 10^{-5} \text{ T}.$$

$$B_2 = \frac{(4\pi \times 10^{-7} \text{ T·m/A})(5.0 \text{ A})}{2\pi(0.10 \text{ m})} = 1.0 \times 10^{-5} \text{ T}.$$

So the net magnetic field is $B = B_1 + B_2 = 4.0 \times 10^{-5}$ T $+ 1.0 \times 10^{-5}$ T $= 5.0 \times 10^{-5}$ T, into the page.

(b) Assume the contributions from both wires are the same at a distance x from the 20-A wire. Then the

distance from the 5.0 A wire is $(0.20 \text{ m} - x)$. $B_1 = B_2$, so $\dfrac{\mu_0 I_1}{2\pi x} = \dfrac{\mu_0 I_2}{(0.20 \text{ m} - x)}$.

Therefore, $I_1(0.20 \text{ m} - x) = I_2 x$, or $(20 \text{ A})(0.20 \text{ m} - x) = (5.0 \text{ A})x$.

Solving, $x = 0.16$ m from the 20 A wire.

---

4. Magnetic Materials (Section 19.4)

Ferromagnetic materials are those in which the electron spin "lock" together to create **magnetic domains** where the magnetic fields of individual electrons add constructively. These materials are easily magnetized. The magnetic permeability of these materials is many times that of free space, reflecting the fact that they enhance an external magnetic field. In an external magnetic field, the domains parallel to the field grow at the expense of other domains and the orientation of some domains may become more aligned with the field.

Common ferromagnetic materials are iron, nickel, and cobalt. Iron is commonly used in the cores of electromagnets. This type of iron is termed "soft" iron because the domains become unaligned and the iron unmagnetized when the external field is removed. "Hard" iron retains some magnetism after a field is removed, and this type of iron is used to make "permanent" magnets by heating the iron above the **Curie temperature** and cooling in a strong magnetic field. Above the Curie temperature, the magnetic domains become thermally disordered and the iron loses its "permanent" magnetic field.

The magnetic permeability is defined as $\mu = K_m \mu_0$, where K_m is the magnetic analog of the dielectric constant and is called the *relative permeability*. $K_m = 1$ for free space (vacuum).

5. Magnetic Forces on Current-Carrying Wires (Section 19.5)

A magnetic field can exert a force (F) on a current-carrying wire. The magnitude of the magnetic force is given by $F = ILB\sin\theta$, where L is the length of the wire, I the current in the wire, and θ the angle between the current direction and the magnetic field vector. The direction of the force is determined by the right-hand rule: when the fingers of the right hand are pointed in the direction of the conventional current and then turned or curled toward the vector **B**, the extended thumb points in the direction of **F**.

Forces exist between two parallel current-carrying wires. This is because the magnetic field produced by the current in one wire exerts a force on the other wire. The force per unit length can be calculated from $\frac{F}{L} = \frac{\mu_0 I_1 I_2}{2\pi d}$, where d is the distance between the parallel wires and I_1 and I_2 the currents in the wires, respectively. According to the right-hand rules (source plus force), the forces are attractive if the currents are in the same direction and repulsive if the currents are in opposite directions (can you show why?).

A magnetic field can exert torque on a current-carrying loop. The magnitude of the torque is equal to $\tau = NIAB\sin\theta$, where N is the number of turns (loops), I the current in the loop, A the area enclosed by the loop, and θ the angle between the normal to the loop and the magnetic field vector. The quantity IA is often referred to as the **magnetic moment**, $m = IA$. The magnitude of the torque can then be expressed as $\tau = NmB\sin\theta$.

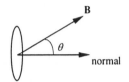

Note: The angle θ in the torque equation is between the *normal to the plane of the loop* and the magnetic field, *not* the angle between the plane of the loop and the magnetic field.

Example 19.6   A wire carries a current of 6.0 A in a direction of 60° with respect to the direction of a magnetic field of 0.75 T. Find the magnitude of the magnetic force on a 0.50 m length of the wire.

Solution: Given: $I = 6.0$ A, $\theta = 60°$, $B = 0.75$ T, $L = 0.50$ m.

Find: F.

$F = ILB\sin\theta = (6.0\text{ A})(0.50\text{ m})(0.75\text{ T})\sin 60° = 1.9$ N.

Example 19.7   A circular loop of wire of radius 0.50 m is in a uniform magnetic field of 0.30 T. The current in the loop is 2.0 A. Find the magnitude of the torque when

(a) the plane of the loop is parallel to the magnetic field,

(b) the plane of the loop is perpendicular to the magnetic field,

(c) the plane of the loop is at 30° to the magnetic field.

Solution: Given: $N = 1$, $r = 0.50$ m, $B = 0.30$ T, $I = 2.0$ A,

(a) $\theta = 90°$,   (b) $\theta = 0°$,   (c) $\theta = 90° - 30° = 60°$.

Find: τ in (a), (b), and (c).

The angle in the torque equation is between the normal to the plane of the loop and the magnetic field, not between the plane of the loop and the magnetic field.

(a) $\tau = NIAB \sin \theta = NA\pi r^2 B \sin \theta = (1)(2.0 \text{ A})(\pi)(0.50 \text{ m})^2 (0.30 \text{ T}) \sin 90° = 0.47 \text{ m·N}.$

(b) $\tau = (1)(2.0 \text{ A})(\pi)(0.50 \text{ m})^2 (0.30 \text{ T}) \sin 0° = 0 \text{ m·N}.$

(c) $\tau = (1)(2.0 \text{ A})(\pi)(0.50 \text{ m})^2 (0.30 \text{ T}) \sin 60° = 0.41 \text{ m·N}.$

Example 19.8   Two long, straight wires separated by a distance of 0.30 m carry currents in the same direction. If the current in one wire is 10 A and the current in the other is 8.0 A, find the magnitude and direction of the forces per unit length (per meter) between the two wires. What if the currents are in opposite directions?

Solution: Given: $I_1 = 10 \text{ A}, \quad I_2 = 8.0 \text{ A}, \quad d = 0.30 \text{ m}.$

Find: $F/L.$

There is a force on each wire because the magnetic field produced by one wire can exert force on the other. According to Newton's third law, the two forces on two wires are equal in magnitude and opposite in direction. The field by wire 1 at wire 2 is $B_1 = \dfrac{\mu_0 I_1}{2\pi d}$,

so the magnitude of the force on wire 2 is $F_2 = I_2 L B_1 = \dfrac{\mu_0 I_1 I_2}{2\pi d} L.$

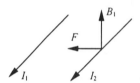

Therefore the force per unit length is

$$\frac{F_2}{L} = \frac{\mu_0 I_1 I_2}{2\pi d} = \frac{(4\pi \times 10^{-7} \text{ T·m/A})(10 \text{ A})(8.0 \text{ A})}{2\pi(0.30 \text{ m})} = 5.3 \times 10^{-5} \text{ N/m}.$$

According to the right-hand rule, the magnetic field due to one wire exerts an attractive force on the other, and vice versa, so the forces are attractive. If the currents are opposite, the magnetic field due to one wire exerts a repulsive force on the other, and vice versa.

---

6.   Applications of Electromagnetism (Section 19.6)

Applications of electromagnetic interactions include the galvanometer, the **dc motor**, the **cathode ray tube (CRT)**, the **mass spectrometer**, and the electronic balance. A galvanometer consists of a small coil. When there is a current in the coil, the magnetic field exerts a torque on the coil, and therefore deflects the coil. The deflection of the coil is directly proportional to the current through the galvanometer. A dc motor is a device that converts electrical energy into mechanical energy. A cathode ray tube (CRT) is a vacuum tube that is used in an oscilloscope, computer monitor, and television. A mass spectrometer is a device used to separate isotopes, or atoms of different masses. It uses the different radii of circular orbits in a magnetic field by particles of the same charge but different mass. In an electronic balance, the magnetic force on a current-carrying wire replaces the known force in a conventional balance.

In a mass spectrometer, a *velocity selector* is used to allow charged particles of only a certain velocity to enter the spectrometer. If a beam of charged particles are not deflected in the presence of a uniform electric field and a magnetic field mutually perpendicular, it must satisfy $v = \dfrac{E}{B_1} = \dfrac{V}{B_1 d}$, where B_1 is the magnetic field in the velocity selector, V the voltage across and d the separations of the plates in the velocity selector. The mass of the particle (m) depends on the radius of the circular orbit (R) in another magnetic field B_2: $m = \dfrac{q d B_1 B_2}{V} R$.

Example 19.9 In a mass spectrometer, a single-charged particle has a speed of 1.0×10^6 m/s and enters a uniform magnetic field of 0.20 T. The radius of the circular orbit is 0.020 m.

(a) What is the mass of the particle?

(b) What is the kinetic energy of the particle?

Solution: Given: $q = 1.6 \times 10^{-19}$ C, $v = 1.0 \times 10^6$ m/s, $B_2 = 0.20$ T, $R = 0.020$ m.

Find: (a) m (b) K.

(a) Since $v = \dfrac{E}{B_1} = \dfrac{V}{B_1 d}$,

$$m = \frac{q d B_1 B_2}{V} R = \frac{q B_2 R}{v} = \frac{(1.6 \times 10^{-19}\ \text{C})(0.20\ \text{T})(0.020\ \text{m})}{1.0 \times 10^6\ \text{m/s}} = 6.4 \times 10^{-27}\ \text{kg.}$$

The particle is likely the helium nucleus, because $4 \times 1.67 \times 10^{-27}$ kg $= 6.4 \times 10^{-27}$ kg.

(b) $K = \frac{1}{2}mv^2 = \frac{1}{2}(6.4 \times 10^{-28}\ \text{kg})(1.0 \times 10^6\ \text{m/s})^2 = 3.2 \times 10^{-16}$ J $= (2.0\ \text{keV})$.

---

*7. The Earth's Magnetic Field (Section 19.7)

The Earth's magnetic field resembles that which would be produced by a large interior bar magnet (with the magnet's south pole near the Earth's north-geographical pole and magnet's north pole near the Earth's south-geographical pole). However, this is not possible because the Earth's interior temperature is above the Curie temperature of ferromagnetic materials. Scientists associate the Earth's magnetic field with motions (electric current) in its liquid outer core.

The magnetic north pole and the geographical south pole do not coincide (nor do the magnetic south pole and geographical north pole), and the magnetic poles "wander" or move about over periods of hundreds of thousands of years. The deviation between the north-south magnetic poles and the south-north geographical poles is called the *magnetic declination*. There is evidence that the Earth's magnetic poles have reversed periodically many times over long geologic time periods.

Charged particles from the Sun and other cosmic rays can be trapped in the Earth's magnetic field, and regions or concentrations of these charged particles are called *Van Allen belts*. It is believed that the recombination of ionized air molecules and electrons that have been ionized by articles from the lower belt give rise to the Aurora Borealis (northern lights) and Aurora Australis (southern lights).

IV.   Mathematical Summary

Magnitude of the Magnetic Force on a Moving Charge Particle	$F = qvB\sin\theta$ (19.3) θ is the angle between **v** and **B**	Computes the magnitude of the magnetic force on a moving charge.
Magnitude of the Magnetic Field near a Long, Straight Current-Carrying Wire	$B = \dfrac{\mu_0 I}{2\pi d}$ (19.4) $\mu_0 = 4\pi \times 10^{-7}$ T·m/A	Calculates the magnitude of the magnetic field due to a long, straight current-carrying wire.
Magnitude of the Magnetic Field at the Center of a Circular Loop of Current-carrying wire	$B = \dfrac{\mu_0 I}{2r}$ (19.5)	Calculates the magnitude of the magnetic field at the center of a circular loop of current-carrying wire.
Magnitude of the Magnetic Field at the Center of a Solenoid (along the axis)	$B = \dfrac{\mu_0 NI}{L}$ (19.6) or $B = \mu_0 nI$ (19.7) where $n = N/L$	Computes the magnitude of the magnetic field along the axis of a long solenoid.
Magnitude of Force on a Straight Current-carrying Wire	$F = ILB\sin\theta$ (19.11)	Computes the magnitude of the magnetic force on a straight current-carrying wire.
Magnitude of the Torque on a Single Current-carrying Loop	$\tau = IAB\sin\theta$ (19.12) where IA is called the magnetic moment, m, of the loop: $m = IA$)	Computes the magnitude of the magnetic torque on a single current-carrying loop.
Magnitude of the Torque on a Current-carrying Coil (of N loops)	$\tau = NIAB\sin\theta$ (19.12)	Computes the magnitude of the magnetic torque on a current-carrying coil of N loops.

V.   Solutions of Selected Exercises and Paired/Trio Exercises

4.   The magnet would attract the unmagnetized iron bar when a pole end is placed at the center of its long side. If the end of the unmagnetized bar were placed at the center of the long side of the magnet, it would not be attracted (field lines perpendicular).

10.   (a) According to the right-hand force rule: $\boxed{\text{(1) negative charge}}$, $\boxed{\text{(2) no charge}}$, $\boxed{\text{(3) positive charge}}$.

   (b) The radius of the circular orbit is proportional to the mass so $\boxed{m_3 > m_1}$.

12.   From $F = qvB \sin \theta$, we have $B = \dfrac{F}{qB \sin \theta} = \dfrac{20 \text{ N}}{(0.25 \text{ C})(2.0 \times 10^2 \text{ m/s}) \sin 90°} = \boxed{0.40 \text{ T}}$

16.   The magnetic field is in the $-z$ direction because electron has negative charge, according to the right-hand force rule.   Since $F = qvB \sin \theta$,

$$B = \frac{F}{qv \sin \theta} = \frac{5.0 \times 10^{-19} \text{ N}}{(1.6 \times 10^{-19} \text{ C})(3.0 \times 10^6 \text{ m/s}) \sin 90°} = \boxed{1.0 \times 10^{-6} \text{ T in } -z \text{ direction}}.$$

18. $\boxed{30° \text{ or } 150°}$.

24. $B = \dfrac{\mu_0 I}{2\pi d} = \dfrac{(4\pi \times 10^{-7} \text{ T·m/A})(2.5 \text{ A})}{2\pi(0.25 \text{ m})} = \boxed{2.0 \times 10^{-6} \text{ T}}$.

28.   (a) The fields by the two wires are equal in magnitude and opposite in direction. So the net field is $\boxed{0}$.

   (b) The fields by the two wires are equal in magnitude and in the same direction. So the net field is

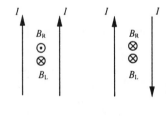

$$B = 2B_R = 2\frac{\mu_0 I}{2\pi d} = \frac{\mu_0 I}{\pi d} = \frac{(4\pi \times 10^{-7} \text{ T·m/A})(4.0 \text{ A})}{\pi(0.25 \text{ m})} = \boxed{6.4 \times 10^{-6} \text{ T}}.$$

29.   The magnitudes at both locations are the same.

   Calculate for the location to the right of I_2. $B = \dfrac{\mu_0 I}{2\pi d}$.

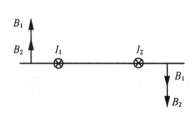

$$B_1 = \frac{(4\pi \times 10^{-7} \text{ T·m/A})(1.5 \text{ A})}{2\pi(0.35 \text{ m})} = 8.57 \times 10^{-7} \text{ T},$$

$$B_2 = \frac{(4\pi \times 10^{-7} \text{ T·m/A})(1.5 \text{ A})}{2\pi(0.15 \text{ m})} = 2.0 \times 10^{-6} \text{ T}.$$

So the net field is $B = B_1 + B_2 = 7.57 \times 10^{-7} \text{ T} + 2.0 \times 10^{-6} \text{ T} = \boxed{2.9 \times 10^{-6} \text{ T}}$.

34. The Earth's magnetic field at the equator is about 10^{-5} T.

From $B = \dfrac{\mu_0 I}{2r}$, we have $I = \dfrac{2rB}{\mu_0} = \dfrac{2(0.10 \text{ m})(10^{-5} \text{ T})}{4\pi \times 10^{-7} \text{ T·m/A}} = \boxed{1.6 \text{ A}}$.

36. The currents are opposite and so the fields by the two loops are also opposite.

$B_1 = \dfrac{\mu_0 I_1}{2r_1} = \dfrac{\mu_0 I_2}{2r_2}$, so $r_2 = \dfrac{I_2}{I_1} r_1 = \dfrac{2.0 \text{ A}}{1.0 \text{ A}} (5.0 \text{ cm}) = \boxed{10 \text{ cm}}$.

40. $d = \sqrt{(a/2)^2 + (a/2)^2} = \dfrac{a}{\sqrt{2}}$.

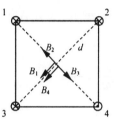

$B_1 = B_2 = B_3 = B_4 = \dfrac{\mu_0 I}{2\pi d} = \dfrac{\mu_0 I}{2\pi(a/\sqrt{2})} = \dfrac{\mu_0 I}{\sqrt{2}\,\pi a}$.

So the net field is $B = B_1 + B_4 = 2\,\dfrac{\mu_0 I}{\sqrt{2}\,\pi a} = \boxed{\dfrac{\sqrt{2}\,\mu_0 I}{\pi a}}$,

at 45° toward the lower-left wire.

47. The electric force provides the centripetal force. $F = \dfrac{kq_1 q_2}{r^2} = \dfrac{ke^2}{r^2} = m\,\dfrac{v^2}{r}$,

so $v = \sqrt{\dfrac{k}{m\,r}}\,e = \sqrt{\dfrac{9.0 \times 10^9 \text{ N·m}^2/\text{C}^2}{(9.11 \times 10^{-31} \text{ kg})(0.053 \times 10^{-9} \text{ m})}} (1.6 \times 10^{-19} \text{ C}) = 2.18 \times 10^6 \text{ m/s}.$

$I = \dfrac{q}{T} = \dfrac{e}{2\pi r/v} = \dfrac{ev}{2\pi r}$. Therefore

$B = \dfrac{\mu_0 I}{2r} = \dfrac{\mu_0\, ev/(2\pi r)}{2r} = \dfrac{\mu_0\, ev}{4\pi r^2} = \dfrac{(4\pi \times 10^{-7} \text{ T·m/A})(1.6 \times 10^{-19} \text{ C})(2.18 \times 10^6 \text{ m/s})}{4\pi(0.053 \times 10^{-9} \text{ m})^2} = \boxed{12 \text{ T}}.$

52. $F = ILB \sin\theta = (5.0 \text{ A})(1.0 \text{ m})(0.30 \text{ T}) \sin 90° = \boxed{1.5 \text{ N, } +y \text{ direction}}$ (taking $+z$ as upward).

57. (a) Since $F = ILB \sin\theta$, $\dfrac{F}{L} = IB \sin\theta = IB \sin 0° = \boxed{0}$.

(b) $\dfrac{F}{L} = (10 \text{ A})(0.40 \text{ T}) \sin 90° = \boxed{4.0 \text{ N/m in } +z}$.

(c) $\dfrac{F}{L} = (10 \text{ A})(0.40 \text{ T}) \sin 90° = \boxed{4.0 \text{ N/m in } -y}$.

(d) $\dfrac{F}{L} = (10 \text{ A})(0.40 \text{ T}) \sin 90° = \boxed{4.0 \text{ N/m in } -z}$.

(e) $\dfrac{F}{L} = (10 \text{ A})(0.40 \text{ T}) \sin 90° = \boxed{4.0 \text{ N/m in } +y}$.

60. $\dfrac{F}{L} = \dfrac{\mu_0 I_1 I_2}{2\pi d} = \dfrac{(4\pi \times 10^{-7}\ \text{T·m/A})(15\ \text{A})^2}{2\pi(0.15\ \text{m})} = \boxed{3.0 \times 10^{-4}\ \text{N/m repulsive}}$. It is repulsive because the

currents are opposite and so the fields in between the wires are in the same direction.

62. $\boxed{1.8 \times 10^{-5}\ \text{N/m repulsive}}$. It is repulsive because the currents are opposite and so the fields in between the

wires are in the same direction.

70. $\theta = 90° - 30° = 60°$ (θ is the angle between the normal to the plane and the field).

$\tau = IAB \sin \theta = (0.25\ \text{A})(0.20\ \text{m}^2)(0.30\ \text{T}) \sin 60° = \boxed{0.013\ \text{m·N}}$.

75. $v = \dfrac{V}{Bd} = \dfrac{E}{B} = \dfrac{8.0 \times 10^3\ \text{V/m}}{0.040\ \text{T}} = \boxed{2.0 \times 10^5\ \text{m/s}}$.

78. $v = \dfrac{V}{Bd} = \dfrac{E}{B} = \dfrac{1.0 \times 10^3\ \text{V/m}}{0.10\ \text{T}} = 1.0 \times 10^4\ \text{m/s}$.

In this circular motion the magnetic force provides the centripetal force. $F = qvB \sin \theta = m\dfrac{v^2}{R}$,

so $m = \dfrac{qBR \sin \theta}{v} = \dfrac{(1.6 \times 10^{-19}\ \text{C})(0.10\ \text{T})(0.012\ \text{m}) \sin 90°}{1.0 \times 10^4\ \text{m/s}} = \boxed{1.9 \times 10^{-26}\ \text{kg}}$.

81. From $K = \tfrac{1}{2}mv^2$, we have $v = \sqrt{\dfrac{2K}{m}} = \sqrt{\dfrac{2(10 \times 10^3\ \text{eV})(1.6 \times 10^{-19}\ \text{J/eV})}{1.67 \times 10^{-27}\ \text{kg}}} = 1.38 \times 10^6\ \text{m/s}$.

In this circular motion the magnetic force provides the centripetal force.

Also $F = qvB \sin \theta = m\dfrac{v^2}{R}$, $B = \dfrac{mv}{qR \sin \theta} = \dfrac{(1.67 \times 10^{-27}\ \text{kg})(1.38 \times 10^6\ \text{m/s})}{(1.6 \times 10^{-19}\ \text{C})(0.50\ \text{m}) \sin 90°} = \boxed{2.9 \times 10^{-2}\ \text{T}}$.

87. $B = \mu_0 \dfrac{N_1}{\ell} I_1 + \mu_0 \dfrac{N_2}{\ell} I_2 = \dfrac{4\pi \times 10^{-7}\ \text{T·m/A}}{0.10\ \text{m}} [(3000)(5.0\ \text{A}) + (2000)(10\ \text{A})] = \boxed{0.44\ \text{T}}$.

92. $v = \dfrac{V}{Bd} = \dfrac{E}{B}$,

$B = \dfrac{E}{v} = \dfrac{100\ \text{V/m}}{2.0 \times 10^2\ \text{m/s}} = \boxed{0.50\ \text{T east}}$ if $\mathbf{v}$ is north and $\mathbf{E}$ is upward.

95. $\tau = IAB \sin \theta = (10\ \text{A})(0.20\ \text{m})(0.30\ \text{m})(0.050\ \text{T}) \sin 90° = \boxed{3.0 \times 10^{-2}\ \text{m·N}}$,

where θ is the angle between the normal to the loop and the field.

VI.   Practice Quiz

1.   An electron moves with a speed of 3.0×10^4 m/s in a uniform magnetic field of 0.20 T. What is the magnitude of the maximum force it can experience?

(a) 2.4×10^{-16} N   (b) 4.8×10^{-16} N   (c) 9.6×10^{-16} N   (d) 1.9×10^{-15} N   (e) zero

2.   A wire of length 2.0 m carrying a current of 0.60 A is oriented in a direction of 35° to a uniform magnetic field of 0.50 T. What is the magnitude of the force it experiences?

(a) zero   (b) 0.30 N   (c) 0.34 N   (d) 0.49 N   (e) 0.60 N

3.   A proton moving along the $+x$ axis enters a region where there is a uniform magnetic field in the $+y$ direction. What is the direction of the magnetic force on the proton, if x axis is to the right, y axis points upward, and $+z$ is out of the page?

(a) $+z$ direction   (b) $-z$ direction   (c) $+y$ direction   (d) $-y$ direction   (e) $-x$ direction

4.   The magnetic field at the center of a solenoid is 25 mT. If the solenoid has 500 turns and is 0.10 m long, what is the current through it?

(a) 0.40 A   (b) 4.0 A   (c) 0.40 mA   (d) 5.0 mA   (e) 8.0 mA

5.   The direction of the force on a current-carrying wire in a magnetic field is described by which of the following?

(a) perpendicular to the current   (b) perpendicular to the magnetic field   (c) parallel to the current
(d) both (a) and (b) are valid   (e) neither (a) nor (b) is valid

6.   A circular loop carrying a current of 2.0 A is in a magnetic field of 0.35 T. The loop has an area of 1.2 m^2 and its plane is oriented at a 53° angle to the field. What is the magnitude of the magnetic torque on the loop?

(a) 46 m·N   (b) 0.67 m·N   (c) 0.51 m·N   (d) 0.10 m·N   (e) zero

7.   Two long parallel wires carry equal currents. The magnitude of the force between the wires is F. The current in each wire is now doubled while the distance between them is halved. What is the magnitude of the new force between the two wires?

(a) F (b) $2F$ (c) $4F$ (d) $8F$ (e) $16F$

8. A proton travels from rest through a voltage of 1.0 kV and then moves into a magnetic field of 0.040 T perpendicular to its velocity. What is the radius of the proton's resulting orbit?

(a) 0.080 m   (b) 0.11 m   (c) 0.14 m   (d) 0.17 m   (e) 0.21 m

9. A thin metal rod 1.0 m long has a mass of 0.050 kg and is in a magnetic field of 0.10 T. What minimum current in the rod is needed in order for the magnetic force to cancel the weight of the rod?

(a) 0.58 A   (b) 1.2 A   (c) 2.5 A   (d) 4.9 A   (e) 9.8 A

10. A stationary electron is in a uniform magnetic field of 0.20 T. What is magnitude of magnetic force on the it?

(a) 1.6×10^{-20} N   (b) 1.6×10^{-21} N   (c) 3.2×10^{-20} N   (d) 3.2×10^{-21} N   (e) zero

Answers to Practice Quiz:

1. c 2. c 3. a 4. b 5. d 6. c 7. d 8. b 9. d 10. e

CHAPTER 20

Electromagnetic Induction

I.   Chapter Objectives

Upon completion of this chapter, you should be able to:

1.   define magnetic flux and explain how induced emf's are created by changing magnetic flux, and calculate the magnitude and predict the polarity of an induced emf.

2.   understand the operation of electrical generators and calculate the emf produced by an ac generator, and explain the origin of back emf and its effect on the behavior of motors.

3.   explain transformer action in terms of Faraday's law, calculate the output of step-up and step-down transformers, and understand the importance of transformers in electric energy delivery systems.

4.   explain the physical nature, origin, and means of propagation of electromagnetic waves, and describe the properties and uses of various types of electromagnetic waves.

II.   Key Terms

Upon completion of this chapter, you should be able to define and/or explain the following key terms:

electromagnetic induction	primary coil
magnetic flux	secondary coil
Faraday's law of induction	step-up transformer
Lenz's law	step-down transformer
alternating current (ac)	eddy currents
ac generator	electromagnetic waves (radiation)
back emf	Maxwell's equations
transformer	radiation pressure

The definitions and/or explanations of the most important key terms can be found in the following section: **III. Chapter Summary and Discussion.**

III. Chapter Summary and Discussion

1. Induced Emf's: Faraday's Law and Lenz's Law (Section 20.1)

Magnetic flux is a measure of the number of magnetic field lines through a particular loop area: $\Phi = BA \cos \theta$, where B is the magnetic field, A the area of the loop, and θ the angle between the magnetic field vector and the normal to the loop area. The SI unit of magnetic flux is T·m^2.

Note: The angle θ is the angle between the magnetic field and the *normal to the plane of the loop*, *not* the angle between the magnetic field and the plane of the loop.

Electromagnetic induction refers to the creation of induced emf's whenever the magnetic flux through a loop or coil changes. The magnitude of induced emf in a conducting coil depends on the time rate change of magnetic flux through the loop, $\mathcal{E} = -N \dfrac{\Delta\Phi}{\Delta t}$, where N is the number of loops. This relationship is known as **Faraday's law of induction**. Since $\Phi = BA \cos \theta$, any change in B, A, or θ will result in an induced emf. Faraday's law can be written in detail as $\mathcal{E} = -N \left[\dfrac{\Delta B}{\Delta t} A \cos \theta + B \dfrac{\Delta A}{\Delta t} \cos \theta + BA \dfrac{\Delta(\cos \theta)}{\Delta t} \right]$. There are three different ways to produce induced emf.

When a conductor of length L moves perpendicular to a magnetic field B with a speed v, the magnitude of the induced emf is called *motional emf* and is given by $\mathcal{E} = BLv$.

The minus sign in Faraday's law of induction gives the polarity of the induced emf, which is found by **Lenz's law:** when a change in magnetic flux induces an emf in a conducting loop, the induced emf produces a current in such a direction as to create a magnetic field that tends to oppose the change in flux. That is, an induced emf gives rise to a current whose magnetic field opposes the change in magnetic flux that produced it.

Example 20.1   A circular loop of radius 0.20 m is rotating in a uniform magnetic field of 0.20 T. Find the magnetic flux through the loop when the plane of the loop and the magnetic field vector
(a) are parallel, (b) are perpendicular, (c) are at 60°.

Solution: Given: $r = 0.10$ m, $B = 0.20$ T,   (a) $\theta = 90°$,   (b) $\theta = 0°$,   (c) $\theta = 90° - 60° = 30°$.
Find: Φ in (a), (b), and (c).

The angle θ in the magnetic flux definition is the angle between the magnetic field and the *normal* to the plane of the loop, *not* the angle between the magnetic field and the plane of the loop.

(a) $\Phi = BA \cos \theta = B\pi r^2 \cos \theta = (0.20 \text{ T})(\pi)(0.20 \text{ m})^2 \cos 90° = 0$.

(b) $\Phi = (0.20\text{ T})(\pi)(0.20\text{ m})^2 \cos 0° = 2.5 \times 10^{-2}\text{ T·m}^2$.

(c) $\Phi = (0.20\text{ T})(\pi)(0.20\text{ m})^2 \cos 30° = 2.2 \times 10^{-2}\text{ T·m}^2$.

Example 20.2 A coil is wrapped with 100 turns of wire on a square frame with sides 18 cm. A magnetic·field is applied perpendicular to the plane of the coil. If the field changes uniformly from 0 to 0.50 T in 8.0 s, find the average value of the magnitude of the induced emf.

Solution: Given: $N = 100,\quad d = 0.18\text{ m},\quad T_i = 0,\quad T_f = 0.50\text{ T},\quad \Delta t = 8.0\text{ s},\quad \theta = 0°$.

Find: $\mathscr{E}$.

From Faraday's law of induction, the magnitude of the induced emf is

$$\mathscr{E} = N\frac{\Delta B}{\Delta t}\, A \cos\theta = (100)\,\frac{0.50\text{ T} - 0}{8.0\text{ s}}\,(0.18\text{ m})^2 \cos 0° = 0.21\text{ V}.$$

Both $\dfrac{\Delta A}{\Delta t}$ and $\dfrac{\Delta(\cos\theta)}{\Delta t}$ are zero because A and θ do not change.

Example 20.3 A square coil of wire with 15 turns and an area of 0.40 m² is placed parallel to a magnetic field of 0.75 T. The coil is flipped so its plane is perpendicular to the magnetic field in 0.050 s. What is the magnitude of the average induced emf?

Solution: For B, A, and θ in this Example, only θ changes.

Given: $N = 15,\quad A = 0.40\text{ m}^2,\quad \Delta t = 0.050\text{ s},\quad B = 0.75\text{ T},\quad \theta_i = 90°,\quad \theta_f = 0°$.

Find: $\mathscr{E}$.

From Faraday's law of induction, the magnitude of the induced emf is

$$\mathscr{E} = NBA\,\frac{\Delta(\cos\theta)}{\Delta t} = (15)(0.75\text{ T})(0.40\text{ m}^2)\,\frac{\cos 0° - \cos 90°}{0.050\text{ s}} = 90\text{ V}.$$

Both $\dfrac{\Delta B}{\Delta t}$ and $\dfrac{\Delta A}{\Delta t}$ are zero because B and A do not change.

Example 20.4 An airplane with a wing span of 50 m flies horizontally with a speed of 200 m/s above the Earth at a location where the downward component of the Earth's magnetic field is 6.0×10^{-5} T. Find the magnitude of the induced emf between the tips of the wing.

Solution: Given: $L = 50\text{m},\quad B = 6.0 \times 10^{-5}\text{ T},\quad v = 200\text{ m/s}$.

Find: $\mathscr{E}$.

The magnitude of the motional emf is $\mathscr{E} = BLv = (6.0 \times 10^{-5}\text{ T})(50\text{ m})(200\text{ m/s}) = 0.60\text{ V}$.

Example 20.5   Find the direction of the induced current through R in the figures shown.

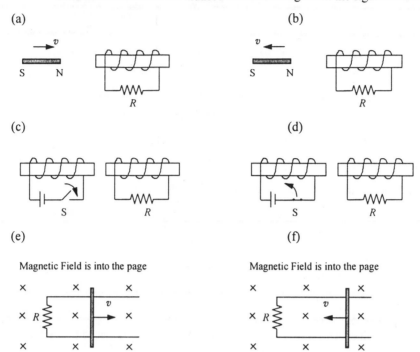

Solution:

We know from Lenz's law that the magnetic field produced by the induced current will tend to oppose or resist the change in flux.

(a) As the north pole of the magnet approaches the coil, the field lines going into the coil increase to the right. The induced current in the coil produces a magnetic field that opposes the increase, so the left end of the coil will be a north pole. By the right-hand source rule, the induced current is from left to right through the resistor.

(b) As the north pole of the magnet moves away from the coil, the field lines going into the coil decrease. The induced current in the coil produces a magnetic field that opposes the decrease, so the left end of the coil will be a south pole. By the right-hand source rule, the induced current is from right to left through the resistor.

Now try to repeat parts (a) and (b) by flipping the magnet, that is, south pole near the coil.

(c) When the switch is closed, the magnetic field due to the coil on the left increases, so the field lines going through the coil on the right increase. The induced current in the coil on the right will produce a magnetic field to oppose that increase, so the left end of the coil on the right will be a south pole. Therefore, the current is from right to left.

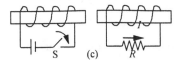 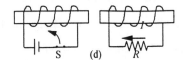

(d) When the switch is opened, the magnetic field due to the coil on the left decreases, so the field lines going through the coil on the right decrease. The induced current in the coil on the right will produce a magnetic field to oppose that decrease, so the left end of the coil on the right will be a north pole. Therefore, the current is from left to right.

(e) As the rod moves to the right, the area of the loop increases, so the magnetic flux through the area also increases. The induced current in the loop will produce a magnetic field to oppose that increase, so the magnetic field due to the induced current is out of the page. By the right-hand source rule, the current is from bottom to top through the rod.

Magnetic Field is into the page   Magnetic Field is into the page

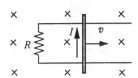 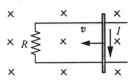

(f) As the rod moves to the left, the area of the loop decreases, so the magnetic flux through the area also decreases into the page. The induced current in the loop will produce a magnetic field to oppose that decrease, so the magnetic field due to the induced current is into the page. The current is from top to bottom through the rod.

---

2. Generators and Back Emf (Section 20.2)

A **generator** is a device to convert mechanical energy into electrical energy. In principle, it is the reverse of an electrical motor. An **ac generator** or alternator produces **alternating current** (ac) meaning that the polarity of the voltage and direction of the current change periodically. A **dc generator** produces direct current (dc). When coils are rotated in a magnetic field (or the field is rotated while the coils are kept fixed), an emf is induced in the coils and can be used to create an electric current. This creation of emf and current is due to the term $\dfrac{\Delta(\cos\theta)}{\Delta t}$ in Faraday's law of induction. The emf thus produced in an ac generator is expressed as $\mathscr{E} = \mathscr{E}_0 \sin \omega t = \mathscr{E}_0 \sin 2\pi f t$, where $\omega = 2\pi f$ is the angular frequency and $\mathscr{E}_0 = NBA\omega$ the maximum emf (N is the number of turns of the coil, B the magnetic field, A the area of the coil).

A **back emf** is a reverse emf created by induction in motors when their armature is rotated in a magnetic field. The direction is opposite to the applied emf due to Lenz's law. The back emf partially cancels the voltage that drives the motor, so the current through the armature when the armature is rotating (operating) is smaller than when it is not (start-up). When a motor is in operating speed, the back emf is given by $\mathcal{E}_b = V - IR$, where V is the applied driving voltage, I the current through the armature, and R the resistance of the armature.

Example 20.6   An ac generator consists of 200 turns of wire of area 0.090 m² and total resistance 12 Ω. The loop rotates in a magnetic field of 0.50 T at a constant angular speed of 60 revolutions per second. Find the maximum induced emf and the maximum current.

Solution: Given: $N = 200$, $A = 0.090$ m², $R = 12$ Ω, $B = 0.50$ T, $f = 60$ Hz.
Find: $\mathcal{E}_o$.

Since the generator is rotating at 60 rev/s, the frequency is 60 cycles/s = 60 Hz.

The angular frequency is $\omega = 2\pi f = 2\pi(60 \text{ Hz}) = 120\pi$ rad/s.

So the maximum emf is $\mathcal{E}_o = NAB\omega = (200)(0.090 \text{ m}^2)(0.50 \text{ T})(120\pi \text{ rad/s}) = 3.4 \times 10^3$ V = 3.4 kV.

Example 20.7   A dc motor of internal resistance 5.0 Ω is connected to a 24 V power supply. The operating current is 2.0 A.

(a) What is the start-up current?

(b) What is the back emf when the motor is running at full speed?

(c) What is the back emf when the motor is running at half speed?

Solution: Given: $R = 5.0$ Ω, $V = 24$ V, $I = 2.0$ A.
Find: (a) I (start-up)   (b) $\mathcal{E}_b$.

(a) At start-up, there is no back emf since the motor is not rotating.

So the current is $I = \dfrac{V}{R} = \dfrac{24 \text{ V}}{5.0 \text{ Ω}} = 4.8$ A.

(b) At full operating speed, the back emf is equal to $\mathcal{E}_b = V - IR = 24 \text{ V} - (2.0 \text{ A})(5.0 \text{ Ω}) = 14$ V.

(c) From Faraday's law or the generator equation, the induced emf (back emf is simply an induced emf, after all) is directly proportional to the angular speed, so at half speed, the back emf is also halved to $\dfrac{14 \text{ V}}{2}$

$= 7.0$ V. The current at this speed is $I = \dfrac{V - \mathcal{E}_b}{R} = \dfrac{24 \text{ V} - 7.0 \text{ V}}{5.0 \text{ Ω}} = 3.4$ A.

---

3. Transformers and Power Transmission (Section 20.3)

A **transformer** is a device that changes the ac voltage and current supplied to it by means of induction. (A transformer cannot work with dc voltage. Why?) It consists of two coils, a **primary coil** (the coil on the input side) and a **secondary coil** (the coil on the output side) and usually an iron core. If the voltage on the secondary is higher than that on the primary, it is a **step-up transformer**. A **step-down transformer** has lower voltage on the secondary side. Whether a transformer is a step-up or step-down transformer is solely determined by the relative number of turns in the primary and secondary coils. A step-up transformer has more turns in the secondary, while a step-down transformer has fewer turns in the secondary. According to the conservation of energy, if the voltage is increased, the current is reduced, and vice versa. A real (in contrast to an ideal one) transformer has $I^2 R$ energy losses in the windings and another mechanism of energy loss caused by **eddy currents** in the transformer core. The number of turns (N), voltage (V), and current (I) ratios are related by $\dfrac{I_p}{I_s} = \dfrac{V_s}{V_p} = \dfrac{N_s}{N_p}$, where the subscript p stands for primary and s for secondary.

In power transmission, the voltage is stepped up before transmission so as to decrease the current and the $I^2 R$ losses (Joule heating) in the transmission lines (see Example 20.8 below). The voltage is then stepped down before the power is supplied to businesses and homes.

Example 20.8 A generator produces 60 A of current at 120 V. The voltage is stepped up to 4500 V by a transformer and transmitted through a power line of total resistance 1.0 Ω. Find:

(a) the number of turns in the secondary coil if the primary coil has 200 turns,

(b) the power lost in the transmission line,

(c) the power that would lost in the transmission line if no transformer is used.

Solution: Given: $I_p = 60$ A, $V_p = 120$ V, $V_s = 4500$ V, $R = 1.0$ Ω, (a) $N_p = 200$.

Find: (a) N_s (b) P_{loss} (c) P_{loss} (no transformer).

(a) From $\dfrac{I_p}{I_s} = \dfrac{V_s}{V_p} = \dfrac{N_s}{N_p}$, we have $N_s = \dfrac{V_s}{V_p} N_p = \dfrac{4500 \text{ V}}{120 \text{ V}} (200) = 7500$ turns.

(b) $I_s = \dfrac{V_p}{V_s} I_p = \dfrac{120 \text{ V}}{4500 \text{ V}} (60 \text{ A}) = 1.6$ A. So the total power loss in the transmission lines is

$P_{loss} = I^2 R = (1.6 \text{ A})^2 (1.0 \text{ Ω}) = 2.6$ W. The total power generated is $P_{tot} = (60 \text{ A})(120 \text{ V}) = 7200$ W.

Therefore the percentage loss is $\dfrac{2.6 \text{ W}}{7200 \text{ W}} = 0.036\%$.

(c) If no transformer is used, the current in transmission is 60 A.

The total power loss is $P_{loss} = (60 \text{ A})^2 (1.0 \text{ Ω}) = 3600$ W.

So the percentage loss is $\dfrac{3600 \text{ W}}{7200 \text{ W}} = 50\%$!

---

4. Electromagnetic Waves (Section 20.4)

An **electromagnetic wave** (or electromagnetic **radiation**) consists of mutually perpendicular, time-varying electric and magnetic fields that propagate at a constant speed in vacuum (the speed of light, $c = 3.00 \times 10^8$ m/s). **Maxwell's equations** are a set of four equations that describe all magnetic and electric field phenomena. Radiation carries energy and momentum and thus can exert force and therefore **radiation pressure** (radiation force per unit area).

The different types of electromagnetic radiation differ only in frequency f, and thus in wavelength λ, because they all travel at the same speed c, in a vacuum and $c = \lambda f$. The major types of electromagnetic radiation in order of increasing frequency or decreasing wavelength are power waves, radio and TV waves, microwaves, infrared radiation, visible light, ultraviolet radiation, X-rays, and gamma rays.

Example 20.9   A Doppler radar indicates that a transmitted pulse is returned by clouds as an echo 40 μs after transmission. How far away are the clouds?

Solution: Given: $\Delta t_{tot} = 40\ \mu s = 40 \times 10^{-6}$ s, $c = 3.00 \times 10^8$ m/s.

Find: d.

Since the time is for the echo (forward and back), the time for one-way distance is $\Delta t_{1/2} = 20 \times 10^{-6}$ s. So the distance is $d = c \Delta t_{1/2} = (3.00 \times 10^8\ \text{m/s})(20 \times 10^{-6}\ \text{s}) = 6.0 \times 10^3$ m = 6.0 km.

Example 20.10   The call number of a radio station indicates the frequency of the carrier wave. For the AM band, the frequency is expressed in kilohertz (kHz) and for the FM band, the frequency is in megahertz (MHz). What are the wavelengths for the carrier waves in the following two radio stations?
(a) WBRN 1460 (AM)   (b) WCKC 107 (FM)

Solution: Given: $c = 3.00 \times 10^8$ m/s,   (a) $f = 1460$ kHz,   (b) $f = 107$ MHz.

Find: λ in (a) and (b).

(a) From $c = \lambda f$, we have $\lambda = \dfrac{c}{f} = \dfrac{3.00 \times 10^8\ \text{m/s}}{1460 \times 10^3\ \text{Hz}} = 205$ m.

(b) $\lambda = \dfrac{3.00 \times 10^8\ \text{m/s}}{107 \times 10^6\ \text{Hz}} = 2.80$ m.

The length of the antenna of a radio station is determined by wavelength of the emitted radio waves. By knowing the wavelength we want, we can construct the antenna to the right length.

IV. Mathematical Summary

Magnetic Flux	$\Phi = BA\cos\theta$ (20.1)	Defines the magnetic flux through an area. θ is the angle between the normal to the plane of the loop and the magnetic field vector
Faraday's Law of Induction	$\varepsilon = -N\dfrac{\Delta\Phi}{\Delta t}$ (20.2) $= -N\left(\dfrac{\Delta B}{\Delta t}\right)(A\cos\theta)$ $-NB\left(\dfrac{\Delta A}{\Delta t}\right)(\cos\theta)$ $-NBA\left(\dfrac{\Delta(\cos\theta)}{\Delta t}\right)$ (20.3)	Calculates the induced emf as the change in magnetic flux divided by the change in time.
Generator emf	$\varepsilon = \varepsilon_0\sin\omega t$ (20.4) $\varepsilon = \varepsilon_0\sin 2\pi ft$ (20.5) where $\varepsilon_0 = NBA\omega$	Expresses the generated emf as a function of time.
Back emf in a motor	$\varepsilon_b = V - IR$ (20.6)	Computes the back emf in an electric motor.
Currents, Voltages, and Turn Ratios for a Transformer	$\dfrac{I_p}{I_s} = \dfrac{V_s}{V_p} = \dfrac{N_s}{N_p}$ (20.9)	Relates the current, voltage, and turn ratio in a transformer.

V. Solutions of Selected Exercises and Paired/Trio Exercises

3. (a) When the bar magnet enters the coil, the needle deflects to one side and when the magnet leaves the coil, the needle reverses direction.

   (b) $\boxed{\text{No}}$, by Lenz's law, it is repelled moving downward toward the loop, and attracted as it leaves the loop.

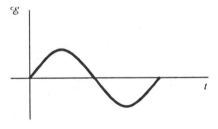

8. $\Phi = BA\cos\theta$, where θ is the angle between the field and the normal to the loop.

   (a) $\Phi = BA\cos 90° = \boxed{0}$.

   (b) $\Phi = (0.30\text{ T})(0.015\text{ m}^2)\cos(90° - 37°) = \boxed{2.7\times10^{-3}\text{ T·m}^2}$.

   (c) $\Phi = (0.30\text{ T})(0.015\text{ m}^2)\cos 0° = \boxed{4.5\times10^{-3}\text{ T·m}^2}$.

11. The length of the third side of the right triangle is $\sqrt{(0.500 \text{ m})^2 - (0.400 \text{ m})^2} = 0.300 \text{ m}$.

$\Phi = BA \cos \theta = (0.550 \text{ T})(\tfrac{1}{2})(0.300 \text{ m})(0.400 \text{ m}) \cos 0° = \boxed{3.3 \times 10^{-2} \text{ T·m}^2}$.

16. $\mathscr{E} = -N\dfrac{\Delta\Phi}{\Delta t} = -N\dfrac{A\Delta B}{\Delta t} = -(60)\dfrac{5.0 \text{ Wb} - 35 \text{ Wb}}{0.10 \text{ s}} = 1.8 \times 10^4 \text{ V}$.

So $R = \dfrac{\varepsilon}{I} = \dfrac{1.8 \times 10^4 \text{ V}}{3.6 \times 10^3 \text{ A}} = \boxed{5.0 \ \Omega}$.

18. $\boxed{0.35 \text{ T}}$.

22. $v = 320 \text{ km/h} = 88.9 \text{ m/s}$. From Example 20.3 in the text book (page 640), the magnitude is

$\mathscr{E} = -BLv = -(5.0 \times 10^{-5} \text{ T})(30 \text{ m})(88.9 \text{ m/s}) = -0.13 \text{ V}$, i.e., $\boxed{0.13 \text{ V}}$.

27. $\mathscr{E} = -N\dfrac{\Delta\Phi}{\Delta t} = -N\dfrac{A\Delta B}{\Delta t} = -(1)\dfrac{(\pi)(0.20 \text{ m})^2 (0 - 14 \times 10^{-3} \text{ T})}{0.25 \text{ s}} = 7.04 \times 10^{-3} \text{ V}$.

$R = \dfrac{\rho L}{A} = \dfrac{(1.70 \times 10^{-8} \ \Omega\text{·m})(2\pi)(0.20 \text{ m})}{\pi[(0.8118 \times 10^{-3} \text{ mm})/2]^2} = 0.0413 \ \Omega$.

$E = Pt = \dfrac{\mathscr{E}^2}{R}t = \dfrac{(7.04 \times 10^{-3} \text{ V})^2}{0.0413 \ \Omega}(0.25 \text{ s}) = \boxed{3.0 \times 10^{-4} \text{ J}}$.

32. (a) After half a period the voltage will be maximum (magnitude, the voltage is actually negative) again.

$t = \dfrac{T}{2} = \dfrac{1}{2f} = \dfrac{1}{2(60 \text{ Hz})} = \boxed{1/120 \text{ s}}$.

(b) After one-quarter of a period the voltage will be zero.

$t = \dfrac{T}{4} = \dfrac{1}{4(60 \text{ Hz})} = \boxed{1/240 \text{ s}}$.

(c) After one period, the value returns. $t = \dfrac{1}{60 \text{ Hz}} = \boxed{1/60 \text{ s}}$.

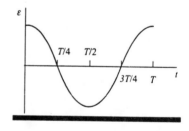

38. $\mathscr{E}_0 = NBA\omega = 2\pi NBAf = 2\pi(20)(0.800 \text{ T})(\pi)(0.10 \text{ m})^2 (60 \text{ Hz}) = \boxed{1.9 \times 10^2 \text{ V}}$.

This maximum value (positive or negative) is attained every half of a period.

$t = \dfrac{T}{2} = \dfrac{1}{2f} = \dfrac{1}{2(60 \text{ Hz})} = \boxed{1/120 \text{ s}}$.

40. $\boxed{\text{Student B's}}$ generates the greater maximum emf.

42. (a) From $\mathscr{E}_b = V - IR$, we have $I = \dfrac{V - \mathscr{E}_b}{R} = \dfrac{12 \text{ V} - 10 \text{ V}}{0.40 \text{ }\Omega} = \boxed{5.0 \text{ A}}$.

(b) When initially starting up, there is no back emf.

So $I = \dfrac{12 \text{ V}}{0.40 \text{ }\Omega} = \boxed{30 \text{ A}}$.

48. (a) $\boxed{\text{Step-up}}$ since $N_s > N_p$.

(b) $\dfrac{I_p}{I_s} = \dfrac{V_s}{V_p} = \dfrac{N_s}{N_p} = \dfrac{450}{75} = \boxed{6:1}$

(c) $\dfrac{V_p}{V_s} = \dfrac{N_p}{N_s} = \dfrac{75}{450} = \boxed{1:6}$.

53. (a) Since $\dfrac{I_p}{I_s} = \dfrac{V_s}{V_p} = \dfrac{N_s}{N_p}$, $I_s = \dfrac{N_p}{N_s} I_p = \dfrac{840}{120}(2.50 \text{ A}) = \boxed{17.5 \text{ A}}$.

(b) $V_s = \dfrac{110}{840}(120 \text{ V}) = \boxed{15.7 \text{ V}}$.

56. (a) From $\dfrac{I_p}{I_s} = \dfrac{V_s}{V_p} = \dfrac{N_s}{N_p}$, we have $N_s = \dfrac{V_s}{V_p} N_p = \dfrac{20 \text{ V}}{120 \text{ V}}(300) = \boxed{50 \text{ turns}}$.

(b) $I_p = \dfrac{50}{300}(0.50 \text{ A}) = \boxed{8.3 \times 10^{-2} \text{ A}}$.

61. (a) At the area substation, $\dfrac{N_s}{N_p} = \dfrac{V_s}{V_p} = \dfrac{100\,000 \text{ V}}{200\,000 \text{ V}} = \boxed{1:2}$;

at the distributing station, $\dfrac{N_s}{N_p} = \dfrac{7200 \text{ V}}{100\,000 \text{ V}} = \boxed{1:14}$;

at the utility pole, $\dfrac{N_s}{N_p} = \dfrac{240 \text{ V}}{7200 \text{ V}} = \boxed{1:30}$.

(b) At the area substation, $\dfrac{I_s}{I_p} = \dfrac{V_p}{V_s} = \dfrac{200\,000 \text{ V}}{100\,000 \text{ V}} = \boxed{2.0}$;

at the distributing station, $\dfrac{I_s}{I_p} = \dfrac{100\,000 \text{ V}}{7200 \text{ V}} = \boxed{14}$;

at the utility pole, $\dfrac{I_s}{I_p} = \dfrac{7200 \text{ V}}{240 \text{ V}} = \boxed{30}$.

(c) Overall, $\dfrac{I_s}{I_p} = \dfrac{200\,000 \text{ V}}{240 \text{ V}} = \boxed{833}$.

68. (a) $f = \dfrac{c}{\lambda} = \dfrac{3.00 \times 10^8 \text{ m/s}}{2.0 \text{ m}} = \boxed{1.5 \times 10^8 \text{ Hz}}$.

(b) $f = \dfrac{3.00 \times 10^8 \text{ m/s}}{25 \text{ m}} = \boxed{1.2 \times 10^7 \text{ Hz}}$.

(c) $f = \dfrac{3.00 \times 10^8 \text{ m/s}}{75 \text{ m}} = \boxed{4.0 \times 10^6 \text{ Hz}}$.

73. For AM: $L = \dfrac{\lambda}{4} = \dfrac{c/f}{4} = \dfrac{c}{4f} = \dfrac{3.00 \times 10^8 \text{ m/s}}{4(0.53 \times 10^6 \text{ Hz} + 1.7 \times 10^6 \text{ Hz})/2} = \boxed{67 \text{ m}}$.

For FM, $L = \dfrac{3.00 \times 10^8 \text{ m/s}}{4(88 \times 10^6 \text{ Hz} + 108 \times 10^6 \text{ Hz})/2} = \boxed{0.77 \text{ m}}$.

77. From $\Phi = BA \cos\theta = B\pi r^2 \cos\theta$,

we have $r = \sqrt{\dfrac{\Phi}{\pi B \cos\theta}} = \sqrt{\dfrac{0.12 \text{ T·m}^2}{\pi (0.150 \text{ T}) \cos 45°}} = \boxed{0.60 \text{ m}}$.

81. (a) $\boxed{\text{Up}}$.

(b) $I = \dfrac{\mathcal{E}}{R} = \dfrac{BLv}{R} = \dfrac{(0.250 \text{ T})(0.50 \text{ m})(2.0 \text{ m/s})}{10\ \Omega} = 2.5 \times 10^{-2} = \boxed{25 \text{ mA}}$.

87. $I_s = \dfrac{P_s}{V_s} = \dfrac{6.6 \times 10^3 \text{ W}}{220 \text{ V}} = 30 \text{ A}$;　$I_p = \dfrac{V_s}{V_p} I_s = \dfrac{220 \text{ V}}{20\,000 \text{ V}} (30 \text{ A}) = 0.33 \text{ A}$.

So the currents are $\boxed{0.33 \text{ A and 30 A}}$.

VI. Practice Quiz

1. According to Lenz's law, the direction of an induced current in a conductor will be that which tends to produce which of the following effects?

(a) enhance the effect which produces it　　(b) oppose the effect which produces it

(c) produce the greatest voltage　　(d) produce a greater heating effect

(e) none of the above

2. A dc motor with an internal resistance of 10 Ω operating on a 12 V source produces a back emf of 9.0 V. What is the current through the motor?

(a) zero　(b) 0.30 A　(c) 0.90 A　(d) 1.2 A　(e) 2.1 A

3.	A uniform magnetic field of 0.50 T passes perpendicularly through the plane of a wire loop 0.30 m^2 in area. What is the magnetic flux through the loop?

(a) 1.7 Wb   (b) 0.80 Wb   (c) 0.60 Wb   (d) 0.15 Wb   (e) zero

4.	A 0.50-meter long wire is moved perpendicularly to a magnetic field of 0.30 T at a speed of 12 m/s. What emf is induced across the ends of the wire?

(a) zero   (b) 0.15 V   (c) 1.8 V   (d) 3.6 V   (e) 6.0 V

5.	A bar magnet falls through a loop of wire with the south pole entering first. When the south pole enters the wire, the induced current will be (as viewed from above)

(a) clockwise.   (b) counterclockwise.   (c) zero.   (d) to right of loop.   (e) to left of loop.

6.	A coil consists of five loops of wire, each with an area of 0.20 m^2. It is oriented with its plane perpendicular to a magnetic field that increases uniformly from 1.0×10^{-2} T to 2.5×10^{-2} T in a time of 5.0×10^{-3} s. What is the induced emf in the coil?

(a) zero   (b) 0.60 V   (c) 2.0 V   (d) 3.0 V   (e) 5.0 V

7.	You are designing a generator with a maximum emf 24 V. If the generator coil has 100 turns and a cross-sectional area of 0.030 m^2, what would be the required magnetic field if the frequency is 60 Hz?

(a) zero   (b) 0.021 T   (c) 0.042 T   (d) 0.067 T   (e) 0.13 T

8.	A transformer consists of a 500-turn primary coil and a 2000-turn secondary coil. If the voltage in the secondary is 4.8 V, what is the primary voltage?

(a) 0.30 V   (b) 0.60 V   (c) 1.2 V   (d) 2.4 V   (e) 19 V

9.	An electromagnetic wave is made up of which of the following time-varying quantities?

(a) electrons only   (b) electric fields only   (c) magnetic fields only

(d) both electric and magnetic fields   (e) neither electric fields nor magnetic fields

10.	What is the frequency of a radio wave signal transmitted at a wavelength of 40 m?

(a) 7.5 Hz   (b) 750 Hz   (c) 7.5 kHz   (d) 75 KHz   (e) 7.5 MHz

Answers to Practice Quiz:

1.b 2.b 3.d 4.c 5.a 6.c 7.b 8.c 9.d 10.e

CHAPTER 21

AC Circuits

I. Chapter Objectives

Upon completion of this chapter, you should be able to:

1. specify how voltage, current, and power vary with time in an ac circuit, understand the concepts of rms and peak values, and learn how resistors respond under ac conditions.

2. explain the behavior of capacitors in ac circuits and calculate their effect on ac current (capacitive reactance).

3. explain what an inductor is, explain the behavior of inductors in ac circuits, and calculate the effect of inductors on ac circuits (inductive reactance).

4. calculate the currents and voltages when various reactive circuit elements are present in ac circuits, use phase diagrams to calculate overall impedance and rms currents, and understand and use the concept of the power factor in ac circuits.

5. understand the concept of resonance in ac circuits and calculate the resonance frequency of an RLC circuit.

II. Key Terms

Upon completion of this chapter, you should be able to define and/or explain the following key terms:

peak voltage	phase diagram
peak current	phasors
rms (or effective) current	impedance
rms (or effective) voltage	phase angle
capacitive reactance	inductive circuit
inductance	capacitive circuit
inductor	power factor
henry (H)	resonance frequency
inductive reactance	

The definitions and/or explanations of the most important key terms can be found in the following section: **III. Chapter Summary and Discussion.**

III. Chapter Summary and Discussion

1. Resistance in an AC Circuit (Section 21.1)

In an ac circuit, voltage and current vary sinusoidally (or voltages and currents are sine or cosine functions) with time. The instantaneous voltage is given by $V = V_0 \sin \omega t = V_0 \sin 2\pi ft$, where V_0 is the **peak voltage** which is the maximum voltage value attained during a cycle of oscillation, and $\omega = 2\pi f$ is the angular frequency.

In an ac circuit with one resistor, the current through that resistor is equal to $I = \dfrac{V}{R} = \dfrac{V_0 \sin 2\pi ft}{R}$

$= \dfrac{V_0}{R} \sin 2\pi ft = I_0 \sin 2\pi ft$, where I_0 is the **peak current** (maximum current value). The voltage across and current through the resistor are in step, or *in phase*, with each other, both reaching zero, minima, and maxima at the same time.

The voltage and current in ac circuits are often given in terms of the **rms voltage** and **rms current**, unless otherwise specified. These are statistical averages of the **effective** *ac voltage and current*. The rms values are less than the peak values, and are related to the peak values by $V_{rms} = \dfrac{V_0}{\sqrt{2}} \approx 0.707\, V_0$ and $I_{rms} = \dfrac{I_0}{\sqrt{2}} \approx 0.707\, I_0$.

The rms voltage and current are related by $V_{rms} = I_{rms} R$ and the average power dissipated by a resistor is equal to $\overline{P} = I_{rms} V_{rms} = I_{rms}^2 R$. Appliances using ac voltage are rated in terms of their average power.

Example 21.1   A computer power supply rated at 200 W is connected to a 120-volt outlet.

(a) What are the rms and peak currents through the power supply?

(b) What is the resistance of the power supply?

Solution: Given: $\overline{P} = 200$ W, $V_{rms} = 120$ V.

Find: (a) I_{rms} and I_0 (b) R.

(a) From $\overline{P} = V_{rms} I_{rms}$, we have $I_{rms} = \dfrac{\overline{P}}{V_{rms}} = \dfrac{200\text{ W}}{120\text{ V}} = 1.67$ A.

The peak current is $I_0 = \sqrt{2}\, I_{rms} = \sqrt{2}\,(1.67\text{ A}) = 2.36$ A.

(b) $R = \dfrac{V_{rms}}{I_{rms}} = \dfrac{120\text{ V}}{1.67\text{ V}} = 71.9\ \Omega$.

Example 21.2   A heating element of resistance 10 Ω is connected to a 120-volt ac power supply.

   (a) What is the peak current through the resistor?

   (b) What is the average power dissipated in the resistor?

Solution: Given: $R = 10\ \Omega$, $V_{rms} = 120$ V.

         Find: (a) I_0 (b) $\overline{P}$.

(a) $I_{rms} = \dfrac{V_{rms}}{R} = \dfrac{120\ V}{10\ \Omega} = 12$ A.

So the peak current is $I_0 = \sqrt{2}\ I_{rms} = \sqrt{2}$ (12 A) = 17 A.

(b) $\overline{P} = I_{rms}V_{rms} = $ (12 A)(120 V) $= 1.44 \times 10^3$ W $= 1.4$ kW.

Note: For convenience and simplicity, the rms subscripts will be omitted from now on — but keep in mind that rms values are given for ac quantities unless otherwise specified.

2.   Capacitive Reactance (Section 21.2)

   A capacitor in an ac circuit limits, or impedes the current, but does not completely prevent the flow of charge as it does in dc circuits. This impeding effect is expressed in terms of **capacitive reactance** (X_C) given by

$X_C = \dfrac{1}{2\pi fC} = \dfrac{1}{\omega C}$. The SI unit of capacitive reactance is ohm (Ω). Capacitive reactance is frequency dependent

(inversely proportional to frequency) and is analogous to resistance in the sense that the voltage across, and current through, the capacitor are related by a more general form of Ohm's law, $V = IX_C$.

   In a purely capacitive ac circuit (capacitor is the only circuit element), the *current leads the voltage* by 90° or a quarter (1/4) cycle. This means the voltage and current are *not* in step, or *not* in phase. When the current reaches maximum, the voltage is zero, and when the current reaches zero the voltage is maximum. Alternatively, we can say that the voltage lags the current by 90°. A capacitor simply charges and discharges in an ac circuit and does not dissipate power or consume energy.

Example 21.3   A capacitor of 10.0 μF is connected to an ac source of 120 V and 60 Hz.

   (a) What is the capacitive reactance?

   (b) What is the rms current?

   (c) What power does the capacitor dissipate?

Solution: Given: $C = 10.0 \ \mu F = 10.0 \times 10^{-6} \ F$, $V = 120 \ V$, $f = 60 \ Hz$ (assume exact).

Find: (a) X_C (b) I (c) P.

(a) $X_C = \dfrac{1}{2\pi f C} = \dfrac{1}{2\pi (60 \ Hz)(10.0 \times 10^{-6} \ F)} = 265 \ \Omega$.

(b) From Ohm's law, $I = \dfrac{V}{X_C} = \dfrac{120 \ V}{265 \ \Omega} = 0.453 \ A$.

(c) An ideal capacitor ($R = 0$) does not dissipate power, so $P = 0$.

3.   Inductive Reactance (Section 21.3)

An inductor in an ac circuit also limits, or impedes the current, but does not completely prevent the flow of charge. This impeding effect is expressed in terms of **inductive reactance** (X_L) given by $X_L = 2\pi f L = \omega L$, where L is the **inductance** of the inductor and has an SI unit of **henry** (H). The SI unit of inductive reactance (X_L) is the ohm (Ω). Inductive reactance is also frequency dependent (directly proportional to frequency) and is analogous to resistance in the sense that the voltage across, and current through, the inductor are related by a more general form of Ohm's law, $V = I X_L$.

In a purely inductive ac circuit (inductor is the only circuit element), the *voltage leads the current* by 90° or a quarter (1/4) cycle. This means the voltage and current are *not* in step, or *not* in phase. When the voltage reaches maximum, the current is zero, and when the voltage reaches zero the current is maximum. Alternatively, the current lags the current by 90°. A inductor simply stores and releases magnetic energy in an ac circuit and does not dissipate power or consume energy.

Note: The phase relationship of current and voltage for purely inductive and purely capacitive circuits are opposite. A phrase that may help you remember the relationship is *EL̲I* the *IC̲E* man. With *E* representing voltage and *I* representing current, *EL̲I* indicates that with an inductance (*L*) the voltage leads the current (*I*). Similarly, *IC̲E* tells you that with a capacitance (*C*) the current leads the voltage.

Example 21.4   An inductor of 25.0 mH is connected to an ac source of 120 V and 60 Hz. Find the inductive reactance and the rms current in the circuit.

Solution: Given: $L = 25.0 \ mH = 25.0 \times 10^{-3} \ H$, $V = 120 \ V$, $f = 60 \ Hz$.

Find: X_L and I.

The inductive reactance is $X_L = 2\pi f L = 2\pi (60 \ Hz)(25.0 \times 10^{-3} \ H) = 9.42 \ \Omega$.

From Ohm's law, the rms current is equal to $I = \dfrac{V}{X_L} = \dfrac{120 \ V}{9.42 \ \Omega} = 12.7 \ A$.

---

4. Impedance: RLC Circuits (Section 21.4)

Impedance (Z) is a generalization of opposition to current which includes not only resistance, but also capacitive and inductive reactances. Because of the phase difference between the voltage and currents on different circuit elements, a **phase diagram** is convenient to use. In a phase diagram, the resistance and reactance of the circuit are given vectorlike properties and their magnitudes are represented as arrows called **phasors**. On a set of x-y axes, the resistance is used as a reference and is plotted on the positive x axis, since the voltage-current phase difference for a resistor is zero ($\phi = 0$). The reactances are plotted based on their phase differences. For example, the voltage lags the current by 90° in a capacitor, so the capacitive reactance is plotted on the $-y$ axis; the voltage leads the current by 90° for a inductor, so the inductive reactance is plotted on the positive y axis.

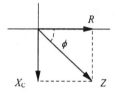 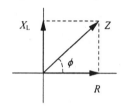

From the phase diagram, the impedance in an RLC series circuit is given by

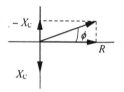

$$Z = \sqrt{R^2 + (X_L - X_C)^2} = \sqrt{R^2 + \left(2\pi f L - \frac{1}{2\pi f C}\right)^2}.$$ The angle ϕ is called the **phase angle**

of the circuit and is equal to $\tan \phi = \dfrac{X_L - X_C}{R}$. If $X_L > X_C$, the angle ϕ is positive and the circuit is said to be an **inductive circuit**; if $X_C > X_L$, the angle ϕ is negative and the circuit is said to be a **capacitive circuit**.

If the circuit is an RC circuit (no inductor), then $X_L = 0$, $Z = \sqrt{R^2 + X_C^2}$ and $\tan \phi = -\dfrac{X_C}{R}$; if the circuit is an RL circuit (no capacitor), then $X_C = 0$, $Z = \sqrt{R^2 + X_L^2}$ and $\tan \phi = \dfrac{X_L}{R}$. The general form of Ohm's law $V = IZ$ is valid for all such circuits.

In an RLC series circuit, *there are no power losses associated with pure capacitor and inductor. Only the resistor dissipates power.* Therefore, the power in an ac circuit can be expressed as $P = I^2 R = I^2 Z \dfrac{R}{Z} = I^2 Z \cos \phi$ $= IV \cos \phi$, where the term $\cos \phi$ is called the **power factor** and is defined as $\cos \phi = \dfrac{R}{Z}$ from the phase diagram. It is a measure of how close the circuit is to expending the maximum power ($\cos 0° = 1$).

Example 21.5 An RLC series circuit consists of a resistor of 100 Ω, a capacitor of 5.00 μF, and an inductor of

0.500 H. The circuit is connected to a power supply of 120 V operating at 60 Hz.

(a) What is the impedance?

(b) What is the phase angle? Is the circuit inductive or capacitive?

(c) What is the rms current?

(d) What is the rms voltage across each element?

(e) What is the power dissipated in the circuit?

Solution: Given: $R = 100\ \Omega$, $C = 5.00\ \mu F = 5.00 \times 10^{-6}$ F, $L = 0.500$ H, $V = 120$ V, $f = 60$ Hz.

Find: (a) Z (b) I (c) V across each element (d) P.

(a) $X_L = 2\pi f L = 2\pi(60\ \text{Hz})(0.500\ \text{H}) = 188\ \Omega$, $X_C = \dfrac{1}{2\pi f C} = \dfrac{1}{2\pi(60\ \text{Hz})(5.00 \times 10^{-6}\ \text{F})} = 531\ \Omega$.

So the impedance is $Z = \sqrt{R^2 + (X_L - X_C)^2} = \sqrt{(100\ \Omega)^2 + (188\ \Omega - 531\ \Omega)^2} = 357\ \Omega$.

(b) $\tan \phi = \dfrac{X_L - X_C}{R} = \dfrac{188\ \Omega - 531\ \Omega}{100\ \Omega} = -3.43$. So $\phi = -73.7°$.

Since the phase angle is negative, the circuit is a capacitive circuit.

(c) $I = \dfrac{V}{Z} = \dfrac{120\ \text{V}}{357\ \Omega} = 0.336$ A.

(d) $V_R = IR = (0.336\ \text{A})(100\ \Omega) = 33.6$ V, $V_C = IX_C = (0.336\ \text{A})(531\ \Omega) = 178$ V, and

$V_L = IX_L = (0.336\ \text{A})(188\ \Omega) = 63.2$ V.

Note: $V_R + V_C + V_L = 33.6\ \text{V} + 178\ \text{V} + 63.2\ \text{V} = 275\ \text{V} > 120$ V. Is this a violation of the conservation of

energy? How can the sum of the individual voltages be greater than the total voltage? (Hint: the voltages

calculated here are rms, or effective voltages, and the voltages, at any instant, are *not* in phase!)

(e) Only the resistor dissipates energy. $P = I^2 R = (0.336\ \text{A})^2 (100\ \Omega) = 11.3$ W.

Or, $P = IV\cos \phi = (0.336\ \text{A})(120\ \text{V}) \cos 73.7° = 11.3$ W.

5.   Circuit Resonance (Section 21.5)

When $X_L = X_C$ (the capacitive reactance and the inductive reactance cancel each other completely), $\phi = 0$

and $Z = R$ (completely resistive) is at minimum, so the circuit dissipates maximum power. This condition is called

resonance and the frequency at which resonance occurs is called the **resonance frequency** (f_0). At resonance,

$X_L = X_C$, or $2\pi f_0 L = \dfrac{1}{2\pi f_0 C}$, so $f_0 = \dfrac{1}{2\pi \sqrt{LC}}$. Resonant circuits have a wide variety of applications such as

radio and television tuning.

Example 21.6 An RLC series circuit consists of a resistor of 100 Ω, a capacitor of 20.0 μF, and an inductor of 0.550 H. The circuit is connected to a power supply of 120 V with an adjustable (tunable) frequency.

(a) What is the resonance frequency of the circuit?

(b) What is the impedance at resonance?

(c) What is the current in the circuit at resonance?

(d) What is the power dissipated at resonance?

(e) What are the voltages across the elements at resonance?

Solution: Given: $R = 100\ \Omega$, $C = 20.0\mu F = 20.0 \times 10^{-6}$ F, $L = 0.550$ H, $V = 120$ V.

Find: (a) f_o (b) Z (c) I (d) P (e) V for each element.

(a) $f_o = \dfrac{1}{2\pi \sqrt{LC}} = \dfrac{1}{2\pi \sqrt{(20.0 \times 10^{-6}\ F)(0.550\ H)}} = 48.0$ Hz.

(b) At resonance, $X_L = X_C$, so $Z = \sqrt{R^2 + (X_L - X_C)^2} = R = 100\ \Omega$ (it is at minimum).

(c) $I = \dfrac{V}{Z} = \dfrac{120\ V}{100\ \Omega} = 1.20$ A.

(d) $P = I^2 R = (1.20\ A)^2 (100\ \Omega) = 144$ W.

(e) $V_R = IR = (1.20\ A)(100\ \Omega) = 120$ V.

$X_L = 2\pi f L = 2\pi(48.0\ Hz)(0.550\ H) = 166\ \Omega$,   so $V_L = IX_L = (1.20\ A)(166\ \Omega) = 199$ V.

$X_C = \dfrac{1}{2\pi f C} = \dfrac{1}{2\pi(48.0\ Hz)(20.0 \times 10^{-6}\ F)} = 166\ \Omega$,   so $V_C = IX_C = (1.20\ A)(166\ \Omega) = 199$ V.

Note: The voltages across the capacitor and the inductor are out of phase (180°, see phase diagram), so they cancel out.

IV.   Mathematical Summary

Instantaneous Voltage in an ac Circuit	$V = V_o \sin \omega t$ $= V_o \sin 2\pi f t$ (21.1)	Gives the instantaneous voltage in an ac circuit as a function of time.
Instantaneous Current in an ac Circuit	$I = I_o \sin 2\pi f t$ (21.2) where $I_o = V_o/R$	Gives the instantaneous current in an ac circuit as a function of time.
Rms (or Effective) Current	$I_{rms} = \dfrac{I_o}{\sqrt{2}} \approx 0.707\ I_o$ (21.6)	Relates the rms current and the peak current.
Effective Power in an ac circuit	$\overline{P} = I_{rms}^2 R$ (21.7)	Computes the effective (average) power in an ac circuit.

Rms (or Effective) Voltage	$V_{rms} = \dfrac{V_o}{\sqrt{2}} \approx 0.707\, V_o \quad (21.8)$	Relates the rms voltage and the peak voltage.
Relationship between Voltage and Current in an ac Circuit*	$V_{rms} = I_{rms}\, R \qquad (21.9)$	Gives the relationship between the rms voltages and rms current.
Capacitive Reactance	$X_C = \dfrac{1}{2\pi f C} = \dfrac{1}{\omega C} \qquad (21.11)$	Defines the capacitive reactance of a capacitor in an ac circuit.
Relationship between Voltage and Current (capacitor only)	$V = I X_C \qquad (21.12)$	Gives the relationship between the rms voltage and rms current on a capacitor.
Inductive Reactance	$X_L = 2\pi f L = \omega L \qquad (21.14)$	Defines the inductive reactance of an inductor in an ac circuit.
Relationship between Voltage and Current (inductor only)	$V = I X_L \qquad (21.15)$	Gives the relationship between the instantaneous voltage and instantaneous current on an inductor.
Ohm's Law Generalized to ac Circuits	$V = I Z \qquad (21.17)$	Relates rms voltage and rms current in any ac circuit (RC, RL, or RLC series).
Impedance for a series RLC Circuit	$Z = \sqrt{R^2 + (X_L - X_C)^2} \qquad (21.19)$	Defines the impedance of a RLC series circuit.
Phase Angle between Voltage and Current in a series RLC Circuit	$\tan \phi = \dfrac{X_L - X_C}{R} \qquad (21.20)$	Defines the phase angle between the voltage and current in an ac circuit.
Power Factor for an RLC Circuit	$\cos \phi = \dfrac{R}{Z} \qquad (21.22)$	Defines the power factor in an ac circuit.
Power in Terms of Power Factor	$P = I V \cos \phi \qquad (21.23)$ or $P = I^2 Z \cos \phi \qquad (21.24)$	Expresses the power in an ac circuit in terms of the power factor.
Resonance Frequency of a series RLC Circuit	$f_o = \dfrac{1}{2\pi \sqrt{L C}} \qquad (21.25)$	Computes the resonance frequency in an RC or RLC series circuit.

V.   Solutions of Selected Exercises and Paired/Trio Exercises

3.   That means the voltage and current reach maximum at the same time, reach minimum at the same time, and are zero at the same, etc.

8. (a) $V_{rms} = I_{rms} R = (0.75 \text{ A})(5.0 \ \Omega) = \boxed{3.8 \text{ V}}$;

$V_o = \sqrt{2} \ V_{rms} = \sqrt{2} \ (3.75 \text{ V}) = \boxed{5.3 \text{ V}}$.

(b) $\bar{P} = I_{rms} V_{rms} = (0.75 \text{ A})(3.75 \text{ V}) = \boxed{2.8 \text{ W}}$.

14. Since $\bar{P} = \dfrac{V_{rms}^2}{R}$, $R = \dfrac{V_{rms}^2}{\bar{P}} = \dfrac{(120 \text{ V})^2}{100 \text{ W}} = \boxed{144 \ \Omega}$. $I_{rms} = \dfrac{V_{rms}}{R} = \dfrac{120 \text{ V}}{144 \ \Omega} = \boxed{0.833 \text{ A}}$.

16. (a) $\boxed{2.9 \times 10^2 \text{ A}}$.   (b) $\boxed{3.4 \times 10^2 \text{ V}}$.

25. In an ac circuit, a capacitor can oppose current. This is because as the capacitor charges, the voltage across its plates increases, opposing the current. Also, an inductor can oppose current because the induced emf opposes the change in flux and thus opposes the current in the circuit.

28. From $X_C = \dfrac{1}{2\pi f C}$, we have $f = \dfrac{1}{2\pi C X_C} = \dfrac{1}{2\pi (25 \times 10^{-6} \text{ F})(25 \ \Omega)} = \boxed{2.5 \times 10^2 \text{ Hz}}$.

34. From $X_L = 2\pi f L$, we have $L = \dfrac{X_L}{2\pi f} = \dfrac{90 \ \Omega}{2\pi (60 \text{ Hz})} = \boxed{0.24 \text{ H}}$.

38. $X_C = \dfrac{V_{rms}}{I_{rms}} = \dfrac{120 \text{ V}}{0.20 \text{ A}} = 600 \ \Omega$.

Also $X_C = \dfrac{1}{2\pi f C}$, so $C = \dfrac{1}{2\pi X_C f} = \dfrac{1}{2\pi (600 \ \Omega)(60 \text{ Hz})} = 4.4 \times 10^{-6} \text{ F} = \boxed{4.4 \ \mu\text{F}}$.

44. (a) $X_C = \dfrac{1}{2\pi f C} = \dfrac{1}{2\pi (60 \text{ Hz})(25 \times 10^{-6} \text{ F})} = \boxed{1.1 \times 10^2 \ \Omega}$.

$Z = \sqrt{R^2 + (X_L - X_C)^2} = \sqrt{(200 \ \Omega)^2 + (0 - 106 \ \Omega)^2} = 226 \ \Omega = \boxed{2.3 \times 10^2 \ \Omega}$.

(b) $I_{rms} = \dfrac{V_{rms}}{Z} = \dfrac{120 \text{ V}}{226 \ \Omega} = \boxed{0.53 \text{ A}}$.

49. If the frequency f doubles, R remains the same, X_L doubles to 80 Ω because $X_L = 2\pi f L$, and X_C halves to

20 Ω since $X_C = \dfrac{1}{2\pi f C}$.

So $Z = \sqrt{R^2 + (X_L - X_C)^2} = \sqrt{(40 \ \Omega)^2 + (80 \ \Omega - 20 \ \Omega)^2} = \boxed{72 \ \Omega}$.

52. At resonance, $f_0 = \dfrac{1}{2\pi\sqrt{LC}} = \dfrac{1}{2\pi\sqrt{(0.100\ \text{H})(5.00 \times 10^{-6}\ \text{F})}} = \boxed{225\ \text{Hz}}$.

54. $\boxed{1.7 \times 10^{-11}\ \text{F to } 1.8 \times 10^{-10}\ \text{F}}$.

58. (a) $Z = \sqrt{R^2 + (X_L - X_C)^2} = \sqrt{(10\ \Omega)^2 + (120\ \Omega - 120\ \Omega)^2} = 10\ \Omega$.

$I_{\text{rms}} = \dfrac{V_{\text{rms}}}{Z} = \dfrac{220\ \text{V}}{10\ \Omega} = 22\ \text{A}$.   So $(V_{\text{rms}})_\text{R} = I_{\text{rms}}\,R = (22\ \text{A})(10\ \Omega) = \boxed{220\ \text{V}}$.

(b) $(V_{\text{rms}})_\text{L} = I_{\text{rms}}\,X_L = (22\ \text{A})(120\ \Omega) = \boxed{2.64 \times 10^3\ \text{V}}$.

(c) $(V_{\text{rms}})_\text{C} = I_{\text{rms}}\,X_C = (22\ \text{A})(120\ \Omega) = \boxed{2.64 \times 10^3\ \text{V}}$.

61. (a) $X_L = 2\pi f L = 2\pi(60\ \text{Hz})(0.450\ \text{H}) = 170\ \Omega$,

$X_C = \dfrac{1}{2\pi f C} = \dfrac{1}{2\pi(60\ \text{Hz})(5.00 \times 10^{-6}\ \text{F})} = 531\ \Omega$.

$Z = \sqrt{R^2 + (X_L - X_C)^2} = \sqrt{(25.0\ \Omega)^2 + (170\ \Omega - 531\ \Omega)^2} = \boxed{362\ \Omega}$.

(b) $f_0 = \dfrac{1}{2\pi\sqrt{LC}} = \dfrac{1}{2\pi\sqrt{(0.450\ \text{H})(5.00 \times 10^{-6}\ \text{F})}} = 106\ \text{Hz}$;

$\boxed{\text{no}}$, the circuit is not in resonance and the resonance frequency is $\boxed{106\ \text{Hz}}$.

69. $\tan \phi = \dfrac{X_L - X_C}{R} = \dfrac{38\ \Omega}{50\ \Omega} = 0.76$, so $\phi = \boxed{37°}$.

71. From $X_C = \dfrac{1}{2\pi f C}$ and $X_L = 2\pi f L$, we can see that the capacitor opposes current more strongly at lower frequencies and the inductor opposes current more strongly at high frequencies. In Fig. 21.18a, the inductor in series with R_L filters out the high frequency current and so only the low frequency current reaches R_L. In Fig. 21.18b, the capacitor in series with R_L filters out the low frequency current and so only the high frequency current reaches R_L.

VI. Practice Quiz

1. The inductive reactance in an ac circuit will change by what factor when the frequency is tripled?

(a) 1/3   (b) 1/9   (c) 1   (d) 9   (e) 3

2.	What is the peak voltage in an ac circuit with an rms voltage of 120 V?

(a) zero	(b) 84.9 V	(c) 120 V	(d) 170 V	(e) 240 V

3.	What is the impedance of an ac circuit with a resistance of 12.0 Ω, an inductive reactance of 15.0 Ω, and a capacitive reactance of 10.0 Ω?

(a) 37.0 Ω	(b) 27.7 Ω	(c) 11.6 Ω	(d) 21.9 Ω	(e) 13.0 Ω

4.	A RLC series circuit has a resistance of 12.0 Ω, an inductive reactance of 15.0 Ω, and a capacitive reactance of 10.0 Ω. If an rms voltage of 120 V is applied, what is the power output?

(a) 435 W	(b) 598 W	(c) 1022 W	(d) 1108 W	(e) 1200 W

5.	What is the phase difference between the voltages of the inductor and capacitor in a RLC series circuit.

(a) 45°	(b) 90°	(c) 180°	(d) 360°	(e) zero

6.	What is the resonant frequency for a circuit containing an inductor of 10 mH and a capacitor of 20 μF?

(a) zero	(b) 0.36 kHz	(c) 1.3 kHz	(d) 2.2 kHz	(e) 0.80 MHz.

7.	An ac series circuit has an impedance of 60 Ω and a resistance of 30 Ω. What is the power factor?

(a) 0.50	(b) 0.71	(c) 1.0	(d) 1.4	(e) 2.0

8.	Resonance occurs in an ac series circuit when

(a) resistance equals capacitive reactance.	(b) resistance equals inductive reactance.

(c) capacitive reactance equals zero.	(d) capacitive reactance equals inductive reactance.

(e) inductive reactance equals zero.

9.	A RLC series circuit consists of a resistor of 100 Ω, a capacitor of 10.0 μF, and an inductor of 0.250 H. The circuit is connected to a power supply of 120 V and 60 Hz. What is the power dissipated in the circuit?

(a) zero	(b) 37 W	(c) 63 W	(d) 73 W	(e) 0.14 kW

10.	What is the phase angle and power factor for a RLC series circuit containing a resistor of 50 Ω, a capacitor of 10 μF, and an inductor of 0.45 H, when connected to a power supply with frequency of 60 Hz?

(a) −62° and 0.47	(b) 62° and 0.47	(c) −38° and 0.79	(d) 38° and 0.79	(e) zero and 1.0

Answers to Practice Quiz:

1. e 2. b 3. e 4. c 5. c 6. b 7. a 8. d 9. b 10. a

CHAPTER 22

Geometrical Optics: Reflection and Refraction of Light

I.   Chapter Objectives

Upon completion of this chapter, you should be able to:

1.   define and explain the concept of wave fronts and rays.

2.   explain the law of reflection and distinguish between regular (specular) and irregular (diffuse) reflections.

3.   explain refraction in terms of Snell's law and the index of refraction, and give examples of refractive phenomena.

4.   describe internal reflection and give examples of fiber optic applications.

5.   explain dispersion and some of its effects.

II.   Key Terms

Upon completion of this chapter, you should be able to define and/or explain the following key terms:

wave front	refraction
plane wave front	Huygen's principle
ray	angle of refraction
geometrical optics	Snell's law
reflection	index of refraction
angle of incidence	critical angle
angle of reflection	total internal reflection
law of reflection	fiber optics
regular (specular) reflection	dispersion
irregular (diffuse) reflection	

The definitions and/or explanations of the most important key terms can be found in the following section:

III. Chapter Summary and Discussion.

III.   Chapter Summary and Discussion

1.   Wave Fronts and Rays (Section 22.1)

A **wave front** is the line (in two dimensions) or surface (in three dimensions) defined by adjacent portions of a wave that are in phase. For example, a point light source emits spherical wave fronts because the points having the same phase angle are on the surface of a sphere. For a parallel beam of light, the wave front is a **plane wave front**. The distance between adjacent wave fronts is the wavelength.

A **ray** is a line drawn perpendicular to a series of wave fronts and pointing in the direction of propagation. For a spherical wave, the rays are radially outward, and for a plane wave, they are parallel to each other. The use of wave fronts and rays in describing optical phenomena such as reflection and refraction is called **geometrical optics**.

2.   Reflection (Section 22.2)

The **law of reflection** states that the **angle of incidence** (the angle between the incident ray and the normal) equals the **angle of reflection** (the angle between the reflected ray and the normal), $\theta_i = \theta_r$, where θ_i is the angle of incidence and θ_r the angle of reflection. The incident ray, the normal, and the reflected ray are always in the same plane.

Note: All angles are measured *from the normal (line perpendicular) to the reflecting surface.*

Regular (specular) reflection occurs from smooth surfaces, with the reflected rays parallel to each other. **Irregular (diffuse) reflection** occurs from rough surfaces, with the reflected rays being at different angles.

Example 22.1   Two mirrors make an angle of 90° with each other. A ray is incident on mirror M_1 at an angle of 30° to the normal. Find the direction of the ray after it is reflected from mirror M_1 and M_2.

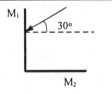

Solution:

All angles should be measured from the normal to the reflecting surface. The angle of incidence of the ray at M_1 is 30°. According to the law of reflection, the angle of reflection at M_1 is also 30°. From geometry, the angle of incidence at M_2 is $90° - 30° = 60°$. Therefore, the angle of reflection at M_2 is also 60°, and the ray emerges parallel to the original incident ray.

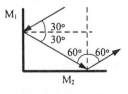

---

3. Refraction (Section 22.3)

Refraction refers to the change in direction of a wave at a boundary where it passes from one medium into another as a result of different wave speeds in different media. This phenomenon can be geometrically explained by **Huygens' principle**: every point on an advancing wave front can be considered to be a source of secondary waves, or wavelets, and the line or surface tangent to all these wavelets defines a new position of the wave front.

Snell's law relates the angle of incidence, θ_1, and angle of refraction θ_2, (the angle between the refracted ray and the normal to the boundary) to the wave speeds in the respective media: $\dfrac{\sin \theta_1}{\sin \theta_2} = \dfrac{v_1}{v_2}$. The **index of refraction** of a medium is defined as the ratio of the speed of light in vacuum to its speed in that medium, $n = \dfrac{c}{v}$. Snell's law can be conveniently expressed in terms of the indices of refraction, $n_1 \sin \theta_1 = n_2 \sin \theta_2$. If the second medium is more optically dense ($n_2 > n_1$), then $\theta_1 > \theta_2$, or the refracted ray is bent toward the normal; if less dense ($n_2 < n_1$), then $\theta_2 > \theta_1$ and bending is away from the normal.

When light travels from one medium to another, the frequency remains constant, but the speed and wavelength change. In terms of wavelength, the index of refraction can be rewritten as $n = \dfrac{\lambda}{\lambda_m}$, where λ is the wavelength in vacuum and λ_m the wavelength in the medium.

Note: Since the geometrical representation of light is used here, a diagram is very helpful (if not necessary) in solving problems. Again, all angles are measured from the normal to the interface boundary.

Example 22.2   A light ray travels through an air–fused quartz interface at an angle of 30° to the normal. Find the speed of light in the quartz and the angle of refraction.

Solution: Given: $n_2 = 1.46$ (fused quartz, from Table 22.1),

$n_1 \approx 1.00$ (air), $\theta_1 = 30°$.

Find: v and θ_2.

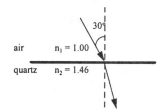

From $n = \dfrac{c}{v}$, we have $v_2 = \dfrac{c}{n_2} = \dfrac{3.00 \times 10^8 \text{ m/s}}{1.46} = 2.05 \times 10^8 \text{ m/s}$.

From Snell's law, $n_1 \sin \theta_1 = n_2 \sin \theta_2$,

$\sin \theta_2 = \dfrac{n_1 \sin \theta_1}{n_2} = \dfrac{(1.00) \sin 30°}{1.46} = 0.342$.   So $\theta_2 = 20.0°$

Example 22.3 A beam of light traveling in air is incident on a slab of transparent material. The incident beam and the refracted beam make angles of 40° and 26° to the normal, respectively. Find the speed of light in the transparent material.

Solution: Given: $n_1 \approx 1.00$ (air), $\theta_1 = 40°$, $\theta_2 = 26°$.

Find: v_2.

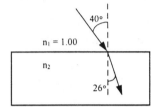

We first find the index of refraction of the transparent material.

From Snell's law, $n_1 \sin \theta_1 = n_2 \sin \theta_2$,

we have $n_2 = \dfrac{n_1 \sin \theta_1}{\sin \theta_2} = \dfrac{(1.00) \sin 40°}{\sin 26°} = 1.47$.

Therefore, from $n = \dfrac{c}{v}$, the speed of light in the material is

$$v_2 = \frac{c}{n_2} = \frac{3.00 \times 10^8 \text{ m/s}}{1.47} = 2.04 \times 10^8 \text{ m/s}.$$

Example 22.4 A light ray from a He-Ne laser has a wavelength of 632.8 nm and travels from air to crown glass.

(a) What is the frequency of the light in air?

(b) What is the frequency of the light in glass?

(c) What is the wavelength of light in glass?

Solution: Given: $n_2 = 1.52$ (crown glass from Table 22.1)

$n_1 = 1.00$ (air), $\lambda = 632.8$ nm $= 632.8 \times 10^{-9}$ m.

Find: (a) f (b) f_m (c) λ_m.

(a) From $c = \lambda f$, we have $f = \dfrac{c}{\lambda} = \dfrac{3.00 \times 10^8 \text{ m/s}}{632.8 \times 10^{-9} \text{ m}} = 4.74 \times 10^{14}$ Hz.

(b) The frequency is a constant (same for all media). So $f_m = f = 4.74 \times 10^{14}$ Hz.

(c) From $n = \dfrac{\lambda}{\lambda_m}$, $\lambda_m = \dfrac{\lambda}{n} = \dfrac{632.8 \text{ nm}}{1.52} = 416$ nm.

---

4. Total Internal Reflection and Fiber Optics (Section 22.4)

At a certain **critical angle** (θ_c), the angle of refraction for a ray going from a medium of greater optical density to a medium of lesser optical density ($n_1 > n_2$) is 90° and the refracted ray is along the media boundary. For any angle of incidence $\theta_1 > \theta_c$, the **total internal reflection** (no refracted light) occurs and the surface acts like a mirror. By Snell's law, the critical angle can be calculated in terms of the indices of refraction of the two media,

$\sin \theta_c = \dfrac{n_2}{n_1}$ (for $n_1 > n_2$). If the second medium is air, then $\sin \theta_c = \dfrac{1}{n}$. **Fiber optics** uses the principle of total internal reflection. Signals can travel a long distance without losing much intensity due to the lack of refraction.

Note: Total internal reflection occurs only if the second medium is less dense than the first *and* if the angle of incidence exceeds the critical angle.

Example 22.5   A diver is 1.5 m beneath the surface of a still pond of water. At what angle must the diver shine a beam of light toward the surface in order for a person on a distant bank to see it?

Solution: Given: $\theta_2 = 90°$, $n_2 = 1.00$ (air), $n_1 = 1.33$ (water).

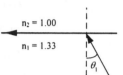

Find: θ_1.

When the refracted light is along the boundary, the angle in water is equal to the

critical angle. $\theta_1 = \theta_c = \sin^{-1} \dfrac{n_2}{n_1} = \sin^{-1} \dfrac{1.00}{1.33} = \sin^{-1} 0.752 = 48.8°$.

Or from Snell's law, $n_1 \sin \theta_1 = n_2 \sin \theta_2$, we have $\sin \theta_1 = \dfrac{n_2 \sin \theta_2}{n_1} = \dfrac{(1.00) \sin 90°}{1.33} = 0.752$,

so $\theta_1 = 48.8°$.

For the light to reach the distant person, it has to shine at or below 48.8°, so there is no total internal reflection. Do you know why some fish-preying birds like pelicans stay very low before they try to catch fish?

Example 22.6   A 45°–45° prism is a wedge-shaped object in which the two acute angles are 45°. It is very useful for changing the direction of light rays in optical devices. If a light ray is traveling through a glass prism according to the diagram shown, what is the minimum index of refraction of the glass?

Solution: Given: $n_2 = 1.00$ (air), $\theta_c = 45°$.   Find: n_1 (minimum).

There is no refracted ray beyond the glass–air boundary, so the light must be internally reflected. For total internal reflection to occur, the index of reflection n_1 must be greater than that of air. The minimum index of reflection corresponds to an angle of incidence of 45°.

$\sin \theta_c = \dfrac{n_2}{n_1}$,   so the minimum n_1 is $\dfrac{n_2}{\sin \theta_c} = \dfrac{1.00}{\sin 45°} = 1.41$.

---

5.   Dispersion (Section 22.5)

Dispersion is the separation of multi-wavelength light into its component wavelengths when the light is refracted. This is caused by the fact that different wavelengths have slightly different indices of refraction and therefore different speeds. In most materials (so called normal dispersion), longer wavelengths have smaller indices of refraction. According to Snell's law, different wavelengths will have different angles of refraction and are therefore separated. A rainbow is produced by refraction, dispersion, and total internal reflection within water droplets.

Example 22.7   A beam of white light strikes a piece of glass at a 70° angle (measured from the normal). A red light of wavelength 680 nm and a blue light of wavelength 430 nm emerge from the boundary after being dispersed. The index of refraction for the red light is 1.4505 and the index of refraction for the blue light is 1.4693.

(a) Which color of light is refracted more?

(b) What is the angle of refraction for each color?

(c) What is the angular separation between the two colors?

Solution: Given: $n_1 = 1.0000$ (air), $n_{2r} = 1.4505$, $n_{2b} = 1.4693$, $\theta_1 = 70°$.

Find: (b) θ_{2r} and θ_{2b} (c) $\Delta\theta$.

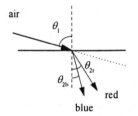

(a) Since the index of refraction of the blue light is greater, its angle of refraction is smaller (bent more toward the normal), and therefore it is refracted more.

(b) We use Snell's law, $n_1 \sin\theta_1 = n_2 \sin\theta_2$,

for red: $\sin\theta_{2r} = \dfrac{n_1 \sin\theta_1}{n_{2r}} = \dfrac{(1.0000)\sin 70°}{1.4505} = 0.64784$, so $\theta_{2r} = 40.379°$;

for blue: $\sin\theta_{2b} = \dfrac{(1.0000)\sin 70°}{1.4693} = 0.63955$, so $\theta_{2b} = 39.758°$.

(c) $\Delta\theta = \theta_{2r} - \theta_{2b} = 40.379° - 39.758° = 0.621°$.

IV.   Mathematical Summary

Law of Reflection	$\theta_i = \theta_r$ (22.1)	Relates the angles of incidence and reflection.
Snell's Law	$\dfrac{\sin\theta_1}{\sin\theta_2} = \dfrac{v_1}{v_2}$ (22.2) or $n_1\sin\theta_1 = n_2\sin\theta_2$ (22.5)	Relates the angles of incidence and refraction, and the speeds of light in the media (or indices of refraction).

Index of Refraction	$n = \dfrac{c}{v} = \dfrac{\lambda}{\lambda_m}$ (22.3, 4)	Defines the index of refraction of a material.
Critical Angle at Boundary between Two Materials	$\sin \theta_c = \dfrac{n_2}{n_1}$ (22.6) where $n_1 > n_2$	Computes the critical angle between two materials for total internal reflection.
Critical Angle at Material–Air Boundary	$\sin \theta_c = \dfrac{1}{n}$ (22.7)	Computes the critical angle between material–air boundary for total internal reflection.

V.   Solutions of Selected Exercises and Paired/Trio Exercises

5.   They are visible because of the diffuse reflections by the particulate matter in the air.

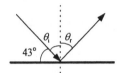

9. $\theta_r = \theta_i = 90° - 43° = \boxed{47°}$

so $\theta_i = \theta_r = \tan^{-1} 0.833 = \boxed{40°}$.

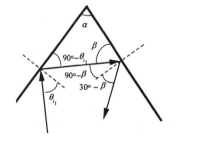

12.   According to the law of reflection, the angle of reflection from the second mirror is $\boxed{20°}$.

16.   According to the law of reflection,

$\beta = 180° - [\alpha + (90° - \theta_{i_1})] = 90° - \alpha + \theta_{i_1}$.

So   the angle of reflection from the second mirror is

$\theta_{r_2} = 90° - \beta = \alpha - \theta_{i_1}$.

(a) $\theta_{r_2} = 70° - 35° = \boxed{35°}$.

(b) $\theta_{r_2} = 115° - 60° = \boxed{55°}$.

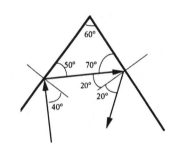

25. $\boxed{\text{The laser}}$ has a better chance to hit the fish. The fish appears to the hunter at a location different from its true location due to refraction. The laser beam obeys the same law of refraction and retraces the light the hunter sees the fish. The arrow goes into the water in a near-straight line path.

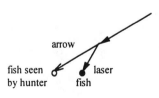

28. $n_1 \sin \theta_1 = n_2 \sin \theta_2$, ☞ $\sin \theta_2 = \dfrac{n_1 \sin \theta_1}{n_2} = \dfrac{(1) \sin 60°}{1.33} = 0.651.$ So $\theta_2 = \boxed{41°}$.

30. $\boxed{1.34}$.

36. $\dfrac{v_B}{v_A} = \dfrac{c/n_B}{c/n_A} = \dfrac{n_A}{n_B} = \dfrac{4/3}{5/4} = \boxed{\dfrac{16}{15}}$.

40. The first refraction is at the air-glass interface and the second refraction is at the glass-water interface.

From Snell's law, $n_1 \sin \theta_1 = n_2 \sin \theta_2$, we have $\sin \theta_2 = \dfrac{n_1 \sin \theta_1}{n_2} = \dfrac{(1) \sin 40°}{1.50} = 0.429.$

So $\theta_2 = 25.4°$.

The critical angle at the glass-water interface is $\theta_c = \sin^{-1} \dfrac{n_3}{n_2} = \sin^{-1} \dfrac{1.33}{1.50} = \sin^{-1} 0.887 = 62.5°$.

Therefore the angle of incidence at the glass-water interface is smaller than the critical angle and thus there is no total internal reflection and $\boxed{\text{yes}}$, the fish is illuminated.

46. The setting is at zero altitude or 90° from the normal, so the angle should equal to the critical angle of

$\theta_c = \sin^{-1} \dfrac{1}{n} = \sin^{-1} \dfrac{1}{1.33} = \sin^{-1} 0.752 = \boxed{48.8°}$.

53. $\theta_2 = \tan^{-1} \dfrac{0.50 \text{ m}}{0.75 \text{ m}} = \tan^{-1} 0.667 = 33.7°$.

From Snell's law, $n_1 \sin \theta_1 = n_2 \sin \theta_2$,

we have $\sin \theta_1 = \dfrac{n_2 \sin \theta_2}{n_1} = \dfrac{(1.33) \sin 33.7°}{1} = 0.738.$

Therefore $\theta_1 = 47.5°$.

Thus $d = (1.8 \text{ m}) \tan \theta_1 = (1.8 \text{ m}) \tan 47.5° = \boxed{2.0 \text{ m}}$.

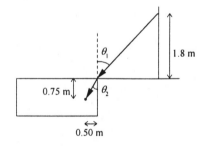

55. (a) We use Snell's law, $n_1 \sin \theta_1 = n_2 \sin \theta_2$.

For refraction at the prism-prism interface, $(1.60) \sin 45° = (1.40) \sin \theta_2$,

so $\theta_2 = \sin^{-1} 0.808 = 53.9°$.

For the prism-air interface,

$\theta_3 = 180° - [135° + (90° - \theta_2)] = \theta_2 - 45° = 53.9° - 45° = 8.9°$.

So $(1.40) \sin 8.9° = (1) \sin \theta$,

therefore $\theta = \sin^{-1} 0.217 = \boxed{12.5°}$.

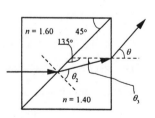

(b) $\theta_c = \sin^{-1} \dfrac{n_2}{n_1} = \sin^{-1} \dfrac{1.40}{1.60} = 61.0°.$

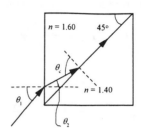

This is the angle required at the interface.

For the air-prism interface,

$\theta_2 = 180° - [135° + (90° - \theta_c)] = \theta_c - 45° = 61.0° - 45° = 16.0°.$

So   (1) $\sin \theta_1 = (1.60) \sin 16°,$

therefore $\theta_1 = \sin^{-1} 0.441 = \boxed{26.2°}$

---

58.   It is due to the $\boxed{\text{different speeds of different frequencies in the material}}$. This, in turn, causes the different

indices of refraction, therefore the different angles of refraction.

---

62.   From Snell's law, $n_1 \sin \theta_1 = n_R \sin \theta_R = n_B \sin \theta_B,$ so $\sin \theta_R = \dfrac{n_1 \sin \theta_1}{n_R} = \dfrac{(1) \sin 37°}{1.515} = 0.3972.$

Therefore $\theta_R = 23.406°.$ $\sin \theta_B = \dfrac{(1) \sin 37°}{1.523} = 0.3952.$ Thus $\theta_B = 23.275°.$

Finally $\Delta\theta = 23.406° - 23.275° = \boxed{0.131°}.$

---

65.   (a) We use Snell's law, $n_1 \sin \theta_1 = n_2 \sin \theta_2.$

For the first air-prism interface,   (1) $\sin 80.0° = (1.400) \sin \theta_2,$

So $\theta_2 = \sin^{-1} 0.7034.$ Therefore $\theta_2 = 44.70°.$

For the second prism-air interface,

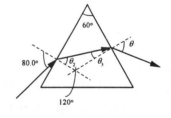

$\theta_3 = 180° - (120° + \theta_2) = 60° - \theta_2 = 60° - 44.70° = 15.30°.$

So   (1.400) $\sin 15.30° = (1) \sin \theta,$

therefore $\theta = \sin^{-1} 0.3693 = \boxed{21.7°}.$

(b) For blue light, $\theta_2 = \sin^{-1} \dfrac{\sin 80.0°}{1.403} = 44.58°.$ $\theta_3 = 60° - 44.58° = 15.42°.$

So $\theta = \sin^{-1} [(1.403) \sin 15.42°] = 21.90°.$

Therefore $\Delta\theta = 21.90° - 21.68° = \boxed{0.22°}.$

(c) For blue light, $\theta_2 = \sin^{-1} \dfrac{\sin 80.0°}{1.405} = 44.50°.$ $\theta_3 = 60° - 44.50° = 15.50°.$

So $\theta = \sin^{-1} [(1.405) \sin 15.50°] = 22.05°.$

Therefore $\Delta\theta = 22.05° - 21.68° = \boxed{0.37°}.$

70.   (a) From Snell's law, $n_1 \sin\theta_1 = n_2 \sin\theta_2$,

we have $\sin\theta_2 = \dfrac{n_1 \sin\theta_1}{n_2} = \dfrac{(1)\sin 40°}{1.52} = 0.423$.

So $\theta_2 = \boxed{25°}$.

(b) $v = \dfrac{c}{n} = \dfrac{3.00 \times 10^8 \text{ m/s}}{1.52} = \boxed{1.97 \times 10^8 \text{ m/s}}$.

(c) $\lambda_m = \dfrac{\lambda}{n} = \dfrac{550 \text{ nm}}{1.52} = \boxed{362 \text{ nm}}$.

VI.   Practice Quiz

1.   If a material has a speed of light of 2.13×10^8 m/s, what is its index of refraction?

(a) 0.710   (b) 1.07   (c) 1.41   (d) 2.13   (e) 5.13

2.   A light ray in air is incident on an air to glass interface boundary at an angle of 45° and is refracted in the glass at an angle of 27° with the normal. What is the index of refraction of the glass?

(a) 0.642   (b) 1.16   (c) 1.41   (d) 1.56   (e) 2.20

3.   An optical fiber is made of clear plastic with index of refraction of $n = 1.50$. What is the minimum angle of incidence so total internal reflection can occur?

(a) 23.4°   (b) 32.9°   (c) 38.3°   (d) 40.3°   (e) 41.8°

4.   A certain kind of glass has an index of refraction of 1.65 for blue light and an index of refraction of 1.61 for red light. If a beam of white light (containing all colors) is incident at an angle of 30°, what is the angle between the refracted red and blue light?

(a) 0.22°   (b) 0.35°   (c) 0.45°   (d) 1.90°   (e) 1.81°

5.   A ray of white light, incident upon a glass prism, is dispersed into its various color components. Which one of the following colors experiences the least refraction?

(a) orange   (b) yellow   (c) red   (d) blue   (e) green

6.   Which one of the following describes what will generally happen to a light ray incident on a glass-to-air boundary?

(a) total reflection   (b) total refraction   (c) partial reflection, partial refraction

(d) either (a) or (c)   (e) either (b) or (c)

7. Light enters water from air. The angle of refraction will be

(a) greater than or equal to the angle of incidence.   (b) less than or equal to the angle of incidence.

(c) equal to the angle of incidence.   (d) greater than the angle of incidence.

(e) less than the angle of incidence.

8. A monochromatic light source emits a wavelength of 633 nm in air. When passing through a liquid, the wavelength reduces to 487 nm. What is the index of refraction of the liquid?

(a) 0.769   (b) 1.30   (c) 1.41   (d) 1.62   (e) 2.11

9. A fiber optic cable ($n = 1.50$) is submerged in water. What is the critical angle for light to stay inside the cable?

(a) 27.6°   (b) 41.8°   (c) 45.0°   (d) 62.5°   (e) 83.1°

10. An oil film ($n = 1.47$) floats on a water ($n = 1.33$) surface. If a ray of light is incident on the air–oil boundary at an angle of 37° to the normal, what is the angle of refraction at the oil–water boundary?

(a) 17.9°   (b) 24.2°   (c) 26.9°   (d) 33.0°   (e) 37.0°

Answers to Practice Quiz:

1. c 2. d 3. e 4. c 5. c 6. d 7. e 8. b 9. b 10. c

CHAPTER 23

Mirrors and Lenses

I.   Chapter Objectives

Upon completion of this chapter, you should be able to:

1.   describe the characteristics of plane mirrors and explain apparent right-left reversals.

2.   distinguish between converging and diverging spherical mirrors, describe images and their characteristics, and determine these image characteristics from ray diagrams and the spherical mirror equation.

3.   distinguish between converging and diverging lenses, describe images and their characteristics, and find image characteristics using ray diagrams and the thin-lens equation.

4.   describe some common lens aberrations and explain how they can be reduced or corrected.

*5.   describe the lens maker's equation and explain how its application differs from that of the thin-lens equation.

II.   Key Terms

Upon completion of this chapter, you should be able to define and/or explain the following key terms:

plane mirror	magnification factor (for spherical mirror)
virtual image	spherical aberration (for mirror)
real image	lens
lateral magnification	converging (biconvex) lens
spherical mirror	diverging (biconcave) lens
concave (converging) mirror	parallel ray (for lens)
convex (diverging) mirror	chief (central) ray (for lens)
center of curvature	focal ray (for lens)
radius of curvature	thin-lens equation
focal point	magnification factor (for lens)
focal length	spherical aberration (for lens)
parallel ray (for mirror)	chromatic aberration
chief (radial) ray (for mirror)	astigmatism
focal ray (for mirror)	*lens maker's equation
spherical mirror equation	*diopters

The definitions and/or explanations of the most important key terms can be found in the following section:

III. Chapter Summary and Discussion.

III.   Chapter Summary and Discussion

1.   Plane Mirrors (Section 23.1)

The images of objects formed by optical systems (mirrors and/or lenses) can be either real or virtual. A **real image** is one formed by light rays that converge at and pass through the image location, and can be seen or formed on a screen. A **virtual image** is one for which light rays *appear* to emanate from the image, but do not actually do so. Virtual images cannot be seen or formed on screens.

A **plane mirror** is a mirror with a flat surface. A mirror forms an image based on the law of reflection. The characteristics of the images formed by a plane mirror are always virtual, upright, and unmagnified ($M = +1$). Also, the image formed by a plane mirror appears to be at a distance behind the mirror that is equal to the distance of the object in front of the mirror and has right-left or front-back reversal.

Example 23.1   A curved arrow is placed in front of a plane mirror as shown. Sketch the image of the arrow formed by the plane mirror.

Solution:

The image is virtual, upright, and unmagnified. The object distance (distance from object to mirror) is equal to the image distance (distance from image to mirror) for images formed by a plane mirror. By sketching the images of each individual point on the object, we obtain the image.

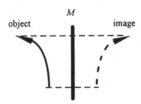

---

2.   Spherical Mirrors (Section 23.2)

A **spherical mirror** is a section of a sphere. Either the outside (convex) surface or the inside (concave) surface of the spherical section may be the reflecting surface. A **concave mirror** is called a **converging mirror** and a **convex mirror** is called a **diverging mirror**. These terms refer to the reflections of rays parallel to a mirror's *optical axis*, which is a line through the center of the spherical mirror that intersects the mirror at the *vertex* of the spherical section.

The **center of curvature** (C) of a spherical mirror is the point on the optic axis that corresponds to the center of the sphere of which the mirror forms a section. The **radius of curvature** (R) is the distance from the vertex to the center of curvature. The **focal point** (F) of a spherical mirror is the point at which parallel rays converge or appear to diverge. The **focal length** (f) of a spherical mirror is the distance from the focal point to the vertex of the spherical section. It is equal to one-half of the radius of curvature, $f = R/2$.

The images formed by spherical mirrors can be studied from geometry (**ray diagrams**). The ray diagram is as important as the free-body diagram in the application of Newton's laws. Three important rays are used to determine the images: a **parallel ray** is a ray incident along a path parallel to the optic axis and reflected through (or appear to go though) the focal point F; a **chief (radial) ray** is a ray incident through (or appear to go though) the center of curvature C and reflected back along its incident path though C; a **focal ray** is a ray which passes through (or appear to go through) the focal point and is reflected parallel to the optic axis.

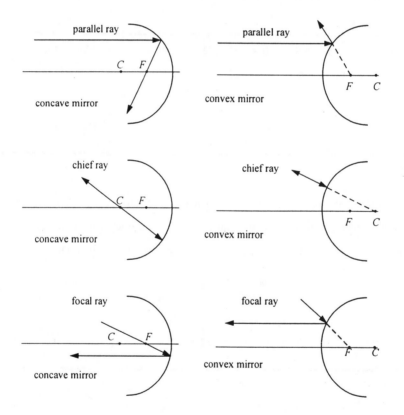

The images of objects can be *upright, inverted, magnified, unmagnified* or *reduced in size*. All these characteristics can be determined from the **lateral magnification**, $M = \dfrac{\text{image height}}{\text{object height}} = \dfrac{h_i}{h_o}$. For real objects (these are the only ones which we will deal with), if M is negative, the image is real, inverted; if M is positive, the image is virtual and upright; if $|M| > 1$, the image is magnified; if $|M| = 1$, the image is unmagnified; and if $|M| < 1$, the image is reduced. These image characteristics are very important in geometrical optics.

Example 23.2   An object is placed 30 cm in front of a concave mirror of radius 20 cm. Determine the characteristics of the image with a ray diagram.

Solution:

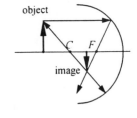

The focal length of the mirror is $f = \dfrac{R}{2} = \dfrac{20 \text{ cm}}{2} = 10$ cm.

We use the parallel ray (parallel in, through focal point out) and chief ray (through center of curvature in, through center of curvature out). These two rays originate from the tip of the object and cross after being reflected by the mirror.

The point where they cross is the image point, so the image is real (because the two rays converged), inverted (the image is upside down), and reduced (the image is smaller than the object), as determined from the ray diagram.

Try to draw the focal ray from the tip of the object here. Will it cross the image point after being reflected?

The images formed by spherical mirrors can also be studied from mathematical equations. The **spherical mirror equation** is given by $\dfrac{1}{d_o} + \dfrac{1}{d_i} = \dfrac{1}{f}$, or $d_i = \dfrac{d_o f}{d_o - f}$, where d_o is the object distance (from the object to the vertex), d_i the image distance (from the image to the vertex), and f the focal length. The lateral magnification can be written in terms of the object and image distances as $M = \dfrac{h_i}{h_o} = -\dfrac{d_i}{d_o}$.

Note: d_i and f should be used as algebraic quantities, that is, they have signs (positive or negative). For a real object (positive d_o), if a mirror is concave (converging), f is positive; if a mirror is convex (diverging), f is negative; if the image is real (formed on the same side of the mirror as the object), d_i is positive; if the image is virtual (formed behind the mirror), d_i is negative.

Example 23.3   Find the location and describe the characteristics of the image formed by a concave mirror of radius 20 cm if the object distance is (a) 30 cm,   (b) 20 cm,   (c) 15 cm,   and   (d) 5.0 cm.

Solution: Given: $R = 20$ cm,   (a) $d_o = 30$ cm,   (b) $d_o = 20$ cm,   (c) $d_o = 15$ cm,   (d) $d_o = 5.0$ cm.

Find: d_i and characteristics in (a), (b), (c), and (d).

The focal length of the mirror is $f = \dfrac{R}{2} = \dfrac{20 \text{ cm}}{2} = 10$ cm.

(a) The object distance is greater than two focal lengths (or the radius of curvature), $d_o > 2f = R$.

The image distance is $d_i = \dfrac{d_o f}{d_o - f} = \dfrac{(30 \text{ cm})(10 \text{ cm})}{30 \text{ cm} - 10 \text{ cm}} = 15$ cm.

The lateral magnification is equal to $M = -\dfrac{d_i}{d_o} = -\dfrac{15 \text{ cm}}{30 \text{ cm}} = -\dfrac{1}{2}$.

So the image is *real* (d_i positive), *inverted* (M is negative), and *reduced* ($|M| < 1$) for $d_o > 2f = R$.

(b) The object distance is equal to two focal lengths (or the radius of curvature), $d_o = 2f = R$.

$$d_i = \frac{(20 \text{ cm})(10 \text{ cm})}{20 \text{ cm} - 10 \text{ cm}} = 20 \text{ cm} \quad \text{and} \quad M = -\frac{20 \text{ cm}}{20 \text{ cm}} = -1.$$

So the image is *real* (d_i positive), *inverted* (M is negative), and *unmagnified* ($|M| = 1$) for $d_o = 2f = R$.

(c) The object distance is greater than one focal length and smaller than two focal lengths, $f < d_o < 2f = R$.

$$d_i = \frac{(15 \text{ cm})(10 \text{ cm})}{15 \text{ cm} - 10 \text{ cm}} = 30 \text{ cm} \quad \text{and} \quad M = -\frac{30 \text{ cm}}{15 \text{ cm}} = -2.$$

So the image is *real* (d_i positive), *inverted* (M is negative), and *magnified* ($|M| > 2$) for $f < d_o < 2f = R$.

(d) The object distance is smaller than one focal length, $d_o < f = R/2$.

$$d_i = \frac{(5.0 \text{ cm})(10 \text{ cm})}{5.0 \text{ cm} - 10 \text{ cm}} = -10 \text{ cm} \quad \text{and} \quad M = -\frac{-10 \text{ cm}}{5.0 \text{ cm}} = -2.$$

So the image is *virtual* (d_i negative), *upright* (M is positive), and *magnified* ($|M| > 1$) for $d_o < f = R/2$.

Hence a concave mirror can form a variety of images depending on the object distance.

Example 23.4   When a person's face is 40 cm in front of a cosmetic mirror (concave mirror), the erect image is three times the size of the object. What is the focal length of this mirror?

Solution: Given: $d_o = 40$ cm, $M = +3.0$.   Find: f.

Since the image is erect and three times the size, the lateral magnification is equal to +3.0 ($M = +3.0$).

From the spherical mirror equation, $\dfrac{1}{d_o} + \dfrac{1}{d_i} = \dfrac{1}{f}$, we first need to know the image distance d_i before we can find the focal length f. Since $M = -\dfrac{d_i}{d_o}$, $\; d_i = -M d_o = -(+3.0)(40 \text{ cm}) = -120$ cm.

Therefore, $\dfrac{1}{f} = \dfrac{1}{40 \text{ cm}} + \dfrac{1}{-120 \text{ cm}} = \dfrac{1}{60 \text{ cm}}$,   or $f = 60$ cm.

Example 23.5   What are the image characteristics of an object 30 cm from a convex mirror of radius 120 cm?

Solution: Given: $d_o = 30$ cm, $R = -50$ cm.   Find: Image characteristics.

For a convex mirror, the radius and focal length are negative. $f = \dfrac{R}{2} = \dfrac{-120 \text{ cm}}{2} = -60$ cm.

From the spherical mirror equation, $d_i = \dfrac{d_o f}{d_o - f} = \dfrac{(30 \text{ cm})(-60 \text{ cm})}{30 \text{ cm} - (-60 \text{ cm})} = -20$ cm.

Since d_i is negative, so the image is virtual.

The lateral magnification is equal to $M = -\dfrac{d_i}{d_o} = -\dfrac{-20 \text{ cm}}{30 \text{ cm}} = +2/3$.

Since $M > 0$ and $|M| < 1$, the image is upright and reduced.

3.   Lenses (Section 23.3)

A lens forms an image based on the law of refraction (Snell's law). A spherical **biconvex lens** is a **converging lens** and a spherical **biconcave lens** is a **diverging lens**. For these lenses, the focal length is *not* equal to one-half the radius of curvature of the surface. Only thin lenses are studied here to ignore the lateral displacement of the ray after refracted from the two lens surfaces.

The images formed by thin lenses can also be studied from geometry (**ray diagrams**). The general rules for drawing ray diagrams for lenses are similar to those for spherical mirrors with some modifications. Three important rays are used to determine the images: a **parallel ray** is a ray incident along a path parallel to the optic axis and refracted through (or appears to go though) the focal point F. a **chief (central) ray** is a ray incident through (or appears to go though) the center of the lens and refracted undeviated; a **focal ray** is a ray which passes through (or appears to go through) the focal point and is refracted parallel to the optic axis.

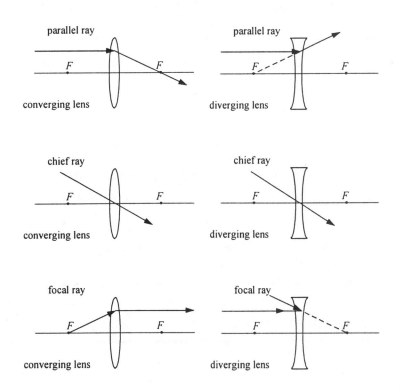

Example 23.6   An object is placed 5.0 cm in front of a thin lens of focal length 15 cm. Determine the characteristics of the image with a ray diagram.

Solution:

We use the chief ray (through center of lens in, undeviated out) and the focal ray (through focal point in, parallel out). These two rays originate from the tip of the object and *appear* to cross after being refracted by the lens. The point where they appear to cross is the image point, so the image is virtual (because the two rays do not cross), upright (the image is right side up), and magnified (the image is larger than the object), as determined from the ray diagram.

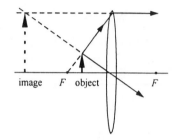

Try to draw the parallel ray from the tip of the object here. Will it cross the image point after being refracted?

The images formed by thin lenses can also be studied from mathematical equations. The **thin-lens equation** is identical in form to the spherical mirror equation, $\frac{1}{d_o} + \frac{1}{d_i} = \frac{1}{f}$, or $d_i = \frac{d_o f}{d_o - f}$, where d_o is the object distance (from the object to the center of the lens), d_i the image distance (from the image to the center of the lens), and f the focal length. The lateral magnification is also defined the same way as for spherical mirrors,

$$M = \frac{h_i}{h_o} = -\frac{d_i}{d_o}.$$

Note: d_i and f should be used as algebraic quantities, that is, they have signs (positive or negative). For real objects (positive d_o), if a lens is convex (converging), f is positive; if a lens is concave (diverging), f is negative; if the image is real (formed on the opposite side of the lens from the object), d_i is positive; if the image is virtual (formed on the same side as the object), d_i is negative. Also, a concave mirror is converging but a concave lens is diverging; a convex mirror is diverging but a convex lens is converging.

Example 23.7   Find the location and describe the characteristics of the image formed by a convex lens of focal length 10 cm if the object distance is (a) 30 cm,   (b) 20 cm,   (c) 15 cm,   and   (d) 5.0 cm.

Solution:

Since the thin-lens equation and lateral magnification for thin lenses are identical to the spherical mirror equation and the lateral magnification for spherical mirrors, where $2f$ is used instead of R, the numerical answers of image distances and image characteristics to this Example are identical to those for Example 23.3. The only difference is on which side of the lens the images are formed. In mirrors, real images are formed on the same side of the mirror as the objects (due to reflection) but in thin lenses, real images are formed on the opposite side of the lens from the objects (due to refraction).

Example 23.8   A biology student uses a converging thin lens to examine a small worm. When the worm is 4.0 cm in front of the lens the upright image is twice the size of the worm. What is the focal length of this lens?

Solution: Given: $d_o = 4.0$ cm, $M = +2.0$.   Find: f.

Since the image is upright and twice the size, the lateral magnification is $+2.0$, ($M = +2.0$).

From the thin lens equation, $\dfrac{1}{d_o} + \dfrac{1}{d_i} = \dfrac{1}{f}$, we first need to know the image distance d_i before we can find

the focal length f. Since $M = -\dfrac{d_i}{d_o}$, $d_i = -Md_o = -(+2.0)(4.0 \text{ cm}) = -8.0$ cm.

Therefore, $\dfrac{1}{f} = \dfrac{1}{4.0 \text{ cm}} + \dfrac{1}{-8.0 \text{ cm}} = \dfrac{1}{8.0 \text{ cm}}$,   or $f = 8.0$ cm.

Example 23.9   Determine the location and describe the characteristics of the image of an object placed at 30 cm in front of a concave (diverging lens) with a focal length of 10 cm.

Solution: Given: $f = -10$ cm, $d_o = 30$ cm.

Find: d_i and image characteristics.

The focal length of a diverging lens is negative.

From the thin lens equation, $d_i = \dfrac{d_o f}{d_o - f} = \dfrac{(40 \text{ cm})(-10 \text{ cm})}{30 \text{ cm} - (-10 \text{ cm})} = -10$ cm.

Since the image distance is negative, the image is virtual.

The lateral magnification is $M = -\dfrac{d_i}{d_o} = -\dfrac{-10 \text{ cm}}{30 \text{ cm}} = 0.33$, so the image is upright and reduced.

Many optical systems such as telescopes and compound microscopes use more than one lens. When two or more lenses are used in combination, the overall image produced may be determined by considering the lenses individually in sequence. That is, the image formed by the first lens is the object for the second lens, and so on. If the first lens produces an image in front of the second lens, the image is treated as a real object for the second lens. If, however, the lenses are close enough together that the image from the first lens is not formed in front of the second lens, then a modification must be made in the sign convention. In this case, the image from the first lens is treated as a *virtual* object for the second lens, and the object distance for it is taken to be *negative* in the lens equation. The total magnification (M_{total}) of a compound lens system is the product of the lateral magnifications of all the component lenses. For a two lens system, $M_{total} = M_1 \times M_2$.

4. Lens Aberrations (Section 23.4)

Rays passing through a lens may not exactly follow the particular rays in a ray diagram because of aberrations. **Spherical aberration** occurs when parallel rays passing through different regions of a lens do not converge or come together on a common image plane. (The thin lens equation is an approximation for rays very close to the optical axis.) **Chromatic aberration** results from the fact that different wavelengths (different colors) have different indices of refraction, and so different angles of refraction, resulting in different image planes for different colors. **Astigmatism** results when a cone (circular cross-section) of light from an off-axis source falls on a lens surface and forms an elliptically illuminated area, giving rise to two images. All these aberrations can be minimized, if not eliminated, by compounding lens systems. For example, some of the common 50-millimeter camera lenses are actually made up of seven lenses to reduce aberrations.

*5. Lens Maker's Equation (Section 23.5)

The **lens maker's equation** is a general equation for determining the focal length of a spherical lens with sides having different (or equal) radii of curvature. In general, the focal length of a lens in air ($n_{air} = 1$) can be calculated from $\dfrac{1}{f} = (n-1)\left(\dfrac{1}{R_1} - \dfrac{1}{R_2}\right)$, where n is the index of refraction of the lens material, and R_1 and R_2 are the radii of the first (front) and second (back) surfaces. The sign convention for the radii are as follows: if the center of curvature C is on the side of the lens from which light emerges or on the back side of the lens, R is positive; if C is on the side of the lens from which light incidents or on the front side of the lens, R is negative. If the lens of a material with an index of refraction n is in a fluid with an index of refraction n_m, the lens maker's equation is modified to $\dfrac{1}{f} = \left(\dfrac{n}{n_m} - 1\right)\left(\dfrac{1}{R_1} - \dfrac{1}{R_2}\right)$.

The power of a lens is defined as $P = \dfrac{1}{f}$, where f is the focal length in meters. The unit of lens power is expressed in **diopters** (D).

Example 23.10   A double convex lens made of glass ($n = 1.52$) has a radius of curvature of 50.0 cm on the front side and 40.0 cm on the back side. Find the power and the focal length of the lens.

Solution: Given: $n = 1.52$, $R_1 = 50.0$ cm, $R_2 = -40.0$ cm.

Find: f.

The center of curvature of the front surface is on the back side of the lens, so its radius is positive. The center of curvature of the back surface is on the front side of the lens, so its radius is negative, according to the sign convention.

From the lens maker's equation, we have

$$P = \frac{1}{f} = (n-1)\left(\frac{1}{R_1} - \frac{1}{R_2}\right) = (1.52 - 1)\left(\frac{1}{0.50 \text{ m}} - \frac{1}{-0.40 \text{ m}}\right) = 2.34 \text{ m}^{-1} = 2.34 \text{ D}.$$

So $f = \dfrac{1}{P} = \dfrac{1}{2.34 \text{ D}} = 0.427 \text{ m} = 42.7 \text{ cm}.$

IV.   Mathematical Summary

Focal Length of Spherical Mirror	$f = \dfrac{R}{2}$ (23.2)	Defines the focal length of a spherical mirror in terms of its radius of curvature.
Spherical Mirror Equation	$\dfrac{1}{d_o} + \dfrac{1}{d_i} = \dfrac{1}{f}$ (23.3) or $d_i = \dfrac{d_o f}{d_o - f}$ (23.4)	Relates object distance, image distance, and focal length of a spherical mirror.
Thin Lens Equation (where $f \neq R/2$)	$\dfrac{1}{d_o} + \dfrac{1}{d_i} = \dfrac{1}{f}$ or $d_i = \dfrac{d_o f}{d_o - f}$ (23.6)	Relates object distance, image distance, and focal length of a thin lens.
Magnification Factor (spherical mirror and thin lens)	$M = -\dfrac{d_i}{d_o}$ (23.5, 23.7)	Defines the lateral magnification as a ratio of the size of the image to the size of the object.
Total Magnification with a Two-Lens System	$M_{\text{total}} = M_1 \times M_2$ (23.8)	Relates the total magnification of a two lens system to the magnifications of the individual lenses.
*Lens Maker's Equation	$\dfrac{1}{f} = (n-1)\left(\dfrac{1}{R_1} - \dfrac{1}{R_2}\right)$ (23.9)	Calculates the focal length of a thin lens.
*Lens Power in Diopters (f in meters)	$P = \dfrac{1}{f}$ (23.10)	Defines the power of a lens in terms of its focal length.

Sign Convention for Spherical Mirrors and Lenses	Positive f for concave mirror and convex lens (converging); negative f for convex mirror and concave lens (diverging). d_o always positive. Positive d_i for real image; negative d_i for virtual image. Positive M for upright image; negative M for inverted image.	Defines the sign convention for focal length f, object distance d_o, image distance d_i, and the lateral magnification M.
*Sign Convention for Lens Maker's Equation	$+R$ for C on back side of lens $-R$ for C on front side of lens $R = \infty$ for a plane (flat) surface $+f$ for converging lens $-f$ for diverging lens	Defines the sign convention for the radii of curvature of a lens.

V.   Solutions of Selected Exercises and Paired/Trio Exercises

3.   (a) Reflections from the window are seen clearly against a dark background. During the day, light passes both ways through the pane, and although some is reflected, it is difficult to see the reflection due to the light coming through. At night, there is little light coming through the pane, so the reflections are seen much more clearly.

(b) The two images are due to reflections on both sides of the pane glass, producing two similar images.

(c) This works on a combination of half-silvering and bright light on one side and dark on the other. For example, at night people inside the house cannot see things outside because there is little light coming through the window from the outside and they see their own reflections.

9.   (a) Image distance equals object distance.

So the distance from object to image is 40 cm + 40 cm = $\boxed{0.80 \text{ m}}$.

(b) The image has the same height as the object. So it is $\boxed{5.0 \text{ cm}}$.

(c) The image is unmagnified and upright. So the magnification is $\boxed{+1.0}$.

12. The image formed by the wall mirror is 0.90 m + 0.90 m = 1.80 m behind the woman or 1.80 m + 0.30 m = 2.10 m behind the hand mirror. The image formed by the hand mirror is then 2.10 m in front of the hand mirror or 2.10 m + 0.30 m = $\boxed{2.40 \text{ m}}$ in front of the woman.

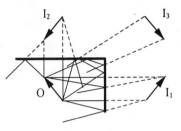

16. The image I_3 is a degenerated image, i.e., two images, one is the image of I_1 by the top mirror and the other is the image of I_2 by the side mirror on top of each other. This happens when the mirrors are at 90° to each other. In Fig 23.23b, the mirrors are not at 90° to each other. So the degeneracy is lifted and the two images are not on top of each other.

24. $d_o = 20$ cm, $\quad f = \dfrac{R}{2} = \dfrac{30 \text{ cm}}{2} = 15$ cm. $\quad d_i = \dfrac{d_o f}{d_o - f} = \dfrac{(20 \text{ cm})(15 \text{ cm})}{20 \text{ cm} - 15 \text{ cm}} = \boxed{60 \text{ cm}}$.

$M = -\dfrac{d_i}{d_o} = -\dfrac{60 \text{ cm}}{20 \text{ cm}} = -3.0$. So $\quad h_i = M h_o = -3.0 (3.0 \text{ cm}) = -9.0$ cm.

It is = $\boxed{9.0 \text{ cm}}$ tall.

26. $d_i = \boxed{-20 \text{ cm}}$; $h_i = \boxed{\tfrac{2}{3} h_o}$.

29. (a)

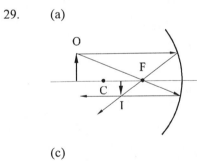

(b)

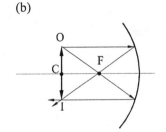

(c)

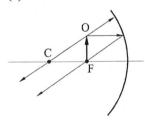

(d)

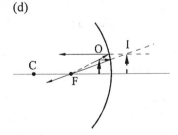

30. Since $\quad d_o < f, \quad d_i = \dfrac{d_o f}{d_o - f} < 0$. Also $\quad M = -\dfrac{d_i}{d_o} = -\dfrac{f}{d_o - f} = \dfrac{f}{f - d_o} > +1$.

So the image is virtual (negative d_i), upright (positive M), and magnified ($|M| > 1$).

37.   (a) $f = \dfrac{R}{2} = \dfrac{30 \text{ cm}}{2} = 15$ cm, $d_o = 20$ cm.

$d_i = \dfrac{d_o f}{d_o - f} = \dfrac{(20 \text{ cm})(15 \text{ cm})}{20 \text{ cm} - 15 \text{ cm}} = \boxed{60 \text{ cm}}$.

$M = -\dfrac{d_i}{d_o} = -\dfrac{60 \text{ cm}}{20 \text{ cm}} = \boxed{-3.0, \text{ real and inverted}}$.

(b)

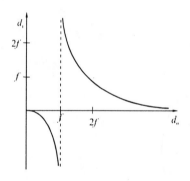

40.   (a) $d_i = \dfrac{d_o f}{d_o - f} = \dfrac{f}{1 - f/d_o}$. $|M| = \dfrac{d_i}{d_o} = \dfrac{f}{d_o - f}$.

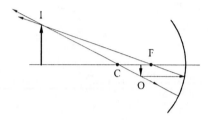

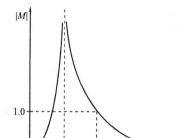

(b) $d_i = \dfrac{d_o(-f)}{d_o + f} = \dfrac{-f}{1 + f/d_o}$. $|M| = \dfrac{d_i}{d_o} = \dfrac{-f}{d_o + f}$.

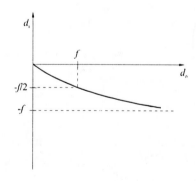

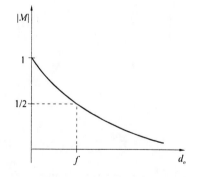

43. $f = \dfrac{R}{2} = \dfrac{20 \text{ cm}}{2} = 10 \text{ cm}$, $M = \pm 2.0$, the + is for a virtual image and the − is for a real image.

$d_i = -M d_o = \mp 2.0 d_o$. $\dfrac{1}{f} = \dfrac{1}{d_o} + \dfrac{1}{d_i}$,   so $\dfrac{1}{10 \text{ cm}} = \dfrac{1}{d_o} + \dfrac{1}{\mp 2.0 d_o}$,

or $\dfrac{2 \mp 1}{2 d_o} = \dfrac{1}{10 \text{ cm}}$.

Solving, $d_o = \dfrac{10 \text{ cm}}{2}(2 \mp 1) = \boxed{5.0 \text{ cm or 15 cm}}$.

45. $\boxed{\text{Yes}}$ it is possible. One is a real image and the other is a virtual image.

$f = \dfrac{R}{2} = \dfrac{40 \text{ cm}}{2} = 20 \text{ cm}$, $M = \pm 3.0$, the + is for a virtual image and the − is for a real image.

$d_i = -M d_o = \mp 3.0 d_o$. $\dfrac{1}{f} = \dfrac{1}{d_o} + \dfrac{1}{d_i}$,   so $\dfrac{1}{20 \text{ cm}} = \dfrac{1}{d_o} + \dfrac{1}{\mp 3.0 d_o}$,

or $\dfrac{3 \mp 1}{3 d_o} = \dfrac{1}{20 \text{ cm}}$.   Solve for $d_o = \dfrac{20 \text{ cm}}{3}(3 \mp 1) = \boxed{13 \text{ cm or 27 cm}}$.

52. $\dfrac{1}{f} = \dfrac{1}{d_o} + \dfrac{1}{d_i} = \dfrac{1}{30 \text{ cm}} + \dfrac{1}{15 \text{ cm}} = \dfrac{1}{10 \text{ cm}}$, so $f = \boxed{10 \text{ cm}}$.

55.   (a) $f = -18.0 \text{ cm}$, $d_o = 10 \text{ cm}$.

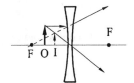

$d_i = \dfrac{d_o f}{d_o - f} = \dfrac{(10 \text{ cm})(-18.0 \text{ cm})}{10 \text{ cm} - (-18.0 \text{ cm})} = \boxed{-6.4 \text{ cm}}$.

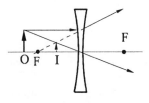

$M = -\dfrac{d_i}{d_o} = -\dfrac{-6.43 \text{ cm}}{10 \text{ cm}} = \boxed{0.64, \text{ virtual and upright}}$.

(b) $d_i = \dfrac{(25 \text{ cm})(-18.0 \text{ cm})}{25 \text{ cm} - (-18.0 \text{ cm})} = \boxed{-10.5 \text{ cm}}$.

$M = -\dfrac{-10.5 \text{ cm}}{25 \text{ cm}} = \boxed{0.42, \text{ virtual and upright}}$.

58.   (a) $\dfrac{1}{f} = \dfrac{1}{d_o} + \dfrac{1}{d_i} = \dfrac{1}{6.0 \text{ cm}} + \dfrac{1}{400 \text{ cm}} = 0.169 \text{ m}^{-1}$, so $f = \boxed{5.9 \text{ cm}}$.

(b) $M = -\dfrac{d_i}{d_o} = -\dfrac{400 \text{ cm}}{6.0 \text{ cm}} = -66.7$.

So $h_i = M h_o = -66.7 (1.0 \text{ cm}) = -66.7 \text{ cm} = \boxed{67 \text{ cm, inverted}}$.

60. $\boxed{-2.5 \text{ cm}}$.

68. For the objective, $\quad d_{i1} = \dfrac{d_{o1}f_o}{d_{o1} - f_o} = \dfrac{(0.30 \text{ cm})(0.28 \text{ cm})}{0.30 \text{ cm} - 0.28 \text{ cm}} = 4.2 \text{ cm}.$

The image by the objective is the object for the eyepiece.

For the eyepiece, $\quad d_{o2} = 7.0 \text{ cm} - d_{i1} = 7.0 \text{ cm} - 4.2 \text{ cm} = 2.8 \text{ cm}.$

$d_{i2} = \dfrac{d_{o2}f_e}{d_{o2} - f_e} = \dfrac{(2.8 \text{ cm})(3.3 \text{ cm})}{2.8 \text{ cm} - 3.3 \text{ cm}} = -18 \text{ cm}.$

So the image is a $\boxed{\text{virtual image 18 cm to the left of eyepiece}}$.

73. (a) $\dfrac{1}{f} = (n - 1)\left(\dfrac{1}{R_1} - \dfrac{1}{R_2}\right)$ assumes $n_{\text{air}} = 1$. If the index of refraction of the surrounding is not air, the lens

maker's equation needs to be modified as $\dfrac{1}{f} = (n_1/n_2 - 1)\left(\dfrac{1}{R_1} - \dfrac{1}{R_2}\right)$, where n_1 is the index of refraction of

the material and n_2 is the index of refraction of the surrounding.

So $\quad \dfrac{f'}{f} = \dfrac{1.6/1.33 - 1}{1.6/1 - 1} = 0.34$, i.e., the $\boxed{\text{focal length decreases by a factor of 0.34}}$.

(b) $\dfrac{f'}{f} = \dfrac{1.3/1.33 - 1}{1.3/1 - 1} = -0.075$, that is,

$\boxed{\text{diverging lens becomes converging lens and vice versa and } f \text{ decreases by a factor of } 0.075}$.

76. $P = \dfrac{1}{f} = (n - 1)\left(\dfrac{1}{R_1} - \dfrac{1}{R_2}\right), \quad \text{☞} \quad \left(\dfrac{1}{R_1} - \dfrac{1}{R_2}\right) = \dfrac{P}{n - 1} = \dfrac{1.5 \text{ D}}{1.6 - 1} = 2.5 \text{ D}.$

So $\quad \dfrac{1}{R_2} = \dfrac{1}{R_1} - 2.5 \text{ D} = \dfrac{1}{0.20 \text{ m}} - 2.5 \text{ D} = 2.5 \text{ D}.$

Therefore $R_2 = \dfrac{1}{2.5 \text{ D}} = 0.40 \text{ m} = \boxed{40 \text{ cm}}.$

83. The image formed by the converging lens is at the mirror. This image is the object for the diverging lens. If the mirror is at the focal point of the diverging lens, the rays refracted after the diverging lens will be parallel to the axis. These rays will be reflected back parallel to the axis by the mirror and will form another image at the mirror. This second image is now the object for the converging lens. By reversing the rays, a sharp image is formed on the screen located where the original object is. Therefore the distance from the diverging lens to the mirror is the focal length of the diverging lens.

88. For the first lens, $d_{o1} = 15$ cm, $f_1 = 10$ cm. $d_{i1} = \dfrac{d_{o1} f_1}{d_{o1} - f_1} = \dfrac{(15 \text{ cm})(10 \text{ cm})}{15 \text{ cm} - 10 \text{ cm}} = 30$ cm.

$M_1 = -\dfrac{d_{i1}}{d_{o1}} = -\dfrac{30 \text{ cm}}{15 \text{ cm}} = -2.0.$ The image of the first lens is the object for the second lens.

For the second lens, $d_{o2} = d - d_{i1} = 60$ cm $- 30$ cm $= 30$ cm, where d is the distance between the lenses.

So $d_{i2} = \dfrac{d_{o2} f_2}{d_{o2} - f_2} = \dfrac{(30 \text{ cm})(20 \text{ cm})}{30 \text{ cm} - 20 \text{ cm}} = \boxed{60 \text{ cm to right of } L_2}.$

$M_2 = -\dfrac{d_{i2}}{d_{o2}} = -\dfrac{60 \text{ cm}}{30 \text{ cm}} = -2.0.$ Therefore $M_{total} = M_1 M_2 = (-2.0)(-2.0) = \boxed{4.0, \text{ real and upright}}.$

89. (a) Since the index of refraction of the lens is greater than that of air, the angle of refraction is smaller than the angle of incidence at the air–lens interface and greater than the angle of incidence at the lens–air interface. So both refractions bend the incident light toward the axis.

(a)   (b)

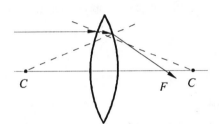   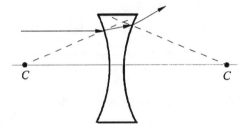

(b) For the same reason, the rays bend away from the axis due to the opposite curvatures of the surfaces.

93. (a)

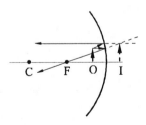

(b) $d_o = 20$ cm, $M = +1.5$ (virtual). $d_i = -M d_o = -1.5 (20 \text{ cm}) = -30$ cm.

$\dfrac{1}{f} = \dfrac{1}{d_o} + \dfrac{1}{d_i} = \dfrac{1}{20 \text{ cm}} + \dfrac{1}{-30 \text{ cm}} = \dfrac{1}{60 \text{ cm}},$ ☞ $f = \boxed{60 \text{ cm}}.$

VI. Practice Quiz

1. Which one of the following describes the image of a concave mirror when the object distance from the mirror is greater than twice the focal length?
    (a) virtual, upright, and magnified   (b) real, inverted, and reduced
    (c) virtual, upright, and reduced   (d) real, inverted, and magnified
    (e) virtual, inverted, and magnified

2. A concave mirror with a radius of curvature of 20 cm creates a real image 30 cm from the mirror. What is the object distance?
    (a) 20 cm   (b) 15 cm   (c) 7.5 cm   (d) 5.0 cm   (e) 2.5 cm

3. A person's face is 30 cm in front of a concave shaving mirror. If the image is an upright image 1.5 times as large as the object, what is the mirror's focal length?
    (a) 12 cm   (b) 20 cm   (c) 50 cm   (d) 70 cm   (e) 90 cm

4. Which of the following describes the image of a convex mirror?
    (a) virtual, inverted, and magnified   (b) real, inverted, and reduced
    (c) virtual, upright, and reduced   (d) virtual, upright, and magnified
    (e) virtual, upright, and unmagnified

5. An object is placed at a distance of 40 cm from a thin lens. If a virtual image forms at a distance of 50 cm from the lens, on the same side as the object, what is the focal length of the lens?
    (a) 200 cm   (b) 90 cm   (c) 75 cm   (d) 45 cm   (e) 22 cm

6. If you stand 2.5 ft in front of a plane mirror, how far away would your image in the mirror?
    (a) 2.5 ft   (b) 5.0 ft   (c) 7.5 ft   (d) 10 ft   (e) 20 ft

7. An object is placed at a distance of 30 cm from a thin convex lens. The lens has a focal length of 10 cm. What are the values, respectively, of the image distance and lateral magnification?
    (a) 65 cm, 2.0   (b) 15 cm, 2.0   (c) 25 cm, 1.0   (d) 60 cm, −0.50   (e) 15 cm, −0.50

8. The ray diagram on the right is an example of a

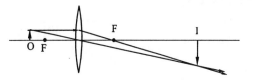

    (a) camera

    (b) overhead projector

    (c) magnifying glass

    (d) microscope

    (e) telescope

9. What causes spherical aberration?

    (a) ray is too far from the optic axis.

    (b) different color has the same index of refraction.

    (c) a circular cone of beam becomes an elliptical beam.

    (d) too many lenses are used in the optical system.

    (e) different wavelengths have different wave speeds.

10. A biconvex lens is formed by using a piece of glass ($n = 1.52$). The radius of the front surface is 30 cm and the radius of the back surface is 40 cm. What is the focal length of the lens?

    (a) 2.3 m,   (b) −33 cm   (c) 33 cm   (d) 3.0 m   (e) −3.0 m

Answers to Practice Quiz:

1. b 2. b 3. e 4. c 5. a 6. b 7. e 8. b 9. a 10. c

CHAPTER 24

Physical Optics: The Wave Nature of Light

I.   Chapter Objectives

Upon completion of this chapter, you should be able to:

1.   explain how Young's experiment demonstrates the wave nature of light and compute the wavelength of light from experimental results.

2.   describe how thin films produce colorful displays and give some examples of practical applications of thin-film interference.

3.   define diffraction and give examples of diffractive effects.

4.   explain light polarization and give examples of polarization, both in the environment and in commercial applications.

5.   define scattering and explain why the sky is blue and sunsets are red.

II.   Key Terms

Upon completion of this chapter, you should be able to define and/or explain the following key terms:

physical (wave) optics	polarizing (Brewster) angle
Young's double-slit experiment	birefringence
thin-film interference	dichroism
optical flats	transmission axis (polarization direction)
Newton's rings	optical activity
diffraction	LCD (liquid crystal display)
diffraction grating	scattering
Bragg's law	Rayleigh scattering
polarization	

The definitions and/or explanations of the most important key terms can be found in the following section: **III. Chapter Summary and Discussion.**

III.   Chapter Summary and Discussion

1.   Young's Double-Slit Experiment (Section 24.1)

Physical (wave) optics treats light as a wave in the study of some light phenomena, such as interference, diffraction, and polarization. These effects can not be successfully explained with geometric optics, in which light is treated as several rays that follow straight line paths.

Young's double-slit experiment not only demonstrates the wave nature of light, but also allows the measurement of the wavelength of light. In this experiment, a light source is incident on two closely spaced narrow slits. The light waves emanating from these two slits can be considered as two in-phase spherical sources and they interfere when they arrive at a screen. On the screen, the interference pattern consists of equally spaced bright fringes separated by equally spaced dark fringes.

As for sound waves (Chapter 14), the condition for interference is determined by the *path-length difference* (ΔL) of the two waves, or the difference in distance traveled. If $\Delta L = n\lambda$ for $n = 0, 1, 2, 3, \ldots$, the interference is constructive (bright fringe); if $\Delta L = \dfrac{m\lambda}{2}$ for $m = 1, 3, 5, \ldots$, the interference is destructive (dark fringe). For two slits separated by a distance d, the path difference is $\Delta L = d \sin\theta$, so the condition for constructive interference in Young's double-slit experiment is $d \sin\theta = n\lambda$ for $n = 0, 1, 2, 3, \ldots$, where n is called the order number. The zeroth-order fringe ($n = 0$) corresponds to the central maximum, the first-order fringe ($n = 1$) is the first bright fringe on either side of the central maximum (there are two first-order fringes), and so on.

For a small angle θ, we have $\sin\theta \approx \tan\theta \approx \dfrac{y}{L}$, where y is the distance from the central maximum on the screen and L is the distance from the slits to the screen.   The distance of the nth bright fringe (y_n) from the central maximum on either side is $y_n \approx \dfrac{nL\lambda}{d}$, the wavelength of light is then $\lambda \approx \dfrac{y_n d}{nL}$ for $n = 1, 2, 3, \ldots$, and the separation between adjacent bright fringes is $y_{n+1} - y_n = \dfrac{L\lambda}{d}$. (The dark fringes are separated by this distance also.)

Example 24.1   Light of wavelength 632.8 nm falls on a double-slit and the third-order bright fringe is seen at an angle of 6.5°. Find the separation between the double slits.

Solution: Given: $\lambda = 632.8 \text{ nm} = 632.8 \times 10^{-9} \text{ m}$, $n = 3$, $\theta_3 = 6.5°$.   Find: d.

From $d \sin\theta = n\lambda$, we have $d = \dfrac{n\lambda}{\sin\theta} = \dfrac{(3)\,\lambda}{\sin\theta_3} = \dfrac{(3)(632.8 \times 10^{-9} \text{ m})}{\sin 6.5°} = 1.7 \times 10^{-5} \text{ m} = 17 \ \mu\text{m}$.

Example 24.2 In a Young's double-slit experiment, if the separation between the two slits is 0.10 mm and the distance from the slits to a screen is 2.5 m, find the spacing between the first-order and second-order bright fringes for light with wavelength of 550 nm.

Solution: Given: $d = 0.10$ mm $= 0.10 \times 10^{-3}$ m, $L = 2.5$ m, $\lambda = 550$ nm $= 550 \times 10^{-9}$ m.

Find: $y_2 - y_1$.

From $y_n = \dfrac{nL\lambda}{d}$, $y_2 - y_1 = \dfrac{(2)L\lambda}{d} - \dfrac{(1)L\lambda}{d} = \dfrac{L\lambda}{d} = \dfrac{(2.5 \text{ m})(550 \times 10^{-9} \text{ m})}{0.10 \times 10^{-3} \text{ m}} = 1.4 \times 10^{-2}$ m $= 1.4$ cm.

2. Thin-Film Interference (Section 24.2)

The waves reflected from the two surfaces of a thin-film can interfere, and this **thin-film interference** depends on the reflective phases of the two reflected waves. Light reflected off a material whose index of refraction is greater than that of the one it is in ($n_2 > n_1$) undergoes a *180° phase change*. If $n_2 < n_1$, there is *no phase change* on reflection. For a particular film, light reflected from the film surfaces may interfere constructively and destructively, depending on the film thickness and any additional phase changes. A practical application of thin-film interference is the nonreflecting coatings for lenses (destructive interference for reflection of certain wavelengths). In a nonreflecting coating, the index of refraction of the film is usually between those of air and glass, so there are 180° phase changes for each reflection off the two surfaces of the film. The minimum thickness of the film is given by $t = \dfrac{\lambda}{4n}$, where n is the index of refraction of the film.

When a perfect spherical lens is placed on an **optical flat**, a very flat (usually as smooth as $\lambda/20$) piece of glass, circular fringes are observed. These fringes are called **Newton's rings**. If a lens has irregularities, the fringes are distorted. This is a simple, yet effective, method to check the quality of lenses in optical industries.

Example 24.3 A transparent material with an index of refraction 1.30 is used to coat a piece of glass with an index of refraction 1.52. What is the minimum thickness of the film in order to minimize the reflected light with a wavelength of 550 nm if the light is incident perpendicularly?

Solution: Given: $\lambda = 550$ nm, n (film) $= 1.30$. Find: t (minimum).

The index of refraction of the film is between those of air and glass, so both reflections from the two surfaces of the film have 180° phase shifts. We can use Equation (24.8).

The minimum film thickness is $t = \dfrac{\lambda}{4n} = \dfrac{550 \text{ nm}}{4(1.30)} = 106$ nm.

3.   Diffraction (Section 24.3)

Diffraction is the deviation or bending of light around objects, edges or corners. Generally, the smaller the size of the opening or object compared to the wavelength of light, the greater the diffraction. The diffraction pattern from a single slit of width w consists of a broader central maximum and some narrower side maxima (the width of the central maximum is twice that of the side maxima). Between two maxima, there is a region of destructive interference (dark fringes). In diffraction, the dark fringes rather than the bright fringes are analyzed.

The condition for the dark fringes is given by $w \sin \theta = m\lambda$ for $m = 1, 2, 3, \ldots$, where θ is the angle for a particular minimum designated by $m = 1, 2, 3, \ldots$ on either side of the central bright fringe (there is no dark fringe corresponding to $n = 0$. Why?) For a small angle approximation, the position of the dark fringes from the center of the central bright fringe on a screen can be calculated from $y_m = \dfrac{mL\lambda}{w}$, where L is the distance from the single slit to the screen. The width of the mth side maximum is the distance between the mth dark and the $(m+1)^{th}$ dark, or $y_{m+1} - y_m$. It is evident from this equation that

- for a given slit width (w), the greater the wavelength (λ), the wider the diffraction pattern (y_m);
- for a given wavelength (λ), the narrower the slit width (w), the wider the diffraction pattern (y_m);
- the width of the central maximum is twice the width of the side maxima.

Example 24.4   Light of wavelength 832.8 nm is incident on a slit of width 0.200 mm. An observing screen is placed 2.50 m from the slit. Find the width of the central maximum and the positions of the third-order dark fringe.

Solution: Given: $\lambda = 632.8$ nm $= 632.8 \times 10^{-9}$ m, $w = 0.200$ mm $= 0.200 \times 10^{-3}$ m, $L = 2.50$ m,
Find: $2y_1$ and y_3.

The central maximum is the region between the first-order dark fringes on either side of this maximum, so its width is simply $2y_1$.

From $y_m = \dfrac{mL\lambda}{w}$,   we have $y_1 = \dfrac{(1)L\lambda}{w} = \dfrac{(1)(2.50 \text{ m})(632.8 \times 10^{-9} \text{ m})}{0.200 \times 10^{-3} \text{ m}} = 7.91 \times 10^{-3}$ m $= 7.91$ mm.

So the width of the central maximum is $2y_1 = 2(7.91 \text{ mm}) = 15.8$ mm.

$y_3 = \dfrac{(3)L\lambda}{w} = \dfrac{(3)(2.50 \text{ m})(632.8 \times 10^{-9} \text{ m})}{0.200 \times 10^{-3} \text{ m}} = 23.7$ mm.

Example 24.5   Light of wavelength 550 nm is incident on a single slit 0.75 mm wide. At what distance from the slit should a screen be placed if the second dark fringe in the diffraction pattern is to be 1.7 mm from the center of the screen?

Solution: Given: $\lambda = 550$ nm $= 550 \times 10^{-9}$ m, $w = 0.75$ mm $= 0.75 \times 10^{-3}$ m, $m = 2$,

$y_2 = 1.7$ mm $= 1.7 \times 10^{-3}$ m.

Find: L.

From $y_m = \dfrac{mL\lambda}{w}$,   we have $L = \dfrac{y_m w}{m\lambda} = \dfrac{(1.7 \times 10^{-3} \text{ m})(0.75 \times 10^{-3} \text{ m})}{(2)(550 \times 10^{-9} \text{ m})} = 1.2$ m.

A **diffraction grating** consists of a larger number of closely spaced narrow slits. The diffraction pattern of a diffraction grating is a combination of multiple slit interference and single slit diffraction. It is very useful for dispersing different wavelengths or colors. (In general, it gives a larger dispersion than a prism.) The sharp (narrow, or well defined) bright fringes for a diffraction grating are given by $d \sin \theta = n\lambda$ for $n = 0, 1, 2, 3, \ldots$, where d is the spacing (or grating's spacing) between adjacent grating slits, which can be obtained from the number of lines per unit length of the grating, $d = \dfrac{1}{N}$. The number of spectral orders produced by a grating depends on the wavelength and on the grating's spacing d. Since $\sin \theta$ cannot be greater than 1, $\sin \theta = \dfrac{n\lambda}{d} \leq 1$, so the order number is therefore limited as follows: $n \leq \dfrac{d}{\lambda}$.

Note: In general, the small angle approximation cannot be used in a diffraction grating because of the grating's larger dispersion power. The bright fringes are usually separated by bigger distances here and, therefore, the angles do not satisfy the small angle approximation.

Example 24.6   Monochromatic light is incident on a grating that is 10.0 cm wide and ruled with 50 000 lines. The third-order maximum is seen at 46.3°. What is the wavelength of the incident light?

Solution: Given: $n = 3$, $\theta_2 = 46.3°$, $N = \dfrac{50\,000 \text{ lines}}{10 \text{ cm}} = 5000$ lines/cm.

Find: λ.

The grating spacing is $d = \dfrac{1}{N} = \dfrac{1}{5000 \text{ lines/cm}} = 2.00 \times 10^{-4}$ cm $= 2.00 \times 10^{-6}$ m.

From $d \sin \theta = n\lambda$, we have $\lambda = \dfrac{d \sin \theta}{n} = \dfrac{(2.00 \times 10^{-6} \text{ m}) \sin 46.3°}{3} = 4.82 \times 10^{-7}$ m $= 482$ nm.

Example 24.7   Monochromatic light of wavelength 632.8 nm is incident normally on a diffraction grating. If the third-order maximum of the diffraction pattern is observed at 32.0°,

   (a) what is the grating's spacing?

   (b) how many total number of visible maxima can be seen?

Solution: Given: $\lambda = 632.8$ nm $= 632.8 \times 10^{-9}$ m, $n = 3$, $\theta_3 = 32.0°$.

   Find: (a) d (b) total number of visible maxima.

   (a) From $d \sin \theta = n\lambda$, we have $d = \dfrac{n\lambda}{\sin \theta} = \dfrac{(3)(632.8 \times 10^{-9}\ \text{m})}{\sin 32.0°} = 3.58 \times 10^{-6}$ m.

   (b) The maximum value of the angle θ is 90° ($\sin 90° = 1$), so the maximum order number is

$n_{max} = \dfrac{d}{\lambda} = \dfrac{3.58 \times 10^{-6}\ \text{m}}{632.8 \times 10^{-9}\ \text{m}} = 5.66.$ Since n is an integer, $n_{max} = 5$.

   Therefore, 11 maxima are seen including the center one ($n = 0$, two for $n = 1$, two for $n = 2$, two for $n = 3$, two for $n = 4$, and two for $n = 5$).

The regular atomic spacing in a crystalline solid acts as a diffraction grating for light of much shorter wavelength than visible light such as X-rays. By measuring the diffraction angle, which is equal to the incidence angle also, of an X-ray beam with known wavelength, the distance between the crystal's internal planes (d) can be determined from **Bragg's law**, $2d\sin \theta = n\lambda$ for $n = 1, 2, 3, \ldots$.

4.   Polarization (Section 24.4)

   Polarization is the preferential orientation of the electromagnetic field vectors that make up a light wave, and is evidence that light is a transverse wave. Light with some partial preferential orientation of the electromagnetic field vectors is said to be *partially polarized*. If the electromagnetic field vectors oscillate in *one* plane (or *one* direction), the light is then *plane (linearly) polarized*. Light can be polarized by reflection, double refraction (**birefringence**), selective absorption (**dichroism**), and scattering.

   Light that is partially reflected and partially refracted is partially polarized. However, when the reflected and the refracted rays form a 90° angle, the reflected ray is linearly polarized or maximum polarization occurs. The angle of incidence for this maximum polarization is called the **polarizing** or **Brewster angle**. The polarizing angle for a material of index of refraction n in air is given by $\tan \theta_p = n$ or $\theta_p = \tan^{-1} n$.

In some materials, the anisotropy of the speed of light with direction (speed of light is different in different directions) gives rise to different indices of refraction in different directions. This property is called birefringence, and such materials are said to be *birefringent* or *double refracting*. When a beam of unpolarized light is incident on a birefringent crystal, it is doubly refracted and separated into two components, or rays. These two rays are linearly polarized with the electromagnetic field vectors in mutually perpendicular directions.

Some crystals, such as tourmaline and herapathite, exhibit the interesting property of absorbing one of the polarized components more than the other. This property is called **dichroism**. If the dichroic material is sufficiently thick, the more strongly absorbed component may be completely absorbed, resulting in a linearly polarized beam with the unabsorbed component. Polaroid films (used in sunglasses) are synthetic polymer materials, which allow light of one polarization direction to pass. This direction is called the **transmission axis**, or **polarization direction**. If two films (the first one is called the polarizer and the second one is called the analyzer) are placed with their transmission axes parallel to each other, light can pass through both films. If the two transmission axes are perpendicular to each other, little or no light can pass. In general, the intensity of the transmitted light is given by $I = I_0 \cos^2 \theta$, where θ is the angle between the transmission axes of the polarizer and analyzer, I_0 the original intensity, and I the intensity after the analyzer. This expression is known as *Malus's law*.

Some transparent materials have the ability to rotate the polarization direction of the linearly polarized light. This is called the **optical activity** and is due to the molecular structure of the material. Some liquid crystals are optically active, and this property forms the basis of the common **liquid crystal display (LCD)**.

Example 24.8   How far above the horizon is the Moon when its image reflected in calm water is completely polarized?

Solution: Given: $n = 1.33$ (water).   Find: $90° - \theta_p$.

The polarizing angle is the angle of incidence (measured from the normal) when the reflected ray is linearly polarized. The angle above the horizon is the angle measured from the surface ($90° - \theta_p$).

From $\tan \theta_p = n$, we have $\theta_p = \tan^{-1} n = \tan^{-1} 1.33 = 53.1°$.

So the angle above the horizon is $90° - 53.1° = 36.9°$.

Example 24.9   Unpolarized light is passed through a polarizer-analyzer combination. The transmission axes of the polarizer and analyzer are at 45° to each other. What percentage of the light gets through the filters?

Solution:

When unpolarized light is incident on the polarizer, only one of the two components can pass; that is, only 50% of the intensity gets through the polarizer. So before the light reaches the analyzer, the intensity is already reduced to half.

From Malus's law, $I = I_0 \cos^2 \theta$, $\dfrac{I}{I_0} = \cos^2 \theta = \cos^2 45° = 0.50$ or 50%.

Therefore, only $\dfrac{50\%}{2} = 25\%$ of the original intensity passes through the polarizer-analyzer.

---

5. Atmospheric Scattering of Light (Section 24.5)

Scattering is the process of particles (like air molecules or dust particles) absorbing light and re-radiating polarized light. The scattering of sunlight by air molecules causes the sky to look blue and sunsets to look red because the shorter wavelength (blue) is scattered more efficiently than long wavelength (red). This is called **Rayleigh scattering** and the scattering intensity is found to be proportional to $\dfrac{1}{\lambda^4}$.

Blue scatters more efficiently than red. In the morning and evening, the blue component of the light from the Sun is scattered more in the denser atmosphere near the Earth, so we see red when we look in the direction of the rising or setting Sun. During the day, we mainly see the blue component from overhead scattering.

IV. Mathematical Summary

Bright Fringe Condition (double-slit interference)	$d\sin\theta = n\lambda$ for $n = 0, 1, 2, \ldots$ (24.3)	Gives the positions of the bright fringes in double-slit interference experiment.
Wavelength Measurement (double-slit interference for small θ only)	$\lambda = \dfrac{y_n d}{nL}$ for $n = 1, 2, 3, \ldots$ (24.7)	Determines the wavelength in double-slit interference experiment.
Nonreflecting Film Thickness (minimum)	$t = \dfrac{\lambda}{4n}$ (for $n_2 > n_1 > n_0$) (24.8)	Gives the minimum thickness of a thin film coating to minimize reflection.
Dark Fringe Condition (single-slit diffraction)	$w\sin\theta = m\lambda$ for $m = 1, 2, 3, \ldots$ (24.9)	Gives the positions of the dark fringes in single slit diffraction experiment.

Lateral Displacement of Dark Fringes (single-slit diffraction)	$y_m = \dfrac{mL\lambda}{w}$ for $m = 1, 2, 3, \ldots$ (24.10)	Gives the lateral position of the mth dark fringe in single-slit diffraction experiment.
Interference Maxima for a Diffraction Grating	$d\sin\theta = n\lambda$ for $n = 0, 1, 2, \ldots$ (24.13) where $d = \dfrac{1}{N}$ and N is the number of lines per unit length	Determines the positions of the bright fringes for a diffraction grating.
Limit of Order Number	$n \leq \dfrac{d}{\lambda}$ (24.14)	Gives the highest order number for a diffraction grating.
Brewster (polarizing) Angle	$\tan\theta_p = n$ (24.16)	Calculates the polarizing or Brewster angle in a medium to air boundary.

V.   Solutions of Selected Exercises and Paired/Trio Exercises

5.   The path-length difference will change because of the airplane. This change in path-length difference results in a change in the condition of interference, i.e., constructive is no longer constructive and destructive is no longer destructive, etc. Therefore the pictures flutter.

8. 0.75 m $= 0.50$ m $+ 0.25$ m $= 1.5(0.50$ m$) = 1.5\lambda$.

So the waves will interfere $\boxed{\text{destructively}}$.

12.   The distance is equal to $y_3 - y_o = 3\Delta y = \dfrac{3L\lambda}{d} = \dfrac{3(1.5 \text{ m})(680 \times 10^{-9} \text{ m})}{0.25 \times 10^{-3} \text{ m}} = \boxed{1.2 \text{ cm}}$.

14. $\boxed{600 \text{ nm (orange-yellow)}}$.

20.   (a) because only the reflection at n_o–n_1 interface has $180°$ phase shift.

25.   (a) $\lambda_n = \dfrac{\lambda}{n} = \dfrac{550 \text{ nm}}{1.5} = 367$ nm. $t = 1.1 \times 10^{-5}$ m $= 30(367 \times 10^{-9} \text{ m}) = \boxed{30\lambda}$.

(b) The path length difference $\Delta L = 2t = 2(30\lambda) = 60\lambda$.

However, the first reflection has a $180°$ phase shift. So they will interfere $\boxed{\text{destructively}}$.

26. $t = \dfrac{\lambda}{4n} = \dfrac{700 \text{ nm}}{4(1.4)} = 125 \text{ nm} = \boxed{1.3 \times 10^{-7} \text{ m}}$.

31.   (a) The two rays for interference are the reflections from the bottom surface of the top plate and the top

        surface from the bottom plate. The reflection from the top surface of the bottom plate has 180° phase shifts,

        so the condition for constructive interference for reflection is $\Delta L = 2t = \dfrac{\lambda}{2}$,

        so $t = \dfrac{\lambda}{4} = \dfrac{632.8 \text{ nm}}{4} = \boxed{158.2 \text{ nm}}$.

        (b) Constructive for transmission is the same as destructive for reflection.

        So $\Delta L = 2t = \lambda$,   ☞ $t = \dfrac{\lambda}{2} = \dfrac{632.8 \text{ nm}}{2} = \boxed{316.4 \text{ nm}}$.

38.   (a) The width of the central maximum is

        $y_1 - y_{-1} = 2\Delta y = \dfrac{2 L \lambda}{w} = \dfrac{2(1.0 \text{ m})(480 \times 10^{-9} \text{ m})}{0.20 \times 10^{-3} \text{ m}} = \boxed{4.8 \text{ mm}}$.

        (b) $y_3 - y_2 = y_4 - y_3 = \Delta y = \dfrac{L \lambda}{w} = \boxed{2.4 \text{ mm}}$.

41.   (a) From $d \sin \theta = m\lambda$, we have $\lambda = \dfrac{d \sin \theta}{m} = (0.025 \text{ m}) \sin 10° = \boxed{4.3 \text{ mm}}$.

        (b) $f = \dfrac{c}{\lambda} = \dfrac{3.00 \times 10^8 \text{ m/s}}{4.34 \times 10^{-3} \text{ m}} = 6.9 \times 10^{10} \text{ Hz}, \boxed{\text{microwave}}$.

46. $d = \dfrac{1}{10\,000 \text{ lines/cm}} = 1.0 \times 10^{-4} \text{ cm} = 1.0 \times 10^{-6} \text{ m}$.

        From $d \sin \theta = m\lambda$, we have $m_{max} = \dfrac{d \sin 90°}{\lambda} = \dfrac{d}{\lambda} = \dfrac{1.0 \times 10^{-6} \text{ m}}{560 \times 10^{-9} \text{ m}} = 1.8$.

        So there are $\boxed{3}$ orders of maxima corresponding to $m = 0$ or ± 1.

48. $\boxed{\text{They do not overlap}}$.

51.   Since $d \sin \theta = m\lambda$, $\theta = \sin^{-1} \dfrac{m\lambda}{d}$.

        For violet, $\theta_{3v} = \sin^{-1} \dfrac{(3)(400 \text{ nm})}{d} = \sin^{-1} \dfrac{1200 \text{ nm}}{d}$.

        For yellow-orange, $\theta_{2y} = \sin^{-1} \dfrac{(2)(600 \text{ nm})}{d} = \sin^{-1} \dfrac{1200 \text{ nm}}{d}$.

        So $\theta_{3v} = \theta_{2y}$, i.e., they overlap.

56. We see the rainbow by the scattering of light from the water droplets. The light is partially polarized in the horizontal direction, so the axis of the analyzer should be in the horizontal direction. We can never block out the polarized light completely because it is only partially polarized.

60. $n = \tan \theta_p = \tan 58° = \boxed{1.6}$.

69. Blue scatters more efficiently than red. In the morning and evening, the blue component of the light from the Sun is scattered more in the denser atmosphere near the Earth, so we see red when we look in the direction of the rising or setting Sun. During the day, we mainly see the blue component from overhead scattering.

73. From $n = \tan \theta_p$, we have $\theta_p = \tan^{-1} n = \tan^{-1} 1.5 = \boxed{56°}$.

77. Since $\Delta y = \dfrac{\lambda L}{d}$, $\Delta \theta = \dfrac{\Delta y}{L} = \dfrac{\lambda}{d} = \dfrac{480 \times 10^{-9} \text{ m}}{0.75 \times 10^{-3} \text{ m}} = \boxed{6.4 \times 10^{-4} \text{ rad}}$.

80. $d = \dfrac{1}{9000 \text{ lines/cm}} = 1.11 \times 10^{-4} \text{ cm} = 1.11 \times 10^{-6} \text{ m}$.

From $d \sin \theta = m\lambda$, we have $m_{max} = \dfrac{d \sin 90°}{\lambda} = \dfrac{d}{\lambda}$.

For red, $m_{max} = \dfrac{1.11 \times 10^{-6} \text{ m}}{700 \times 10^{-9} \text{ m}} = 1.6$. So $m_{max} = \boxed{1 \text{ for red}}$.

For violet, $m_{max} = \dfrac{1.11 \times 10^{-6} \text{ m}}{400 \times 10^{-9} \text{ m}} = 2.8$. So $m_{max} = \boxed{2 \text{ for violet}}$.

VI.   Practice Quiz

1. If a wave from one slit of a Young's double-slit experiment arrives at a point on the screen two wavelengths behind the wave from the other slit, what is observed at that point?

   (a) bright fringe   (b) dark fringe   (c) gray fringe   (d) multi-colored fringe   (e) none of the above

2. A monochromatic light is incident on a Young's double-slit separated by 3.00×10^{-5} m. The resultant bright fringe separation is 2.15×10^{-2} m on a screen 1.20 m from the double slit. What is the separation between the third order bright fringe and the zeroth order fringe?

   (a) 8.60×10^{-2} m   (b) 7.35×10^{-2} m   (c) 6.45×10^{-2} m   (d) 4.30×10^{-2} m   (e) 2.15×10^{-2} m

3. What is the minimum thickness of a nonreflecting coating ($n = 1.35$) on a glass lens ($n = 1.52$) for wavelength 550 nm?

   (a) zero   (b) 102 nm   (c) 204 nm   (d) 90.5 nm   (e) 181 nm

4. What will happen to the width of the central maximum if the width of the slit decreases in a single slit experiment?

   (a) decrease   (b) increase   (c) remain unchanged   (d) does not depend on the separation
   (e) cannot be determined because not enough information is given

5. Light of wavelength 610 nm is incident on a slit 0.20-millimeter wide and the diffraction pattern is produced on a screen that is 1.5 m from the slit. What is the width of the central maximum?

   (a) 0.34 cm   (b) 0.68 cm   (c) 0.92 cm   (d) 1.22 cm   (e) 1.35 cm

6. A beam of unpolarized light in air strikes a flat piece of glass at an angle of incidence of 57.0°. If the reflected beam is completely polarized, what is the index of refraction of the glass?

   (a) 0.54   (b) 0.84   (c) 1.12   (d) 1.54   (e) 1.84

7. What is the process to obtain polarized light in a dichroic material like Polaroid film?

   (a) reflection   (b) refraction   (c) double refraction   (d) selective absorption   (e) scattering

8. When the transmission axes (polarization directions) of two Polaroid sheets are parallel to each other, what is the percentage of the incident light which will pass the two sheets?

   (a) 0%   (b) 25%   (c) 50%   (d) 75%   (e) 100%

9. White light is spread out into spectral hues by a diffraction grating. If the grating has 2000 lines per centimeter, at what angle will red light ($\lambda = 640$ nm) appear in the first-order?

   (a) 0°   (b) 3.57°   (c) 7.35°   (d) 11.2°   (e) 13.4°

10. A helium–neon laser ($\lambda = 632.8$ nm) is used to calibrate a diffraction grating. If the first-order maximum occurs at 20.5°, how many lines are there in a millimeter?

    (a) 138   (b) 185   (c) 276   (d) 455   (e) 552

Answers to Practice Quiz:

1.a 2.c 3.b 4.b 5.c 6.d 7.d 8.e 9.c 10.e

CHAPTER 25

Optical Instruments

I. Chapter Objectives

Upon completion of this chapter, you should be able to:

1. describe the optical workings of the eye and explain some common vision defects and how they are corrected.

2. distinguish between lateral and angular magnifications and describe simple and compound microscopes and their magnifications.

3. distinguish between refractive and reflective telescopes and describe the advantages of each.

4. describe the relationship of diffraction and resolution and state and explain Rayleigh's criterion.

*5. relate color vision and light.

II. Key Terms

Upon completion of this chapter, you should be able to define and/or explain the following key terms:

retina	astronomical telescope
rods	terrestrial telescope
cones	reflecting telescope
nearsightedness	resolution
farsightedness	Rayleigh criterion
astigmatism	resolving power
magnifying glass (simple microscope)	additive primary colors
angular magnification	additive method of color production
compound microscope	complementary colors
objective	subtractive method of color production
eyepiece (ocular)	subtractive primary pigments
refracting telescope	

The definitions and/or explanations of the most important key terms can be found in the following section: **III. Chapter Summary and Discussion.**

III. Chapter Summary and Discussion

1. The Human Eye (Section 25.1)

The crystalline lens in the eye is a converging lens composed of microscopic glassy fibers. Through muscle action, the shape of the lens is adjusted (this adjustment is called accommodation) and sharp images are formed on the **retina**, a light-sensitive surface at the back of the eye. The photo-sensitive **rod** and **cone** cells of the retina are responsible for twilight (black and white) vision and color vision, respectively. The extremes of the range over which distinct vision (sharp focus) is possible are known as the far point and the near point. The far point is the greatest distance at which the eye can see objects clearly, and is for the normal eye taken to be infinity. The near point is the position closest to the eye at which objects can be seen clearly, and depends on the extent the lens can be deformed (thickened) by accommodation, which varies with age even to the normal eye.

There are three common vision defects. **Nearsightedness** is the condition of being able to see nearby objects clearly, but not distant objects. This is caused by the focal length of the crystalline lens being too short (the lens is too converging, image is focused in front of the retina). A person with nearsightedness will have the far point not at infinity, but at a nearer point. This vision defect can be corrected by using a diverging lens. **Farsightedness** is the condition of being able to see distant objects clearly, but not nearby objects. This is caused by the focal length of the crystalline lens being too long (the lens is too diverging, image is focused behind the retina). A person with farsightedness will have the near point not at the normal position, but at some point farther from the eye. This vision defect can be corrected by using a converging lens. **Astigmatism** is caused by a refractive surface, most usually the cornea or crystalline lens, being out of spherical shape. As a result, the eye has different focal lengths in different planes. This condition may be corrected with lenses that have greater curvature in the plane in which the cornea or crystalline lens has deficient curvature.

Example 25.1 A student cannot see clearly objects more than 95.0 cm away. What power of lens should be prescribed if the glass is to be worn 1.00 cm in front of the eye? Is the lens converging or diverging?

Solution: Given: $d_o = \infty$, $d_i = -(0.950 \text{ m} - 0.010 \text{ m}) = -0.940 \text{ m}$. Find: $P = \dfrac{1}{f}$.

The normal eye has the far point at infinity. To correct for nearsightedness, the image of an object at infinity must be formed at 95.0 cm in front of the eye, so the object distance is infinity and the image distance is $-(95.0 \text{ cm} - 1.00 \text{ cm}) = -94.0 \text{ cm}$. The image distance is negative because the image is on the same side as the object (virtual image).

$$P = \frac{1}{f} = \frac{1}{d_o} + \frac{1}{d_i} = \frac{1}{\infty} + \frac{1}{-0.940 \text{ m}} = 0 - 1.06 \text{ m}^{-1} = -1.06 \text{ D}.$$

Since the power of the lens is negative, it is a diverging lens.

---

2. Microscopes (Section 25.2)

A **magnifying glass (simple microscope)** is a single converging lens which allows one to view an object clearly when it is brought closer than the near point. In such a position, an object subtends a greater angle and therefore appears larger, or magnified. The magnification of an object viewed through a magnifying glass is expressed in terms of **angular magnification** (m), which is defined as the ratio of the *angular* size of the object viewed through the magnifying glass (θ) to the angular size of the object viewed without the magnifying glass (θ_0), or $m = \dfrac{\theta}{\theta_0}$. If the image is at the near point (normally 25 cm from the eye), the magnification of the magnifying glass is $m = 1 + \dfrac{25 \text{ cm}}{f}$, where f is the focal length of the converging lens. If the image is formed at infinity (eye in relaxed position), the magnification is then $m = \dfrac{25 \text{ cm}}{f}$. Note that the shorter the focal length, the greater the magnification. A magnifying glass provides limited magnification because for very short focal lengths, the image becomes distorted.

A **compound microscope** consists of a pair of converging lenses, each of which contributes to the magnification. A compound microscope provides greater magnification than can be attained with a single lens. A converging lens having a relatively short focal length ($f_o < 1$ cm) is known as the **objective**. It forms a real, inverted, and magnified image of an object positioned *slightly beyond* its focal point. The other lens, called the **eyepiece** (or **ocular**), has longer focal length (f_e is a few centimeters) and is positioned so that the image formed by the objective falls just *inside* its focal point. This lens forms a magnified virtual image that is viewed by the observer. The total angular magnification of the combination is $M_{\text{total}} = \dfrac{(25 \text{ cm}) L}{f_o f_e}$, where L is the separation between the two lenses, and f_o, f_e, and L are in centimeters. When finding the magnification of a compound microscope, make sure to convert f_o, f_e, and L into the same units, usually centimeters because the near point is usually so expressed.

Example 25.2   A person uses a converging lens of focal length 5.0 cm as a magnifying glass.

(a) What is the maximum possible angular magnification?

(b) What is the magnification if the person's eye is relaxed?

Solution: Given: $f = 5.0$ cm.   Find: (a) m_{max} (b) m_{relaxed}.

(a) The maximum magnification is attained when the image is at the near point.

$$m = 1 + \frac{25 \text{ cm}}{f} = 1 + \frac{25 \text{ cm}}{5.0 \text{ cm}} = 6.0\times.$$

(b) When the eyes are relaxed, the image is at infinity,   so $m = \dfrac{25 \text{ cm}}{f} = 5.0\times.$

Example 25.3 A compound microscope has an objective with a focal length of 4.5 mm and an eyepiece of focal length 5.0 cm. If the two lenses are separated by 25 cm, what is the total angular magnification?

Solution: Given: $f_o = 4.5$ mm $= 0.45$ cm, $f_e = 5.0$ cm, $L = 20$ cm.

Find: M_{total}.

$$M_{total} = \frac{(25 \text{ cm}) L}{f_o f_e} = \frac{(25 \text{ cm})(25 \text{ cm})}{(0.45 \text{ cm})(5.0 \text{ cm})} = 280 \times.$$

3. Telescopes (Section 25.3)

A **refracting telescope** uses a converging lens to collect and converge light from a distant object while a **reflecting telescope** uses a mirror to collect and converge light. The principle of a refracting telescope is similar to that of a compound microscope. The major components are objective and eyepiece lenses. The objective is a large converging lens with a long focal length. The movable eyepiece has a relatively short focal length. The image formed by the objective acts like the object for the eyepiece and a magnified image is seen. If the eyepiece of a refracting telescope is also a converging lens, the final image is inverted and the setup is known as an **astronomical telescope**. If the final image is upright, the setup is known as a **terrestrial telescope**. There are several ways to achieve an upright image. One of these is to use a diverging lens as the eyepiece (Galilean telescope). The separation of the objective and the eyepiece is equal to the sum of the focal lengths of the two lenses, $L = f_o + f_e$.

The angular magnification of a refracting telescope for final image at infinity is given by $m = -\dfrac{f_o}{f_e}$.

Example 25.4 A student constructs an astronomical telescope with a magnification 10. If the telescope has a converging lens of focal length 50 cm,

(a) what should be the focal length of the eyepiece?

(b) what is the resulting length of the telescope?

Solution: An astronomical telescope has an inverted final image, so the magnification is negative.

Given: $m = -10$, $f_o = 50$ cm.

Find: (a) f_e (b) L.

(a) From $m = -\dfrac{f_o}{f_e}$, we have $f_e = -\dfrac{f_o}{m} = -\dfrac{50 \text{ cm}}{-10} = 5.0$ cm.

(b) The length of a refracting telescope is equal to the sum of the focal lengths of the two lenses.

$L = f_o + f_e = 50$ cm $+ 5.0$ cm $= 55$ cm.

4. Diffraction and Resolution (Section 25.4)

The diffraction of light places a limitation on our ability to distinguish objects that are close together when using microscopes or telescopes. In general, images of two sources can be resolved if the center of the central maximum of one falls at or beyond the first minimum (dark fringes) of the other. This generally accepted limiting condition for the **resolution** of two diffracted images is known as the **Rayleigh criterion**. The limiting, or minimum, angle of resolution (θ_{min}) for a slit of width d is given by $\theta_{min} = \dfrac{\lambda}{d}$ (where θ_{min} is a pure number and therefore must be expressed in radians). Thus, the images of two sources will be *distinctly* resolved if the angular separation of the sources is greater than λ/d. For *circular* apertures, the minimum angle of resolution is $\theta_{min} = \dfrac{1.22\,\lambda}{D}$, where D is the diameter of the aperture.

For a microscope, it is convenient to specify the **resolving power**, or the actual separation (s) between two point sources. Since the objects are usually near the focal point of the objective, the minimum distance between two points whose images can be just resolved is $s = f\theta_{min} = \dfrac{1.22\,\lambda f}{D}$.

Note: The term resolving power used here is actually a distance, not power as used in our everyday lives.

Example 25.5   A binary star system in the constellation Orion has an angular separation of 2.5×10^{-5} rad. If the wavelength of the light from the system is $\lambda = 550$ nm, what is the smallest aperture (diameter) telescope that can just resolve the two stars? (Ignore atmospheric blurring.)

Solution: Given: $\theta_{min} = 2.5 \times 10^{-5}$ rad, $\lambda = 550$ nm $= 550 \times 10^{-9}$ m.
            Find: D.

From $\theta_{min} = \dfrac{1.22\,\lambda}{D}$, we have $D = \dfrac{1.22\,\lambda}{\theta_{min}} = \dfrac{1.22(550 \times 10^{-9}\text{ m})}{2.5 \times 10^{-5}\text{ rad}} = 2.7 \times 10^{-2}$ m $= 2.7$ cm.

Due to atmospheric blurring, the diameter of the telescope is much larger than the value predicted here.

Example 25.6   A compound microscope is designed to resolve objects which are 0.010 mm apart. If the focal length of the objective is 4.0 cm and the wavelength of light used is 550 nm, what is the diameter of the aperture of the objective?

Solution: Given: $s = 0.010 \text{ mm} = 0.010 \times 10^{-3} \text{ m}$, $f = 4.0 \text{ cm} = 0.040 \text{ m}$, $\lambda = 550 \text{ nm} = 550 \times 10^{-9} \text{ m}$.

Find: D.

The 0.010 mm distance is the resolving power of the microscope.

From $s = \dfrac{1.22\lambda f}{D}$, we have $D = \dfrac{1.22\lambda f}{s} = \dfrac{1.22(550 \times 10^{-9} \text{ m})(0.040 \text{ m})}{0.010 \times 10^{-3} \text{ m}} = 2.7 \times 10^{-3} \text{ m} = 2.7 \text{ mm}$.

In reality, the aperture of the microscope must be larger than the predicted value to take other laboratory conditions into account, such as air movement, temperature fluctuations, etc.

---

*5.   Color (Section 25.5)

The **additive primary colors** (additive primaries) are red, blue, and green. The mixing of light of the additive primaries is called the **additive method of color production**. When light of the three additive primaries is mixed in proper portion, the mixture appears white to the eye. Pairs of color combinations that appear white to the eye are called **complementary colors**. The complementary color of red is cyan, of blue is yellow, and of green is magenta, etc.

The **subtractive primary pigments** (subtractive primaries) are cyan, yellow, and magenta. A mixture of absorbing pigments results in the subtraction of colors, and one sees the color of light that is not absorbed or subtracted. This is called **subtractive method of color production.** When the three subtractive primaries are mixed in the proper portion, the mixture appears black (all colors are absorbed) to the eye.

IV.   Mathematical Summary

Angular Magnification	$m = \dfrac{\theta}{\theta_o}$ (25.1)	Defines the angular magnification.
Magnification of a Magnifying Glass with Image at Near Point (25 cm)	$m = 1 + \dfrac{25 \text{ cm}}{f}$ (25.3)	Computes the angular magnification of a magnifying glass when the image is at near point (25 cm).
Magnification of a Magnifying Glass with Image at Infinity	$m = \dfrac{25 \text{ cm}}{f}$ (25.4)	Computes the angular magnification of a magnifying glass when the image is at infinity.
Angular Magnification of a Compound Microscope (with L and f's in cm)	$M_{\text{total}} = M_o\, m_e = \dfrac{(25 \text{ cm})L}{f_o f_e}$ (25.5)	Computes the magnification of a compound microscope consisting of an objective and an eyepiece.

Magnification of a Refracting Telescope	$m = -\dfrac{f_o}{f_e}$ (25.6)	Calculates the magnification of a refracting telescope consisting of an objective and an eyepiece.
Minimum Angle of Resolution for a Slit	$\theta_{min} = \dfrac{\lambda}{d}$ (25.7)	Defines the minimum angle of resolution for a slit width d.
Minimum Angle of Resolution for a Circular Aperture	$\theta_{min} = \dfrac{1.22\,\lambda}{D}$ (25.8) diameter D	Defines the minimum angle of resolution for a circular aperture of diameter D.
Resolving Power	$s = f\theta_{min} = \dfrac{1.22\,\lambda f}{D}$ (25.9)	Defines the resolving power of a circular lens of focal length f.

V.   Solutions of Selected Exercises and Paired/Trio Exercises

3.   The eye focuses by changing the shape of its lens to change the focal length according to the lens maker's equation. The focal length is adjusted to form a sharp image. The image distance is fairly constant and is the distance from the lens to the retina. From the thin lens equation, the eye must have short focal length for looking at close objects and so the radius is small; the eye must have long focal length for looking at distant objects and so the radius is large.

7.   (a) $P = \dfrac{1}{f} = \dfrac{1}{0.20\text{ m}} = \boxed{+5.0\text{ D}}$.

      (b) $P = \dfrac{1}{-0.50\text{ m}} = \boxed{-2.0\text{ D}}$.

10.   (a) $\boxed{\text{Nearsighted}}$.

      (b) $d_o = \infty$, $d_i = -12.5$ m (image on object side).

      $P = \dfrac{1}{f} = \dfrac{1}{d_o} + \dfrac{1}{d_i} = \dfrac{1}{\infty} + \dfrac{1}{-12.5\text{ m}} = \boxed{-0.080\text{ D, diverging}}$.

12.   (a) $\boxed{\text{Converging}}$.   (b) $f = \boxed{36\text{ cm}}$.

18.   Top: $d_o = \infty$, $d_i = -500$ cm $= -5.0$ m (image on object side).

      $P = \dfrac{1}{f} = \dfrac{1}{d_o} + \dfrac{1}{d_i} = \dfrac{1}{\infty} + \dfrac{1}{-5.0\text{ m}} = \boxed{-0.20\text{ D}}$.

      Bottom: $d_o = 25$ cm $= 0.25$ m, $d_i = -70$ cm $= -0.70$ m (image on object side).

      $P = \dfrac{1}{0.25\text{ m}} + \dfrac{1}{-0.70\text{ m}} = \boxed{+2.6\text{ D}}$.

21. Nearsightedness: $d_o = \infty$, $d_i = -(220 \text{ cm} - 3.0 \text{ cm}) = -2.17 \text{ m}$ (image on object side).

$$P = \frac{1}{f} = \frac{1}{d_o} + \frac{1}{d_i} = \frac{1}{\infty} + \frac{1}{-2.17 \text{ m}} = -0.46 \text{ D}.$$

Right farsightedness: $d_o = 25 \text{ cm} - 3.0 \text{ cm} = 0.22 \text{ m}$, $d_i = -0.320 \text{ m}$ (image on object side).

$$P_r = \frac{1}{0.22 \text{ m}} + \frac{1}{-0.320 \text{ m}} = 1.42 \text{ D}.$$

Left farsightedness: $d_o = 25 \text{ cm} - 3.0 \text{ cm} = 0.22 \text{ m}$, $d_i = -0.420 \text{ m}$ (image on object side).

$$P_r = \frac{1}{0.22 \text{ m}} + \frac{1}{-0.420 \text{ m}} = 2.16 \text{ D}.$$

So the prescription is $\boxed{\text{right: } +1.42 \text{ D}, -0.46 \text{ D}; \text{ left: } +2.16 \text{ D}, -0.46 \text{ D}}$.

24. A short focal length lens has a very small radius according to the lens maker's equation. The aberration (geometrical optics or small angle approximation is no longer valid if the object is large compared with the size of the lens) will be bigger and bigger as the focal length gets smaller and smaller. This limits the magnification to about $3\times$ to $4\times$.

28. $m = 1 + \dfrac{25 \text{ cm}}{f} = 1 + \dfrac{25 \text{ cm}}{12 \text{ cm}} = \boxed{3.1\times}.$

34. From $M_{total} = \dfrac{(25 \text{ cm})L}{f_o f_e}$, we have $P_o = \dfrac{1}{f_o} = \dfrac{M_{total} f_e}{(0.25 \text{ m})L} = \dfrac{(360)(0.0080 \text{ cm})}{(0.25 \text{ cm})(0.15 \text{ cm})} = \boxed{+77 \text{ D}}.$

37. (a) $M_{total} = \dfrac{(25 \text{ cm})L}{f_o f_e} = \dfrac{(25 \text{ cm})(22 \text{ cm})}{(0.50 \text{ cm})(3.25 \text{ cm})} = 338\times = \boxed{340\times}.$

(b) $m = 1 + \dfrac{25 \text{ cm}}{f_e} = 1 + \dfrac{25 \text{ cm}}{3.25 \text{ cm}} = 8.7\times.$

So the percentage is $\dfrac{340\times}{8.7\times} = 39 = \boxed{3900\%}.$

42. Since $d_o = \dfrac{d_i f_o}{d_i - f_o}$, $M_o = \dfrac{d_i}{d_o} = \dfrac{d_i - f_o}{f_o}.$ So $M_1 = \dfrac{150 \text{ mm} - 16 \text{ mm}}{16 \text{ mm}} = 8.38\times,$

$M_2 = \dfrac{150 \text{ mm} - 4.0 \text{ mm}}{4.0 \text{ mm}} = 36.5\times,$ and $M_3 = \dfrac{150 \text{ mm} - 1.6 \text{ mm}}{1.6 \text{ mm}} = 92.8\times.$

Therefore $M_{max} = (92.8\times)(10\times) = \boxed{930\times}$ and $M_{min} = (8.38\times)(5.0\times) = \boxed{42\times}.$

48. From $L = f_o + f_e$, we have $f_o = L - f_e = 1.5 \text{ m} - 10 \times 10^{-3} \text{ m} \approx 1.49 \text{ m}.$

$m = \dfrac{f_o}{f_e} = \dfrac{1.49 \text{ m}}{10 \times 10^{-3} \text{ m}} = 149\times = \boxed{150\times}.$

51. (a) $m_1 = \dfrac{f_{o1}}{f_{e1}} = \dfrac{90.0 \text{ cm}}{0.84 \text{ cm}} = 107\times,$ $\quad m_2 = \dfrac{85.0 \text{ cm}}{0.77 \text{ cm}} = 110\times.$

So the $\boxed{\text{second}}$ one has a higher magnification.

(b) The resolution depends on the diameter of the objective, so the $\boxed{\text{first}}$ one has a higher resolution.

58. The central maximum of one pattern falls on the first minimum of the other. The angular position of the first minimum is determined by $\quad w \sin\theta \approx w\theta = m\lambda = (1)\,\lambda.$

So $\quad \theta_{min} = \dfrac{\lambda}{w} = \dfrac{680 \times 10^{-9} \text{ m}}{0.55 \times 10^{-3} \text{ m}} = \boxed{1.2 \times 10^{-3} \text{ rad}}.$

62. $\theta_{min} = \dfrac{1.22\lambda}{D} = \dfrac{1.22(550 \times 10^{-9} \text{ m})}{7.0 \times 10^{-3} \text{ m}} = \boxed{9.6 \times 10^{-5} \text{ rad}}.$

64. $\boxed{18 \text{ km}}.$

66. (a) $\theta_{min} = \dfrac{1.22\lambda}{D} = \dfrac{1.22(570 \times 10^{-9} \text{ m})}{0.0250 \text{ m}} = \boxed{2.78 \times 10^{-5} \text{ rad}}.$

(b) $s = f\theta_{min} = (30.0 \text{ mm})(2.78 \times 10^{-5} \text{ rad}) = \boxed{8.34 \times 10^{-4} \text{ mm}}.$

71. Since white is obtained by adding colors, it cannot be obtained by the subtractive method. That method subtracts colors, and the one we see is the one that is not absorbed.

Black objects do not absorb all wavelengths of light. We see the objects because we perceive the extremely faint light as black. (Think of twilight vision.)

77. The far point is at the image location with glasses.

$d_o = \infty, \quad P = \dfrac{1}{f} = -0.15 \text{ D}.$

$\dfrac{1}{d_o} + \dfrac{1}{d_i} = \dfrac{1}{f} = P, \quad \text{so} \quad d_i = f = \dfrac{1}{P} = \dfrac{1}{-0.15 \text{ D}} = -6.7 \text{ m}.$

Therefore the far point is $\boxed{6.7 \text{ m}}.$

82. $m = \dfrac{f_o}{f_e} = \dfrac{50 \text{ cm}}{1.5 \text{ cm}} = 33.3. \quad \theta_o = \dfrac{0.10 \text{ m}}{50 \text{ m}} = 2.0 \times 10^{-3} \text{ rad}.$

So $\quad \theta = m\theta_o = (33.3)(2.0 \times 10^{-3} \text{ rad}) = 0.0666 \text{ rad} = \boxed{3.8°}.$

87. Assume the diameter of a typical iris is 1.0 cm and use 550 nm as the wavelength.

The resolution is $\theta_{min} = \dfrac{1.22\lambda}{D} = \dfrac{1.22(550 \times 10^{-9} \text{ m})}{0.010 \text{ m}} = 6.7 \times 10^{-5}$ rad.

The resolving power on the Earth is then $s = d\theta_{min} = (150 \times 10^{3} \text{ m})(6.7 \times 10^{-5} \text{ rad}) = 10 \text{ m} \approx 33$ ft.

So, in theory, she is able to identify $\boxed{\text{objects as large as typical houses}}$.

VI. Practice Quiz

1. The farthest distance at which the normal eye can see objects clearly is

   (a) the near point.   (b) the far point.   (c) nearsightedness.   (d) farsightedness.   (e) astigmatism.

2. Which one of the following is not a primary color of light?

   (a) red   (b) yellow   (c) blue   (d) green   (e) both (c) and (d)

3. A nearsighted person wears glasses whose lenses have power of -0.15 D. What is the person's far point if the glasses are very close to the eyes?

   (a) 1.5 m   (b) 3.3 m   (c) 6.0 m   (d) 6.7 m   (e) infinity

4. A magnifying glass has a focal length of 5.0 cm. What is the angular magnification if the image is viewed by a relaxed eye?

   (a) 3.0×   (b) 4.0×   (c) 5.0×   (d) 6.0×   (e) 7.0×

5. A compound microscope has a 18-centimeter barrel and an eyepiece with a focal length of 8.0 mm. What is the focal length of the objective to give a total magnification of 240×?

   (a) 0.094 cm   (b) 0.13 cm   (c) 1.5 cm   (d) 1.9 cm   (e) 2.3 cm

6. A person is designing a 10× telescope. If the telescope is limited to a length of 22 cm, what is the approximate focal length of the objective?

   (a) 16 cm   (b) 18 cm   (c) 20 cm   (d) 22 cm   (e) 24 cm

7. The 2.4-meter reflecting Hubble Space Telescope has been placed into Earth orbit by the space shuttle. What angular resolution could this telescope achieve by the Rayleigh criterion if the wavelength is 550 nm?

   (a) 5.2×10^{-6} rad   (b) 4.4×10^{-6} rad   (c) 4.6×10^{-7} rad   (d) 2.8×10^{-7} rad   (e) 2.3×10^{-7} rad

8.   To decrease the minimum angle of resolution a microscope can resolve,

(a) the diameter of the objective should be decreased.

(b) the diameter of the objective should be increased.

(c) the wavelength of light should be increased.

(d) the microscope's magnification should be more powerful.

(e) None of the above.

9.   If a farsighted person wears glasses of prescription +3.2D, what is the person's far point?

(a) 1.2 m   (b) 1.4 m   (c) 1.6 m   (d) 1.8 m   (e) 2.0 m

10.   A refracting telescope has an angular magnification m. If the objective focal length is doubled and the eyepiece focal length is halved, what is the new magnification?

(a) 4m (b) 2m (c) m (d) $m/2$ (e) $m/4$

Answers to Practice Quiz:

1. b 2. b 3. d 4. c 5. e 6. c 7. d 8. b 9. a 10. a

CHAPTER 26

I.   Chapter Objectives

Upon completion of this chapter, you should be able to:

1. summarize the concepts of classical relativity and relative velocities, define inertial and noninertial reference frames, and explain the reasoning behind the ether hypothesis.

2. explain the general concept and operation of the Michelson-Morley experiment, its result, and the effect on the ether concept.

3. explain how the two postulates of relativity imply the relativity of simultaneity and how the relativity of simultaneity leads to length contraction.

4. understand the concepts of time dilation and length contraction and calculate the relationship between time intervals and lengths observed in different inertial frames.

5. understand the relativistically correct expressions for kinetic energy, momentum, and total energy when objects move near the speed of light; understand the equivalence of mass and energy, and use the relativistically correct expressions to calculate energy and momentum in particle interactions.

6. explain the principle of equivalence and specify some of the predictions of general relativity.

*7. understand the necessity for a relativistic velocity addition equation and apply it to simple relative velocity calculations.

II.   Key Terms

Upon completion of this chapter, you should be able to define and/or explain the following key terms:

relativity	constancy of the speed of light	relativistic kinetic energy
inertial reference frame	special theory of relativity	relativistic momentum
principle of classical	event	rest energy
(Newtonian) relativity	time dilation	mass-energy equivalence
ether	proper time	general theory of relativity
interferometer	proper length	principle of equivalence
Michelson-Morley experiment	length contraction	black hole
principle of relativity	twin (clock) paradox	Schwarzschild radius
		event horizon

The definitions and/or explanations of the most important key terms can be found in the following section:
III. Chapter Summary and Discussion.

III.   Chapter Summary and Discussion

1.   Classical Relativity (Section 26.1)

Relativity is a branch of physics that deals with fast moving (close to the speed of light) particles and observations from two different reference frames. Classical (nonrelativistic) physics is an approximation of relativity at low speeds.

An **inertial reference frame** is a nonaccelerating frame in which Newton's first law holds. In an inertial frame, an *isolated* object (one on which there is no net force) is stationary or moves with constant velocity. The **principle of classical (Newtonian) relativity** states: the laws of mechanics are the same in all inertial reference frames. The speed of light in vacuum was predicted by Maxwell equations to be 3.00×10^8 m/s. But, relative to what reference frame does light have this speed? Classically the speed of light measured from different frames of reference would be expected to differ -- possibly to be even greater than 3.00×10^8 m/s if you were approaching the light wave -- by simple vector addition of velocities, so it was reasoned that this speed of light that Maxwell predicted *must* be referenced to a particular frame. This gives rise to the concept of a unique reference frame associated with the proposed medium of transport for electromagnetic waves called the **ether**; an absolute reference frame. As a result, it would seem that Maxwell's equations (which govern light and its speed) did *not* satisfy the Newtonian principle as did the laws of mechanics.

2.   The Michelson-Morley Experiment (Section 26.2)

The **Michelson-Morley experiment** was an attempt to detect the ether by using the interference of light in an extremely sensitive **interferometer**. In Michelson-Morley's interferometer, a beam of light is divided into two beams, the first traveling in a direction perpendicular to the velocity of the Earth and the second traveling in a direction parallel to the velocity of the Earth. If the ether existed, interference fringe *shifts* should be observed when the equipment is rotated 90°, but the experiment yielded *null* results, that is, *no fringe shift at all* was detected. This experiment ruled out the existence of the ether, although attempts were made to explain the null result of the experiment in terms of other theories.

3. The Postulates of Special Relativity and the Relativity of Simultaneity

(Section 26.3)

The above null result, or the failure to detect the ether, was resolved by the **special theory of relativity** formulated by Albert Einstein. His theory had two basic postulates:

Postulate I (**principle of relativity**): all the laws of physics (not just mechanics) are the same in all inertial reference frames.

Postulate II (**constancy of the speed of light**): the speed of light in vacuum c is the same as measured in any inertial reference frame.

The first postulate implies that all inertial reference frames are equivalent, with physical laws being the same in all of them. The second postulate implies that two observers in different inertial reference frames would measure the speed of light to be c independent of the speed of the source and/or the observer.

In the language of relativity, an **event** is a "happening". We need to specify the location *and* time to tell an event. Events that are simultaneous in a particular inertial reference frame may *not* be simultaneous as measured in a different inertial frame. Simultaneity is thus a relative, not an absolute, concept. The nonintuitive results of length contraction and time dilation (discussed next) are just two of the many relativistic results that follow directly from the two postulates and the relativity of simultaneity. They are nonintuitive because we are used to dealing with speeds much slower than light, where lengths and time intervals are absolute.

4. The Relativity of Length and Time: Length Contraction and Time Dilation

(Section 26.4)

Two results of the postulates of the special theory of relativity are time dilation and length contraction. **Time dilation** refers to a phenomenon in which a fast moving clock is measured to run more slowly than a clock at rest in the observer's own frame of reference. Since time intervals are relative and there are two time intervals, the time measured in a frame at rest with respect to the clock is called the **proper time** (t_0). The dilated time (t), measured by a clock moving with respect to the frame, and the proper time are related by the equation

$t = \dfrac{t_0}{\sqrt{1 - (v/c)^2}}$, where v is the relative speed between the two frames and c is the speed of light. Many relativistic

equations can be written more simply if we represent the expression $\dfrac{1}{\sqrt{1 - (v/c)^2}}$ as a factor, gamma (γ):

$\gamma = \dfrac{1}{\sqrt{1 - (v/c)^2}}$. Gamma ($\gamma$) is always greater than or equal to 1. Therefore, the time dilation equation can be

written as $t = \gamma t_0$.

Note: t and t_0 are time intervals (Δt and Δt_0), not instants of time. The Δ's are dropped for simplicity.

Length contraction refers to a phenomenon in which the length of a fast moving object relative to an observer in an inertial reference frame is less (in the dimension of relative motion) than if the object were at rest in the observer's frame. The length measured in a frame at rest with respect to the object is called the **proper length** (L_0). The contracted length (L) and the proper length are related by $L = \dfrac{L_0}{\gamma} = L_0 \sqrt{1 - (v/c)^2}$.

Note: It is very important to identify the proper time and the proper length in problem solving. They are the time and length measured in a frame at rest with respect to the clock and object.

Time dilation gives rise to another popular relativistic topic -- the so called **twin (clock) paradox**. A twin on a space journey relative to the Earth would age more slowly than an Earth-based twin. The twin paradox has been experimentally verified with atomic clocks.

Example 26.1   One 20-year old twin brother takes a space trip with a speed of $0.70c$ for 30 years according to a clock on the spaceship. Upon returning to the Earth, what is his own age and the age of the Earth-based twin?

Solution: Given: $t_0 = 30$ y, $v = 0.70c$.

Find: (a) Age of traveling twin (b) age of Earth-based twin.

The time 30 years is measured by the traveling twin with his own clock, so it is the proper time.

(a) The age of the traveling twin is simply 20 y $+ 30$ y $= 50$ y.

(b) According to the Earth-based twin, 30 years on the spaceship is

$$t = \gamma t_0 = \frac{t_0}{\sqrt{1 - (v/c)^2}} = \frac{(30 \text{ y})}{\sqrt{1 - (0.70)^2}} = 42 \text{ y on the Earth.}$$

So the age of the Earth-based twin is 20 y $+ 42$ y $= 62$ y.

Example 26.2   A spaceship is moving toward you with a speed of $0.75c$. By what percentage will its length change compared to the spaceship's length of 15 m when it is at rest?

Solution: Given: $L_0 = 15$ m, $v = 0.75c$.

Find: L/L_0.

The proper length of the spaceship is 15 m.

From length contraction, $L = \dfrac{L_0}{\gamma} = L_0 \sqrt{1 - (v/c)^2} = (15 \text{ m}) \sqrt{1 - (0.75)^2} = 9.9$ m.

So the spaceship appears to be $\dfrac{9.9 \text{ m}}{15 \text{ m}} = 0.66 = 66\%$ as compared to its length at rest. That means the

length decreases by 34%.

Example 26.3   The closest star to our solar system is Alpha Centauri, which is 4.30 light years away. A spaceship
with a constant velocity of $0.800c$ relative to the Earth travels toward the star.

(a) What distance does the space ship travel according to a passenger on the ship?

(b) How much time would elapse on a clock on board the spaceship?

(c) How much time would elapse on a clock on the Earth?

Solution: Given: $L_0 = 4.30$ ly, $v = 0.800c$.

Find: (a) L (b) t_0 (c) t.

The 4.30 light-year distance is measured by observers on the Earth, so it is the proper length. The proper
time is the time measured with a clock on the spaceship.

(a) From length contraction, $L = \dfrac{L_0}{\gamma} = L_0 \sqrt{1 - (v/c)^2} = (4.30 \text{ ly}) \sqrt{1 - (0.800)^2} = 2.58$ ly.

(b) The proper time (the time measured on the spaceship) is $t_0 = \dfrac{L}{v} = \dfrac{2.58 \text{ ly}}{0.800c} = \dfrac{2.58c \text{ y}}{0.800c} = 3.23$ y.

(c) The dilated time (the time measured by an Earth-based clock) is then

$t = \gamma t_0 = \dfrac{t_0}{\sqrt{1 - (v/c)^2}} = \dfrac{3.23 \text{ y}}{\sqrt{1 - (0.800)^2}} = 5.38$ y.

Or, $\Delta t = \dfrac{L_0}{v} = \dfrac{4.30 \text{ ly}}{0.800c} = \dfrac{4.30c \text{ y}}{0.800c} = 5.38$ y.   Two different ways, the same result.

---

5.   Relativistic Kinetic Energy, Momentum, Total Energy, and Mass-Energy Equivalence (Section 26.5)

Other important quantities in the special theory of relativity are relativistic kinetic energy, momentum, total
energy, and mass—energy Equivalence.

Relativistic kinetic energy is given by $K = \left[\dfrac{1}{\sqrt{1 - (v/c)^2}} - 1 \right] mc^2 = (\gamma - 1)m c^2.$

Relativistic momentum is given by $\mathbf{p} = \dfrac{m\mathbf{v}}{\sqrt{1 - (v/c)^2}} = \gamma m\mathbf{v}.$

Total relativistic energy is equal to $E = \dfrac{mc^2}{\sqrt{1-(v/c)^2}} = \gamma mc^2$. When $v = 0$, $E = E_o = mc^2$, which is called the **rest energy**. E can also be written as $E = K + E_o = K + mc^2 = \gamma E_o$. The expression $E_o = mc^2$ is the famous **mass-energy equivalence** because it points out that mass is also a form of energy. In particular, a particle has rest energy E_o associated with its mass m. For speeds below 10% of the speed of light, the use of nonrelativistic formulas ($K = \frac{1}{2}mv^2$, etc.) is acceptable since it will cause errors of less than 1% error.

Example 26.4   The kinetic energy of a proton is 80% of its total energy.

(a) What is the speed of the proton?

(b) What is the magnitude of the momentum of the proton?

(c) What is the proton's total energy?

Solution: Given: $K = 0.80E$, $m = 1.67 \times 10^{-27}$ kg.

Find: (a) v (b) p.

(a) Since $E = K + E_o = 0.80E + E_o$, $E_o = 0.20E$.   From $E = \gamma E_o$, $\gamma = \dfrac{E}{E_o} = \dfrac{1}{0.20} = 5.0$.

Since $\dfrac{1}{\gamma} = \sqrt{1-(v/c)^2} = 0.20$, $v = c\sqrt{1-(0.20)^2} = 0.98c$.

(b) $p = \gamma mv = (5.0)(1.67 \times 10^{-27} \text{ kg})(0.98)(3.00 \times 10^8 \text{ m/s}) = 2.5 \times 10^{-18}$ kg·m/s.

(c) $E = \gamma E_o = \gamma mc^2 = (5.0)(1.67 \times 10^{-27} \text{ kg})(3.00 \times 10^8 \text{ m/s})^2 = 7.5 \times 10^{-10}$ J $= 4.7$ GeV.

---

6.   The General Theory of Relativity (Section 26.6)

The special theory of relativity applies only to inertial reference frames, *not* noninertial (accelerating) systems. The **general theory of relativity** considers accelerating frames and is essentially a gravitational theory. Its **principle of equivalence** states that an inertial reference frame in a uniform gravitational field is equivalent to a reference frame in the absence of a gravitational field that has a constant acceleration with respect to that inertial frame. It basically means that no experiment performed in a closed system can distinguish between the effects of a gravitational field and the effects of an acceleration.

Some of the predictions of the general theory of relativity are gravitational light bending, gravitational lensing, black holes, and the gravitational red shift. One way a **black hole** can form is from the gravitational collapse of a massive star after it is done producing light. Such an object has a density so great and a gravitational field so intense that nothing (including light) can escape it. The critical radius around a black hole from which light will not be able to escape is given by the **Schwarzschild radius**, $R = \dfrac{2GM}{c^2}$, where G is the universal gravitational constant and M the mass within the radius. The boundary of a sphere of radius R defines what is called the **event horizon**. Any event occurring within this horizon is invisible to an observer outside, since light cannot escape.

Example 26.5   What radius must our Sun have in order for light to not be able to escape from it?

Solution: Given: $M = 2.0 \times 10^{30}$ kg, $G = 6.67 \times 10^{-11}$ N·m²/kg².   Find: R.

$$R = \frac{2GM}{c^2} = \frac{2(6.67 \times 10^{-11} \text{ N·m}^2/\text{kg}^2)(2.0 \times 10^{30} \text{ kg})}{(3.00 \times 10^8 \text{ m/s})^2} = 3.0 \times 10^3 \text{ m} = 3.0 \text{ km.}$$

---

*7.   Relativistic Velocity Addition (Section 26.7)

When dealing with relative velocities for fast moving particles, the relativistic velocity addition must be

used, $u = \dfrac{v + u'}{1 + \dfrac{v\,u'}{c^2}}$. The meanings of the symbols in the equation are:

v = velocity of object 1 with respect to an inertial observer (e.g., on Earth)

u' = velocity of object 2 with respect to object 1

u = velocity of object 2 with respect to an inertial observer (e.g., on Earth).

Example 26.6   A spaceship moves away from the Earth with a speed of $0.80c$. The spaceship then fires a missile with a speed of $0.50c$ relative to the ship. What is the velocity of the missile measured by observers on the Earth if

(a) the missile is fired away from the Earth?

(b) the missile is fired toward the Earth?

Solution: Given: $v = 0.80c$,   (a) $u' = +0.50c$,   (b) $u' = -0.50c$.

Find: u (a) and (b).

Here we assumed away from the Earth as the positive direction for velocity.

(a) $u = \dfrac{v + u'}{1 + \dfrac{v\,u'}{c^2}} = \dfrac{0.80c + 0.50c}{1 + (0.80)(0.50)} = 0.93c$ away from the Earth.

According to the classical theory, $u = v + u' = 0.80c + 0.50c = 1.3c$!

(b) $u = \dfrac{0.80c + (-0.50c)}{1 + (0.80)(-0.50)} = 0.50c$ away from the Earth.

According to the classical theory, $u = 0.80c + (-0.50c) = 0.30c$!

IV. Mathematical Summary

γ (dimensionless relativistic factor)	$\gamma = \dfrac{1}{\sqrt{1-(v/c)^2}}$ (26.7)	Defines the dimensionless relativistic factor.
Time Dilation	$\Delta t = \dfrac{\Delta t_0}{\sqrt{1-(v/c)^2}}$ (26.6) and $t = \dfrac{t_0}{\sqrt{1-(v/c)^2}} = \gamma t_0$ (26.8)	Computes the dilated time in terms of proper time.
Length Contraction	$L = \dfrac{L_0}{\gamma} = L_0\sqrt{1-(v/c)^2}$ (26.10)	Computes the contracted length in terms of proper length.
Relativistic Kinetic Energy	$K = (\gamma - 1)mc^2$ (26.11)	Defines relativistic kinetic energy.
Relativistic Momentum	$\mathbf{p} = \gamma m\mathbf{v}$ (26.12)	Defines relativistic momentum.
Rest Energy	$E_0 = mc^2$ (26.14)	Defines rest energy.
Relativistic Total Energy	$E = K + E_0 = K + mc^2 = \gamma mc^2$ (26.13, 26.15)	Computes total relativistic energy.
Total Energy and Rest Energy	$E = \gamma E_0$ (26.16)	Relates total relativistic energy and rest energy.
Schwarzschild Radius	$R = \dfrac{2GM}{c^2}$ (26.18)	Defines the radius of an event horizon.
*Relativistic Velocity Addition	$u = \dfrac{v + u'}{1 + \dfrac{vu'}{c^2}}$ (26.19)	Computes relativistic velocity addition.

V. Solutions of Selected Exercises and Paired/Trio Exercises

7.   Since the speed of light is constant in all directions through all frames of references and no ether (wind) would be present, there is no extra time difference between the two beams and no fringe shift is observed.

10.   (a) The velocity relative to ground is 200 km/h $+ (-35$ km/h$) = \boxed{165 \text{ km/h}}$.

     (b) The velocity relative to ground is 200 km $+ 25$ km/h $= \boxed{225 \text{ km/h}}$.

12.   (a) Perpendicular to the shore, $t = \dfrac{y}{v_\perp} = \dfrac{50 \text{ m}}{0.50 \text{ m/s}} = \boxed{1.0 \times 10^2 \text{ s}}$.

0.15 m/s

0.45 m/s θ

(b) Along the shore, $x = v_{||}\, t = (0.15 \text{ m/s})(100 \text{ s}) = \boxed{15 \text{ m}}$.

(c) $\theta = \sin^{-1}\left(\dfrac{0.15}{0.50}\right) = \boxed{17° \text{ up stream}}$.

15. $t_2 = \dfrac{2dc}{c^2 - v^2} = \dfrac{2\,d}{c(1 - v^2/c^2)}$.

If d is contracted by $\sqrt{1 - v^2/c^2}$, we have $t_2 = \dfrac{2\,d\sqrt{1 - v^2/c^2}}{c(1 - v^2/c^2)} = \dfrac{2\,d}{c\sqrt{1 - v^2/c^2}} = t_1$.

21.   (a) 300 m/s. An inertial frame O' moving at this velocity will determine the proper time between the two events.

(b) The time interval between these two events observed by another inertial frame is $\Delta t = \dfrac{\Delta t_o}{\sqrt{1 - v^2/c^2}}$,

where Δt_o is the proper time. Δt can never be negative and so the firing of the gun will always precede the hitting of the target.

27.   The proper time for one beat is $\dfrac{1}{80 \text{ beats/min}} = \dfrac{1}{80}$ min/beat.

So $t = \dfrac{t_o}{\sqrt{1 - v^2/c^2}} = \dfrac{1/80 \text{ min/beat}}{\sqrt{1 - 0.85^2}} = \dfrac{1}{42}$ min/beat.

Therefore the number of beats per min is $\boxed{42 \text{ beats/min}}$.

30.   From $t = \dfrac{t_o}{\sqrt{1 - v^2/c^2}}$, we have $v = \sqrt{1 - t_o^2/t^2}\ c = \sqrt{1 - 2.20^2/34.8^2}\ c = \boxed{0.998c}$.

32.   (a) $\boxed{5.7 \text{ y}}$.   (b) $\boxed{7.2 \text{ y}}$.

34.   (a) The height is still 15.0 m because it is perpendicular to the relative velocity.

The base, however, is $L = L_o \sqrt{1 - v^2/c^2} = (40.0 \text{ m})\sqrt{1 - 0.90^2} = 17.44$ m.

So the area is $A = \frac{1}{2}(17.44 \text{ m})(15.0 \text{ m}) = \boxed{131 \text{ m}^2}$.

(b) The angle between the hypotenuse and the base is $\theta = \tan^{-1}\left(\dfrac{15.0}{17.44}\right) = \boxed{40.7°}$.

39.   From $L = L_o \sqrt{1 - v^2/c^2}$, we have $v = c\sqrt{1 - L^2/L_o^2} = c\sqrt{1 - (3 \times 0.3048)^2} = 0.40c$.

So she is traveling at 0.40c in the direction of the lengths of the two sticks .

43.   No , there are no such limits on momentum and energy since $p = \gamma mv$ and $E = \gamma mc^2$.

46. $E_o = mc^2 = (9.11 \times 10^{-31} \text{ kg})(3.00 \times 10^8 \text{ m/s})^2 = 8.20 \times 10^{-14} \text{ J} \approx 0.511 \text{ MeV}$.

$E = \gamma E_o = \dfrac{1}{\sqrt{1 - v^2/c^2}} E_o = \dfrac{1}{\sqrt{1 - 0.600^2}} (0.511 \text{ MeV}) = \boxed{0.639 \text{ MeV}}$.

48.   (a) $\boxed{0.14c}$.   (b) $\boxed{0.86c}$.

52. $E_o = mc^2 = (9.11 \times 10^{-31} \text{ kg})(3.00 \times 10^8 \text{ m/s})^2 = 8.20 \times 10^{-14} \text{ J} \approx 0.511 \text{ MeV}$.

From $E = \gamma E_o = \dfrac{E_o}{\sqrt{1 - v^2/c^2}}$, we have $v = c\sqrt{1 - E_o^2/E^2} = c\sqrt{1 - (0.511)^2/(2.8)^2} = 0.983c$.

So $p = \dfrac{mv}{\sqrt{1 - v^2/c^2}} = \dfrac{(9.11 \times 10^{-31} \text{ kg})(0.983)(3.00 \times 10^8 \text{ m/s})}{\sqrt{1 - 0.983^2}} = \boxed{1.5 \times 10^{-21} \text{ kg·m/s}}$.

55.   (a) P is plotted in units of classical $p_o = mv$ (b) E is plotted in units of rest energy $E_o = mc^2$

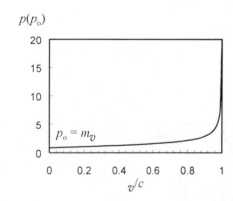

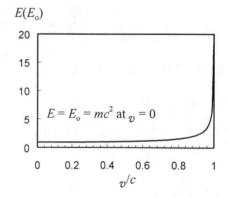

59.   (a) $E_o = mc^2 = (9.11 \times 10^{-31} \text{ kg})(3.00 \times 10^8 \text{ m/s})^2 = 8.20 \times 10^{-14} \text{ J} \approx 0.511 \text{ MeV}$.

$K = E - E_o = E_o(\gamma - 1) = E_o \left(\dfrac{1}{\sqrt{1 - v^2/c^2}} - 1 \right) = (0.511 \text{ MeV}) \left(\dfrac{1}{\sqrt{1 - 0.950^2}} - 1 \right)$

$= \boxed{1.13 \text{ MeV}}$.

(b) $E = K + E_o = 1.13 \text{ MeV} + 0.511 \text{ MeV} = \boxed{1.64 \text{ MeV}}$.

63.   (a) The minimum energy is the rest energy, which for a proton is

$$E_o = mc^2 = (1.67 \times 10^{-27} \text{ kg})(3.00 \times 10^8 \text{ m/s})^2 = 1.502 \times 10^{-10} \text{ J} = 939 \text{ MeV} > 600 \text{ MeV}.$$

(b) $E = K + E_o = 600 \text{ MeV} + 939 \text{ MeV} = 1539 \text{ MeV}.$

Since $E = \dfrac{E_o}{\sqrt{1 - v^2/c^2}}$, $v = c\sqrt{1 - E_o^2/E^2} = c\sqrt{1 - (939)^2/(1539)^2} = \boxed{0.792c}$.

(c) $p = \dfrac{mv}{\sqrt{1 - v^2/c^2}} = \dfrac{(1.67 \times 10^{-27} \text{ kg})(0.792)(3.00 \times 10^8 \text{ m/s})}{\sqrt{1 - 0.792^2}} = \boxed{6.50 \times 10^{-19} \text{ kg·m/s}}.$

68. $R_E = \dfrac{2GM_E}{c^2} = \dfrac{2(6.67 \times 10^{11} \text{ N·m}^2/\text{kg}^2)(6.0 \times 10^{24} \text{ kg})}{(3.00 \times 10^8 \text{ m/s})^2} = 8.9 \times 10^{-3} \text{ m} = \boxed{8.9 \text{ mm}}.$

$R_J = \dfrac{2GM_J}{c^2} = \dfrac{2G(318M_E)}{c^2} = 318R_E = \boxed{2.8 \text{ m}}.$

73. $u = \dfrac{v + u'}{1 + v\,u'/c^2} = \dfrac{0.40c + (-0.15c)}{1 + (0.40)(-0.15)} = \boxed{0.27c}$, same direction as spacecraft

78. $v = 9.5 \times 10^7 \text{ m/s} = 0.317c.$

$E_o = mc^2 = (9.11 \times 10^{-31} \text{ kg})(3.00 \times 10^8 \text{ m/s})^2 = 8.20 \times 10^{-14} \text{ J} \approx 0.511 \text{ MeV}.$

$E = \dfrac{E_o}{\sqrt{1 - v^2/c^2}} = \dfrac{0.511 \text{ MeV}}{\sqrt{1 - 0.317^2}} = \boxed{0.54 \text{ MeV}}.$

82.   The rest energy of an electron or a positron is

$$E_o = mc^2 = (9.11 \times 10^{-31} \text{ kg})(3.00 \times 10^8 \text{ m/s})^2 = 8.20 \times 10^{-14} \text{ J} \approx 0.511 \text{ MeV}.$$

The total energy of the radiation is $0.511 \text{ MeV} + 0.511 \text{ MeV} = \boxed{1.02 \text{ MeV}}$.

VI.   Practice Quiz

1.   As the speed of a particle approaches the speed of light, the rest energy of the particle

(a) increases.   (b) decreases.   (c) remains the same.   (d) approaches zero.   (e) none of the above.

2.   What is the momentum in kg·m/s of an electron when it is moving with a speed of $0.75c$?

(a) 1.0×10^{-22} (b) 2.0×10^{-22} (c) 2.6×10^{-22} (d) 3.1×10^{-22} (e) 4.6×10^{-22}

3. A fast spaceship is traveling with a speed of $0.80c$ relative to an observer. How fast would light travel from the headlights of the ship relative to a that observer?

(a) $0.20c$ (b) $0.80c$ (c) c (d) $0.90c$ (e) $1.8c$

4. A spaceship takes a nonstop journey to a planet and returns in 10 h according to a clock on the spaceship. If the speed of the spaceship is $0.80c$, how much time has elapsed on the Earth?

(a) 6.0 h   (b) 6.3 h   (c) 10 h   (d) 16 h   (e) 17 h

5. The length of a spaceship is 10 m when it is at rest. If the spaceship travels past you with its velocity $(0.70c)$ parallel to its length, what length does it appear to you?

(a) 5.5 m   (b) 7.1 m   (c) 10 m   (d) 14 m   (e) 18 m

6. What is the total energy of an electron moving with a speed of $0.95c$?

(a) 2.6×10^{-13} J   (b) 8.2×10^{-14} J   (c) 1.1×10^{-13} J   (d) 1.2×10^{-14} J   (e) 3.7×10^{-14} J

7. What was the result of the Michelson-Morley experiment?

(a) verified the existence of the ether   (b) detected the interference fringe shifts

(c) proved light is a wave   (d) detected no fringe shift

(e) verified the gravitational light bending

8. A 3.0 GW (3.0×10^{9} W) nuclear power plant loses how much fuel mass in one hour?

(a) 1.2 kg   (b) 0.12 kg   (c) 12 g   (d) 1.2 g   (e) 0.12 g

9. What is the speed of a proton if its total energy is twice its rest energy?

(a) $0.866c$ (b) $0.750c$ (c) $0.707c$ (d) $0.581c$ (e) $0.500c$

10. A space traveler in a ship moves away from the Earth with a speed of $0.30c$ when a missile is fired with a speed of $0.80c$ back towards the Earth relative to the ship. How fast does the missile appear to travel toward the Earth according to the Earth?

(a) $0.40c$ (b) $0.50c$ (c) $0.66c$ (d) $0.89c$ (e) c

Answers to Practice Quiz:

1. c 2. d 3. c 4. e 5. b 6. a 7. d 8. e 9. a 10. c

CHAPTER 27

Quantum Physics

I.   Chapter Objectives

Upon completion of this chapter, you should be able to:

1. define blackbody radiation and use Wien's law, and understand how Planck's hypothesis paved the way for quantum ideas.

2. describe the photoelectric effect, explain how it can be understood by assuming that light energy is carried by particles, and summarize the properties of photons.

3. understand how the photon model of light explains the scattering of light from electrons (the Compton effect) and calculate the wavelength of the scattered light in the Compton effect.

4. understand how the Bohr model of the hydrogen atom explains atomic emission and absorption spectra, calculate the energies and wavelengths of emitted photons for transitions in atomic hydrogen, and understand how the generalized concept of atomic energy levels can explain other atomic phenomena.

5. understand some of the practical applications of the quantum hypothesis—in particular, the laser.

II.   Key Terms

Upon completion of this chapter, you should be able to define and/or explain the following key terms:

thermal radiation	absorption spectrum
blackbody	Balmer series
Wien's displacement law	principal quantum number
Planck's constant	ground state
quantum	excited state
photon	binding energy
photoelectric effect	lifetime
stopping potential	fluorescence
work function	laser
threshold frequency	metastable state
Compton effect	phosphorescent
dual nature of light	stimulated emission
Bohr theory of the hydrogen atom	population inversion
emission spectrum	holography

III. Chapter Summary and Discussion

1. Quantization: Planck's Hypothesis (Section 27.1)

One of the problems scientists had at the end of nineteen century was how to explain **thermal radiation**, the continuous spectra of radiation emitted by hot objects. A **blackbody** is an ideal system that absorbs and emits all radiation that falls on it. The wavelength of maximum radiation (spectral component), λ_{max} is inversely proportional to its absolute temperature, T, and obeys Wien's displacement law, $\lambda_{max} T = 2.90 \times 10^{-3}$ m·K. Classically, the intensity of the blackbody radiation at a particular wavelength is predicted to be proportional to $\frac{1}{\lambda^4}$ (Chapter 11), which leads to what is sometimes called the *ultraviolet catastrophe*—"ultraviolet" because disagreement between theory and experiment occurs for short wavelengths beyond the violet end of the visible spectrum, and "catastrophe" because it predicts the emitted intensity at these short wavelengths will become infinitely large.

Note: Wien's displacement law relates the wavelength of the *most intense* radiation and the absolute temperature (Kelvin), not maximum wavelength and temperature. λ_{max} does not mean it is the maximum or longest wavelength emitted.

Max Planck successfully explained the spectrum of blackbody radiation by proposing a radical hypothesis. According to **Planck's hypothesis**, the energy of a thermal oscillator is *quantized*. That is, an oscillator can have only *discrete*, or particular, amounts of energy, rather than a continuous distribution of energies. The smallest quantum amount of energy is given by $E = hf$, where $h = 6.63 \times 10^{-34}$ J·s is called the **Planck's constant**. This is called a **quantum** of energy. The thermal oscillators in a blackbody can have only integer multiples of this quantum of energy, $E_n = n(hf)$.

Example 27.1 What is the most intense color of light emitted by a giant star of surface temperature 4400 K? What is the color of the star?

Solution: Given: $T = 4400$ K.

Find: λ_{max}.

From Wien's displacement law, $\lambda_{max} T = 2.90 \times 10^{-3}$ m·K,

we have $\lambda_{max} = \dfrac{2.90 \times 10^{-3} \text{ m·K}}{T} = \dfrac{2.90 \times 10^{-3} \text{ m·K}}{4400 \text{ K}} = 659$ nm.

This is in the red end of the visible spectrum. This star emits light in all colors but red is the most intense.

---

2. Quanta of Light: Photons and the Photoelectric Effect (Section 27.2)

A quantum or packet of light energy is referred to as a **photon**, and each photon has an energy $E = hf$, where f is the frequency of light. This idea suggests that light may sometimes behave as discrete quanta, or "particles" of energy, rather than as a wave.

Einstein used the photon theory to explain the **photoelectric effect**, another area in which the classical (wave) description of light was inadequate. Some materials are *photosensitive*, that is, when light strikes their surface, electrons are emitted and a current may be established. However, if the frequency of light is below a certain cutoff value, *no photoelectrons* are emitted no matter how strong the light intensity is and how long light is incident on the material! If the frequency is higher than the certain cutoff value, photoelectrons are emitted instantaneously, no matter how low the light intensity is. The intensity of light apparently has only to do with the number of photoelectrons. Classical wave physics cannot explain these observations.

To study the maximum kinetic energy (K_{max}) of the photoelectrons, a **stopping potential** (V_0) is applied to a beam of electrons. When there is no photocurrent, the maximum kinetic energy is related to the stopping potential by $K_{max} = eV_0$, where eV_0 is the work needed to stop the most energetic electrons. The minimum energy needed to free the electrons from the material is called the **work function** (ϕ_0). According to energy conservation, $hf = K_{max} + \phi_0$, that is, the energy of the absorbed photon goes into the work of freeing the electron, and the rest is carried off by that emitted electron as kinetic energy. The **threshold frequency**, the lowest frequency of light which releases photoelectrons, corresponds to photoelectrons having zero kinetic energy. $hf_0 = 0 + \phi_0$, so $f_0 = \dfrac{\phi_0}{h}$.

Note: We are often given the wavelength of light (in nm) rather than the frequency, and energy. The energy of the photon in units of eV can be calculated quickly from $E = \dfrac{1.24 \times 10^3 \ \text{eV·nm}}{\lambda}$.

Example 27.2   What is the photon energy of visible light having wavelength of 632.8 nm?

Solution: Given: $\lambda = 632.8$ nm $= 632.8 \times 10^{-9}$ m.   Find: E.

The frequency of light is calculated from $f = \dfrac{c}{\lambda} = \dfrac{3.00 \times 10^8 \ \text{m/s}}{632.8 \times 10^{-9} \ \text{m}} = 4.74 \times 10^{14}$ Hz.

So the energy of a photon is $E = hf = (6.63 \times 10^{-34} \ \text{J·s})(4.74 \times 10^{14} \ \text{Hz}) = 3.14 \times 10^{-19}$ J $= 1.96$ eV.

Or, $E = \dfrac{1.24 \times 10^3 \ \text{eV·nm}}{\lambda} = \dfrac{1.24 \times 10^3 \ \text{eV·nm}}{632.8 \ \text{nm}} = 1.96$ eV.

Example 27.3   A metal has a work function of 4.5 eV. Find the maximum kinetic energy of the emitted photoelectrons if the wavelength of light falling on the metal is (a) 300 nm   (b) 250 nm.

Solution: Given: $\phi_0 = 4.5$ eV.   (a) $\lambda = 300$ nm   (b) $\lambda = 250$ nm.

Find: K_{max} in (a) and (b).

(a) From energy conservation, $E = hf = K_{max} + \phi_0$,

so $K_{max} = E - \phi_0 = \dfrac{1.24 \times 10^3 \text{ eV·nm}}{\lambda} - \phi_0 = \dfrac{1.24 \times 10^3 \text{ eV·nm}}{300 \text{ nm}} - 4.5 \text{ eV} = -0.37$ eV.

Since kinetic energy cannot be negative, the result indicates there are no photoelectrons ejected.

(b) $K_{max} = \dfrac{1.24 \times 10^3 \text{ eV·nm}}{250 \text{ nm}} - 4.5 \text{ eV} = 0.46$ eV.

Example 27.4   When light of wavelength 350 nm is incident on a metal surface, the stopping potential of the photoelectrons is measured to be 0.500 V.

(a) What is the work function of the metal?

(b) What is the threshold frequency of the metal?

(c) What is the maximum kinetic energy of the photoelectrons?

Solution: Given: $\lambda = 350$ nm, $V_0 = 0.500$ V or $eV_0 = 0.500$ eV

Find: (a) ϕ_0 (b) f_0 (c) K_{max}.

(a) According to energy conservation, $E = hf = K_{max} + \phi_0 = eV_0 + \phi_0$,

so $\phi_0 = E - eV_0 = \dfrac{1.24 \times 10^3 \text{ eV·nm}}{\lambda} - eV_0 = \dfrac{1.24 \times 10^3 \text{ eV·nm}}{350 \text{ nm}} - 0.500 \text{ eV} = 3.04$ eV.

(b) $f_0 = \dfrac{\phi_0}{h} = \dfrac{(3.04 \text{ eV})(1.60 \times 10^{-19} \text{ J/eV})}{6.63 \times 10^{-34} \text{ J·s}} = 7.34 \times 10^{14}$ Hz.

(c) The maximum kinetic energy is equal to the stopping potential, eV_0, expressed in eV.

$K_{max} = eV_0 = 0.500$ eV.

---

3.   Quantum "Particles": The Compton Effect (Section 27.3)

The **Compton effect** is the increase in wavelength of light scattered by electrons or other charged particles. When a photon collides with an electron, a scattered photon of longer wavelength (lower frequency or lower energy because the electron carries off some energy) emerges. Using the conservation of relativistic momentum and total energy, the shift is given by $\Delta\lambda = \lambda - \lambda_0 = \lambda_C (1 - \cos\theta)$, where $\lambda_C = h/(mc) = 2.43 \times 10^{-12}$ m $= 2.43 \times 10^{-3}$ nm is called the *Compton wavelength* of the electron. Since the Compton shift is very small, it is only significant for X-ray and gamma-ray scattering where the wavelengths are on the order of λ_C.

Note: The Compton wavelength is *not* the wavelength of the electron. It is the characteristic wavelength shift when photons are scattered by an electron. If light is scattered by a proton, the Compton wavelength would be that of the proton and would be much smaller than the electrons (why?).

Einstein's and Compton's successes in explaining electromagnetic phenomena in terms of quanta left scientists with two apparently competing theories of electromagnetic radiation, the wave theory and the photon theory. The two theories gave rise to a description that is called the **dual nature of light**. That is, light apparently behaves sometimes as a wave and at other times as photons or "particles."

Example 27.5   X-rays of wavelength of 0.200 nm are scattered by a metal. The wavelength shift is observed to be 1.50×10^{-12} m at a certain scattering angle measured relative to the incoming X-ray.

(a) What is the scattering angle?

(b) What is the maximum shift possible for the Compton efffect?

Solution: Given: $\lambda = 0.200$ nm $= 2.00 \times 10^{-10}$ m, $\Delta\lambda = 1.50 \times 10^{-12}$ m.

Find: (a) θ (b) $\Delta\lambda_{max}$.

(a) From $\Delta\lambda = \lambda_C (1 - \cos\theta) = (2.43 \times 10^{-12}$ m$)(1 - \cos\theta)$,

$$\cos\theta = 1 - \frac{\Delta\lambda}{\lambda_C} = 1 - \frac{1.50 \times 10^{-12} \text{ m}}{2.43 \times 10^{-12} \text{ m}} = 0.617, \quad \text{so} \quad \theta = 51.9°.$$

(b) Maximum shift occurs when $\theta = 180°$ (or $\cos\theta = -1$).

$$\Delta\lambda_{max} = \lambda_C (1 - \cos 180°) = 2\lambda_C = 4.86 \times 10^{-12} \text{ m}.$$

---

4.   The Bohr Theory of the Hydrogen Atom (Section 27.4)

It is experimentally observed that a hydrogen atom can only emit and absorb light at only certain wavelengths, not at all wavelengths. The small number of visible wavelengths the hydrogen atom can emit and absorb (called the **Balmer series**) are given by an empirical equation $\frac{1}{\lambda} = R\left(\frac{1}{2^2} - \frac{1}{n^2}\right)$ for $n = 3, 4, 5,$ and 6 where

$R = 1.097 \times 10^{-2}$ nm^{-1} is called the *Rydberg constant*. The **Bohr theory of the hydrogen atom** successfully explained the **emission spectrum** (a series of bright lines) and **absorption spectrum** (a series of dark lines superimposed on a continuous spectrum) of the hydrogen atom. In Bohr's theory, he assumed that

- the hydrogen electron orbits the nuclear proton in a circular orbit (analogous to planets orbiting the Sun);
- the angular momentum of the electron is quantized in integral multiples of Planck's constant, h;
- the electron does not radiate energy when it is in certain discrete circular orbits;
- the electron radiates or absorbs energy only when it makes a transition to another orbit.

From these assumptions, Bohr showed that the electron can have only certain sized orbits with certain energies. The energies and the radii of the orbits are given by $E_n = -\dfrac{13.6 \text{ eV}}{n^2}$ and $r_n = 0.0529 n^2$ nm for $n = 1, 2, 3, 4, \ldots$, where n is an example of a *quantum number*, specifically the **principal quantum number**. The $n = 1$ orbit is known as the **ground state** and orbits with $n > 1$ are called the **excited states**. The energy of the electron in any state is E_n, and the energy needed to completely free the electron from the atom in that state is $-E_n$, which is called the **binding energy**.

An electron generally does not remain in an excited state for long. It decays, or makes a transition to a lower energy level, in a short time. The time an electron spends in an excited state is called the **lifetime** of the excited state. If an electron makes a downward transition from n_i state to n_f state, a photon is released and its energy is equal to the energy difference between the final and initial states, $\Delta E = E_f - E_i = 13.6 \left(\dfrac{1}{n_f^2} - \dfrac{1}{n_i^2} \right)$ eV. The

wavelength of the photon is then $\lambda = \dfrac{hc}{\Delta E} = \dfrac{1.24 \times 10^3 \text{ eV·nm}}{\Delta E}$. For transitions with $n_f = 1$, the spectrum series is called the *Lyman series* (all ultraviolet), for $n_f = 2$, the *Balmer series* (visible if $n_i = 3, 4, 5,$ and 6), for $n_f = 3$, the *Paschen series* (infrared), etc.

In **fluorescence**, an electron that has been excited by absorbing a photon returns to the ground state in two or more steps. At each step a photon is emitted each with less energy (longer wavelength) than the absorbed light.

Example 27.6   What are the orbital radius and total energy of an electron for a hydrogen atom in (a) the ground state and (b) the second excited state?

Solution: Given: (a) $n = 1$ (b) $n = 3$.   Find: r and E.

Ground state corresponds to $n = 1$, and $n = 2$ and 3 are for the first and second excited states, respectively.

(a) From $r_n = 0.0529 n$ nm, $r_1 = 0.0529(1)$ nm $= 0.0529$ nm.

From $E_1 = -\dfrac{13.6 \text{ eV}}{n^2}$, $E_1 = -\dfrac{13.6 \text{ eV}}{1^2} = -13.6$ eV.

(b) $r_3 = 0.0529(3)$ nm $= 0.159$ nm,   and $E_3 = -\dfrac{13.6 \text{ eV}}{3^2} = -1.51$ eV.

Example 27.7   The electron of a hydrogen atom makes a transition from the fourth excited state to the first excited state. What are the energy and wavelength of the emitted photon?

Solution: Given: $n_i = 5$, $n_f = 2$. Find: (a) ΔE (b) λ.

Since ground state corresponds to $n = 1$, $n = 2$ and 5 are for the first and fourth excited states.

(a) $\Delta E = 13.6 \left(\dfrac{1}{n_f^2} - \dfrac{1}{n_i^2} \right) \text{eV} = 13.6 \left(\dfrac{1}{2^2} - \dfrac{1}{5^2} \right) \text{eV} = 2.86 \text{ eV}$.

(b) From $\Delta E = \dfrac{1.24 \times 10^3 \text{ eV·nm}}{\lambda}$,

we have $\lambda = \dfrac{1.24 \times 10^3 \text{ eV·nm}}{\Delta E} = \dfrac{1.24 \times 10^3 \text{ eV·nm}}{2.86 \text{ eV}} = 434 \text{ nm}$.

The emitted photon is in the blue/violet region of the visible spectrum.

---

5. A Quantum Success: The Laser (Section 27.5)

The quantum theory of atomic structure gave rise to the development of the **laser**, an acronym which stands for *l*ight *a*mplification by *s*timulated *e*mission of *r*adiation. **Stimulated emission** is the process in which a photon with an energy equal to an allowed transition strikes an atom in an excited state, stimulating the atom to make a transition and emit a photon of the same energy, frequency, traveling direction, and phase as the incoming photon. In this process, one photon goes in and two come out (the incoming one and the one emitted by the atom), resulting in the amplification of that light. Another basic requirement for laser action is **population inversion**, that is, more electrons must occupy a **metastable state**, a state in which an excited electron remains for a relatively longer time, than an excited state of lower energy.

Phosphorescent materials are examples of substances made up of atoms with metastable states. Normally, there are more electrons at lower energy levels. This can be achieved by various "pumping" or energy input processes. Laser beams are very intense, highly directional, coherent (same phase), and monochromatic (same frequency). Depending on the lasing medium, lasers can be classified as gas lasers, solid lasers, semiconductor lasers, dye lasers, X-ray lasers, free-electron lasers, etc. Lasers have a wide variety of applications, ranging from compact disk players to welding and eye surgery.

An interesting application of laser light is the production of three-dimensional images in a process called **holography**. The key to holography is the coherent property of laser light, which gives the light waves a definite spatial relationship to each other. The process does not use lenses as ordinary image-forming processes do, yet it recreates the original scene in three dimensions. Holography has many potential applications in medicine, sciences, and even in our everyday life (holographic three-dimensional television).

IV. Mathematical Summary

Wien's Displacement Law	$\lambda_{max} T = 2.90 \times 10^{-3}$ m·K (27.1)	Relates the absolute temperature and the wavelength of maximum intensity.
Photon Energy	$E = hf$ (27.4) $h = 6.63 \times 10^{-34}$ J·s	Defines the energy of a photon (quantum) in terms of its frequency.
K_{max} and Stopping Potential in the Photoelectric Effect	$K_{max} = eV_o$ (27.5)	Relates the maximum electron kinetic energy and the stopping potential in the photoelectric effect.
Energy Conservation in the Photoelectric Effect	$E = hf = K + \phi$ (27.6)	Applies energy conservation to the photoelectric effect.
Energy Conservation for Least Bound Electron in the Photoelectric Effect	$hf = K_{max} + \phi_o$ (27.7)	Applies energy conservation to the least bound electron in the photoelectric effect.
Work-Function and Threshold Frequency in the Photoelectric Effect	$f_o = \dfrac{\phi_o}{f}$ (27.8)	Gives the minimum light frequency (threshold frequency) to eject photo-electron.
Compton Equation (scattering from free electrons)	$\Delta\lambda = \lambda - \lambda_o = \lambda_C (1 - \cos\theta)$ (27.9) $\lambda_C = h/(mc) = 2.43 \times 10^{-12}$ m $= 0.00243$ nm	Computes the wavelength shift (increase) of light scattered by a free electron or other charged particles.
Bohr Theory Orbit Radius	$r_n = 0.0529 n^2$ nm $n = 1, 2, 3, \ldots$ (27.16)	Calculates the orbit radius of an electron in a hydrogen atom.
Bohr Theory Electron Energy	$E_n = \dfrac{-13.6}{n^2}$ eV $n = 1, 2, 3, \ldots$ (27.17)	Calculates the total energy of an electron in a hydrogen atom.
Emitted Photon energy (in eV)	$\Delta E = 13.6 \left(\dfrac{1}{n_f^2} - \dfrac{1}{n_i^2} \right)$ (27.18)	Computes the energy difference (transition energy) between two orbits of an electron in a hydrogen atom.
Bohr Theory Photon Wavelength	$\lambda = \dfrac{1.24 \times 10^3}{\Delta E \text{ (in eV)}}$ nm (27.19)	Calculates the wavelength of a photon emitted when an electron makes a transition.

V. Solutions of Selected Exercises and Paired/Trio Exercises

5. (a) From Wien's displacement law, $\lambda_{max} T = 2.9 \times 10^{-3}$ m·K,

    we have $T = \dfrac{2.9 \times 10^{-3} \text{ m·K}}{\lambda_{max}} = \dfrac{2.9 \times 10^{-3} \text{ m·K}}{c} f \propto f.$

    (b) $T = \dfrac{2.9 \times 10^{-3} \text{ m·K}}{\lambda_{max}} \propto \dfrac{1}{\lambda_{max}}.$

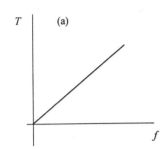
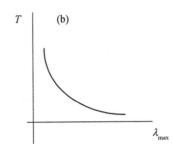

6. From Wien's displacement law, $\lambda_{max} T = 2.9 \times 10^{-3}$ m·K,

    we have $T = \dfrac{2.9 \times 10^{-3} \text{ m·K}}{\lambda_{max}} = \dfrac{2.9 \times 10^{-3} \text{ m·K}}{700 \times 10^{-9} \text{ m}} = \boxed{4.1 \times 10^{3} \text{ K}}.$

11. From Wien's displacement law, $\lambda_{max} T = 2.9 \times 10^{-3}$ m·K,

    we have $\lambda_{max} = \dfrac{2.9 \times 10^{-3} \text{ m·K}}{T} = \dfrac{2.9 \times 10^{-3} \text{ m·K}}{373 \text{ K}} = 7.77 \times 10^{-6}$ m.

    So $f = \dfrac{c}{\lambda_{max}} = \dfrac{3.00 \times 10^{8} \text{ m/s}}{7.77 \times 10^{-6} \text{ m}} = 3.86 \times 10^{13}$ Hz.

    Therefore $E = hf = (6.63 \times 10^{-34} \text{ J·s})(3.86 \times 10^{13} \text{ Hz}) = \boxed{2.56 \times 10^{-20} \text{ J}}.$

17. The energy of radiation is proportional to its intensity, according to the wave theory, and its frequency, according to the particle theory. It takes a certain amount of energy to eject a photoelectron. Since only the frequency, not the intensity, matters in this case, it favors the particle theory.

22. The current is also $\boxed{\text{doubled}}$ since it is directly proportional to the intensity.

25. (a) Since $eV_o = hf + \phi_o$, the slope of the graph is equal to Planck's constant.

    $h = \dfrac{[(3-1) \text{ eV}](1.6 \times 10^{-19} \text{ J/eV})}{(115 - 67) \times 10^{13} \text{ Hz}} = \boxed{6.7 \times 10^{-34} \text{ J·s}}.$

    (b) $\phi_o = hf_o = (6.63 \times 10^{-34} \text{ J·s})(43.9 \times 10^{13} \text{ Hz}) = \boxed{2.9 \times 10^{-19} \text{ J}}.$

30.   (a) $eV_\mathrm{o} = K_\mathrm{max} = hf - \phi_\mathrm{o} = \dfrac{hc}{\lambda} - \phi_\mathrm{o} = \dfrac{1.24 \times 10^3 \ \mathrm{eV \cdot nm}}{300 \ \mathrm{nm}} - 3.5 \ \mathrm{eV} = 0.63 \ \mathrm{eV}.$

So $V = \boxed{0.63 \ \mathrm{V}}.$

(b) $f_\mathrm{o} = \dfrac{\phi_\mathrm{o}}{h} = \dfrac{(3.5 \ \mathrm{eV})(1.6 \times 10^{-19} \ \mathrm{J/eV})}{6.63 \times 10^{-34} \ \mathrm{J \cdot s}} = \boxed{8.4 \times 10^{14} \ \mathrm{Hz}}.$

32.   (a) $\boxed{2.93 \ \mathrm{eV}}.$ (b) $\boxed{1.16 \times 10^{15} \ \mathrm{Hz}}.$

35.   Since $K_\mathrm{max} = hf - \phi_\mathrm{o}$, the slope is equal to Planck's constant.

The slope of the line is $4.1 \times 10^{-15} \ \mathrm{eV/Hz} = 4.1 \times 10^{-15} \ \mathrm{eV \cdot s}.$

$h \approx (4.1 \times 10^{-15} \ \mathrm{eV \cdot s})(1.6 \times 10^{-19} \ \mathrm{J/eV}) = \boxed{6.6 \times 10^{-34} \ \mathrm{J \cdot s}}.$

From the graph, $f_\mathrm{o} \approx 3.7 \times 10^{14} \ \mathrm{Hz}.$

So $\phi_\mathrm{o} = hf_\mathrm{o} = (6.63 \times 10^{-34} \ \mathrm{J \cdot s})(3.7 \times 10^{14} \ \mathrm{Hz}) = 2.45 \times 10^{-19} \ \mathrm{J} = \boxed{1.5 \ \mathrm{eV}}.$

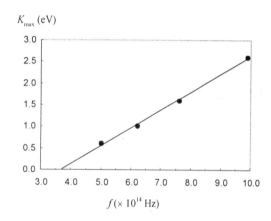

39.   Assume average scattering angle of 90°, for each scattering the wavelength shift is approximately equal to

$\Delta\lambda = \lambda_\mathrm{c} = 0.00243 \ \mathrm{nm}.$ The wavelength change required to go from X–ray to visible light is in the order of

$550 \ \mathrm{nm} - 0.01 \ \mathrm{nm} = 550 \ \mathrm{nm}.$ So after approximately $\dfrac{550 \ \mathrm{nm}}{0.00243 \ \mathrm{nm}}$ scatterings = 200 000 scatterings, X–

ray becomes visible light.

42. $\Delta\lambda = \lambda_\mathrm{c}(1 - \cos\theta) = (0.00243 \ \mathrm{nm})(1 - \cos 30°) = \boxed{3.26 \times 10^{-4} \ \mathrm{nm}}.$

44. $\boxed{0.0055 \ \mathrm{nm}}.$

48. (a) $\lambda_C = \dfrac{h}{m_p c^2} = \dfrac{6.63 \times 10^{-34} \text{ J·s}}{(1.67 \times 10^{-27} \text{ kg})(3.00 \times 10^8 \text{ m/s})} = \boxed{1.32 \times 10^{-15} \text{ m}}$.

(b) $\dfrac{\Delta \lambda_e}{\Delta \lambda_p}\bigg|_{max} = \dfrac{2\lambda_{Ce}}{2\lambda_{Cp}} = \dfrac{2.43 \times 10^{-12} \text{ m}}{1.32 \times 10^{-15} \text{ m}} = \boxed{1.84 \times 10^3}$.

55. (a) $\lambda = \dfrac{1.24 \times 10^3}{\Delta E \text{ (in eV)}} \text{ nm} = \dfrac{1.24 \times 10^3}{10.2} = 122 \text{ nm}$, i.e., $\boxed{\text{ultraviolet}}$.

(b) $\lambda = \dfrac{1.24 \times 10^3}{1.89} \text{ nm} = 656 \text{ nm}$, i.e., $\boxed{\text{visible (red)}}$.

58. (a) $r_n = 0.0529n^2$ nm. $r_3 = (0.0529)(2)^2$ nm $= \boxed{0.212 \text{ nm}}$.

(b) $r_6 = 0.0529(4)^2 = \boxed{0.846 \text{ nm}}$.

(c) $r_{10} = 0.0529(5)^2$ nm $= \boxed{1.32 \text{ nm}}$.

64. (a) $\Delta E = (-13.6 \text{ eV}) \left(\dfrac{1}{n_f^2} - \dfrac{1}{n_i^2} \right)$.

$\Delta E_{52} = (13.6 \text{ eV}) \left(\dfrac{1}{2^2} - \dfrac{1}{5^2} \right) = 2.856 \text{ eV}.$ $\lambda_{52} = \dfrac{1.24 \times 10^3}{\Delta E \text{ (in eV)}} \text{ nm} = \dfrac{1.24 \times 10^3}{2.856} \text{ nm} = \boxed{434 \text{ nm}}$.

$\Delta E_{21} = (13.6 \text{ eV}) \left(\dfrac{1}{1^2} - \dfrac{1}{2^2} \right) = 10.2 \text{ eV}.$ $\lambda_{21} = \dfrac{1.24 \times 10^3}{10.2} \text{ nm} = \boxed{122 \text{ nm}}$.

(b) $\boxed{\text{Yes}}$, 5 to 2 is in the visible region (violet).

67. (a) $E = -\dfrac{(13.6 \text{ eV})Z^2}{n^2}$, where Z is the atomic number (number of protons).

The binding energy is $\Delta E = E_\infty - E_1 = 0 - E_1 = -E_1 = \dfrac{(13.6 \text{ eV})(2)^2}{1^2} = \boxed{54.4 \text{ eV}}$.

(b) The binding energy is $-E_1 = \dfrac{(13.6 \text{ eV})(3)^2}{1^2} = \boxed{122 \text{ eV}}$.

70. $E = -\dfrac{ke^2}{2r} = -\dfrac{(9.0 \times 10^9 \text{ N·m}^2/\text{C}^2)(1.6 \times 10^{-19} \text{ C})^2}{2(0.0529 \times 10^{-9} \text{ m})} = -2.18 \times 10^{-18} \text{ J} = -13.6 \text{ eV}$.

75. In each second $E = Pt = (750 \times 10^3 \text{ W})(1.0 \text{ s}) = 750 \times 10^3 \text{ J}$.

Since $E = nhf$, $n = \dfrac{E}{hf} = \dfrac{750 \times 10^3 \text{ J}}{(6.63 \times 10^{-34} \text{ J·s})(98.9 \times 10^6 \text{ Hz})} = \boxed{1.14 \times 10^{31} \text{ photons}}$.

79.   (a) $\Delta E = (-13.6 \text{ eV}) \left(\frac{1}{n_f^2} - \frac{1}{n_i^2}\right)$. $\Delta E_{13} = (-13.6 \text{ eV}) \left(\frac{1}{3^2} - \frac{1}{1^2}\right) = \boxed{12.1 \text{ eV}}$.

   (b) $\Delta E_{15} = (-13.6 \text{ eV}) \left(\frac{1}{5^2} - \frac{1}{1^2}\right) = \boxed{13.1 \text{ eV}}$.

83.   (a) $hf = K_{max} + \phi_0 = 2.0 \times 10^{-19} \text{ J} + (5.0 \text{ eV})(1.6 \times 10^{-19} \text{ J/eV}) = 1.0 \times 10^{-18}$ J.

   So $f = \dfrac{1.0 \times 10^{-18} \text{ J}}{6.63 \times 10^{-34} \text{ J·s}} = \boxed{1.5 \times 10^{15} \text{ Hz}}$.

   (b) $eV_0 = K_{max} = 2.0 \times 10^{-19} \text{ J} = 1.25 \text{ eV}$.   So $V_0 = \boxed{1.3 \text{ V}}$.

86.   The electrons emitted have a wide range of kinetic energies. When a retarding voltage is applied, only those electrons which have a kinetic energy higher than eV will reach the anode. So the photocurrent decreases. When the stopping voltage ($V = V_0$) is finally reached, no electron will reach the anode and so the photocurrent drops to zero.

VI.   Practice Quiz

1.   Which of the following colors is associated with a blackbody of the lowest temperature?
   (a) red   (b) yellow   (c) orange   (d) green   (e) blue

2.   If the wavelength of a photon is doubled, by what factor does the energy change?
   (a) 4   (b) 2   (c) 1   (d) 1/2   (e) 1/4

3.   A hypothetical atom has two excited states in addition to its ground state. How many different spectral lines are possible?
   (a) 3   (b) 4   (c) 5   (d) 6   (e) 7

4.   The kinetic energy of a photoelectron depends on which one of the following?
   (a) intensity of light   (b) duration of illumination   (c) stopping potential
   (d) wavelength of light   (e) angle of illumination

5.   If the scattering angle in Compton scattering from electron is 45°, what is the wavelength shift?
   (a) 7.11×10^{-13} m   (b) 1.72×10^{-12} m   (c) 2.43×10^{-12} m   (d) 3.44×10^{-12} m   (e) 6.08×10^{-12} m

6. What is the frequency of the most intense radiation from an object with temperature 100°C?

(a) 7.8×10^{-6} Hz   (b) 2.9×10^{-5} Hz   (c) 3.9×10^{13} Hz   (d) 1.0×10^{13} Hz   (e) 1.0×10^{11} Hz

7. A monochromatic light beam is incident on the surface of a metal with work function 2.50 eV. If a stopping potential of 1.0 V is required to make the photocurrent zero, what is the wavelength of light?

(a) 1.42×10^{3} nm   (b) 744 nm   (c) 497 nm   (d) 423 nm   (e) 354 nm

8. The binding energy of the hydrogen atom in its ground state is 13.6 eV. What is the energy of the atom when it is in the $n = 5$ state?

(a) 2.72 eV   (b) −2.72 eV   (c) 0.544 eV   (d) −0.544 eV   (e) −13.6 eV

9. A hydrogen atom in ground state absorbs a photon of energy 12.09 eV. To which state will the electron make a transition?

(a) $n = 1$ (b) $n = 2$ (c) $n = 3$ (d) $n = 4$ (e) $n = 5$

10. The wavelength of a He-Ne laser is 632.8 nm. What is the energy difference between the two energy states involved in producing this laser action?

(a) 0.509 eV   (b) 1.96 eV   (c) 3.14 eV   (d) 4.74 eV   (e) 13.6 eV

Answers to Practice Quiz:

1.a 2.d 3.a 4.d 5.a 6.c 7.e 8.d 9.c 10.b

CHAPTER 28

Quantum Mechanics and Atomic Physics

I.   Chapter Objectives

Upon completion of this chapter, you should be able to:

1.   explain the de Broglie's hypothesis, calculate the "wavelength" of a matter wave, and specify under what circumstances the wave nature of matter will be observable.

2.   understand qualitatively the reasoning that underlies the Schrödinger wave equation and this equation's use in finding particle wave functions.

3.   understand the structure of the periodic table in terms of quantum mechanical electron orbits and the Pauli exclusion principle.

4.   understand the inherent quantum mechanical limits on the accuracy of physical observations.

5.   understand the relationship between particles and antiparticles and the energy requirements for pair production.

II.   Key Terms

Upon completion of this chapter, you should be able to define and/or explain the following key terms:

quantum mechanics	electron configuration
de Broglie hypothesis	electron period
de Broglie (matter) wave	periodic table of elements
wave function	period
Schrödinger's wave equation	group
probability density	Heisenberg uncertainty principle
orbital quantum number	positron
magnetic quantum number	antiparticle
spin quantum number	pair production
shell	pair annihilation
subshell	antimatter
Pauli exclusion principle	

The definitions and/or explanations of the most important key terms can be found in the following section: **III. Chapter Summary and Discussion.**

III. Chapter Summary and Discussion

1. Matter Waves: The de Broglie Hypothesis (Section 28.1)

The **de Broglie hypothesis** associates wavelength with moving material particles, by reverse analogy with the assignment of particle nature to wave of light. From relativity, the momentum of a photon (which has no mass) is $p = \dfrac{E}{c} = \dfrac{hf}{c} = \dfrac{h}{\lambda}$. The de Broglie hypothesis states that whenever a particle has momentum of magnitude, p, there is a wave associated with it. In analogy with a light photon, that wave should have a wavelength of $\lambda = \dfrac{h}{p} = \dfrac{h}{mv}$.

These waves associated with moving particles are called **matter waves** or, more commonly, **de Broglie waves**. Matter waves of particles have been experimentally verified by the *Davisson-Germer* experiment (electron diffraction) and the *Thomson* experiment (also electron diffraction).

Note: For convenience, the wavelength of the matter wave of an electron accelerated from rest through a potential V (in volts) can be calculated with $\lambda = \sqrt{\dfrac{1.50}{V}} \times 10^{-9}$ m $= \sqrt{\dfrac{1.50}{V}}$ nm.

Example 28.1 What is the wavelength of the matter wave associated with

      (a) a ball of mass 0.50 kg moving with a speed of 25 m/s?

      (b) an electron moving with a speed of 2.5×10^7 m/s?

Solution: Given: (a) $m = 0.50$ kg, $v = 25$ m/s. (b) $m = 9.11 \times 10^{-31}$ kg, $v = 2.5 \times 10^7$ m/s.

Find: λ (a) and (b).

(a) $\lambda = \dfrac{h}{p} = \dfrac{h}{mv} = \dfrac{6.63 \times 10^{-34} \text{ J·s}}{(0.50 \text{ kg})(25 \text{ m/s})} = 5.3 \times 10^{-35}$ m.

(b) $\lambda = \dfrac{6.63 \times 10^{-34} \text{ J·s}}{(9.11 \times 10^{-31} \text{ kg})(2.5 \times 10^7 \text{ m/s})} = 2.9 \times 10^{-11}$ m $= 290$ nm.

The wavelength of the ball is much shorter than that of the electron. That is why matter waves are important for small particles like electrons since their wavelengths are comparable to the sizes of the objects they interact with. It is easier to observe interference and diffraction for electrons than the ball and all other everyday objects.

Example 28.2 An electron is accelerated by a potential difference of 120 V.

(a) What is the wavelength of the matter wave associated with the electron?

(b) What is the momentum of the electron?

(c) What is the kinetic energy of the electron?

Solution: Given: $m = 9.11 \times 10^{-31}$ kg, $V = 120$ V. Find: (a) λ (b) p (c) K.

(a) $\lambda = \sqrt{\dfrac{1.50}{V}}$ nm $= \sqrt{\dfrac{1.50}{120}}$ nm $= 0.112$ nm.

(b) From $\lambda = \dfrac{h}{p}$, we have $p = \dfrac{h}{\lambda} = \dfrac{6.63 \times 10^{-34} \text{ J·s}}{0.112 \times 10^{-9} \text{ m}} = 5.92 \times 10^{-24}$ kg·m/s.

(c) $K = \frac{1}{2}mv^2 = \dfrac{p^2}{2m} = \dfrac{(5.92 \times 0^{-24} \text{ kg·m/s})^2}{2(9.11 \times 10^{-31} \text{ kg})} = 1.92 \times 10^{-17}$ J. Here we used $p = mv$.

2. The Schrödinger Wave Equation (Section 28.2)

The **Schrödinger wave equation** is an equation that enables us to calculate matter waves for particles in various systems like atoms. The function satisfying the wave equation is called the **wave function**, ψ, which describes the wave as a function of time and space. The general form of the wave equation can be written as $(K + U)\psi = E\psi$, where K, U, and E are the kinetic, potential, and total energy of the particle, respectively. The square of the wave function (ψ^2) is called the **probability density**, which represents the relative probability of finding a particle in space and time. The interpretation of ψ^2 as the probability of finding a particle at a particular place altered the idea that particles are found in certain definite locations. Now we can only say a particle has a certain probability of being at some location and cannot predict exactly where it is or will be. In some instances, a particle has a finite probability of being even in a classically forbidden region. For example, the classical locations of particles such as the electron orbits in a hydrogen atom are only the *most probable* locations the particles can be.

The solutions to the Schrödinger wave equation lead to many experimentally proven quantum effects (classically forbidden) such as *tunneling* through a *potential energy barrier*. A classical analogy to quantum tunneling would be that a person could run though a concrete wall if he/she tries enough "hits". It takes a lot of energy, say E, to go through the concrete wall and the kinetic energy of a human being is certainly smaller than E. Therefore, classically the person cannot run through the concrete wall because he/she does not have enough energy to overcome the energy barrier of the wall. However, according to quantum mechanics, there is a small probability that the person might appear on the other side of the wall. It is possible for the impossible (in classical sense) to happen if it is tried enough times. The electron tunneling microscope is an application of this quantum effect.

3.   Atomic Quantum Numbers and the Periodic Table (Section 28.3)

When the Schrödinger wave equation is solved for the hydrogen atom, the solution gives three quantum numbers: the principal quantum number n, the orbital quantum number ℓ, the magnetic quantum number m_ℓ. Later, the spin quantum number m_s was added to agree with experimental observations. The **principal quantum number** is the sole determiner of the energy levels of a hydrogen atom as predicted by the Bohr theory (in the absence of external magnetic field). The **orbital quantum number** determines the allowed value of angular momentum for an electron orbit. The **magnetic quantum number** determines the orientation of the plane of the electron orbit with respect to a given axis. The **spin quantum number** is a purely quantum-mechanical concept, that, in a classical analogy, indicates whether the spin angular momentum of an electron is up or down relative to a given axis.

Numerically, the principal quantum number can be any positive integer, $n = 1, 2, 3, \ldots, \infty$. For a given n, the orbital quantum number can be $\ell = 0, 1, 2, 3, \ldots, (n-1)$, or n different values. For example if $n = 2$, $\ell = 0, 1$, or two different values; if $n = 5$, $\ell = 0, 1, 2, 3$, and 4 or five different values. For a given ℓ, the magnetic quantum number can be $m_\ell = 0, \pm1, \pm2, \pm3, \ldots, \pm\ell$, or $2\ell + 1$ different values. The spin quantum number can be either $+\frac{1}{2}$ (spin up) or $-\frac{1}{2}$ (spin down). Each set of unique quantum numbers, (n, ℓ, m_ℓ, m_s), defines a quantum state.

Example 28.3   Write down the possible sets of quantum numbers for each individual quantum state in the $n = 4$ shell.

Solution:

For each n, $\ell = 0, 1, 2, 3, \ldots, n-1$. For each ℓ, $m_\ell = 0, \pm1, \pm2, \pm3, \ldots, \pm\ell$. m_s can be $\pm\frac{1}{2}$.

For $n = 4$, $\ell = 0, 1, 2$, and 3.

For $\ell = 0$, $m_\ell = 0$, and $m_s = +\frac{1}{2}$ or $-\frac{1}{2}$.

So there are two states, $(4, 0, 0, +\frac{1}{2})$ and $(4, 0, 0, -\frac{1}{2})$, for $\ell = 0$.

For $\ell = 1$, $m_\ell = 0$, and ±1.

For $m_\ell = 0$, $m_s = +\frac{1}{2}$ or $-\frac{1}{2}$, so there are two states, $(4, 1, 0, +\frac{1}{2})$ and $(4, 1, 0, -\frac{1}{2})$.

For $m_\ell = +1$, $m_s = +\frac{1}{2}$ or $-\frac{1}{2}$, so there are two states, $(4, 1, 1, +\frac{1}{2})$ and $(4, 1, 1, -\frac{1}{2})$.

For $m_\ell = -1$, $m_s = +\frac{1}{2}$ or $-\frac{1}{2}$, so there are two states, $(4, 1, -1, +\frac{1}{2})$ and $(4, 1, -1, -\frac{1}{2})$.

Therefore, there are six states for $\ell = 1$.

For $\ell = 2$, $m_\ell = 0, \pm1, \pm2$.

For $m_\ell = 0$, $m_s = +\frac{1}{2}$ or $-\frac{1}{2}$, so there are two states, $(4, 2, 0, +\frac{1}{2})$ and $(4, 2, 0, -\frac{1}{2})$.

For $m_\ell = +1$, $m_s = +\frac{1}{2}$ or $-\frac{1}{2}$, so there are two states, $(4, 2, 1, +\frac{1}{2})$ and $(4, 2, 1, -\frac{1}{2})$.

For $m_\ell = -1$, $m_s = +\frac{1}{2}$ or $-\frac{1}{2}$, so there are two states, $(4, 2, -1, +\frac{1}{2})$ and $(4, 2, -1, -\frac{1}{2})$.

For $m_\ell = +2$, $m_s = +\frac{1}{2}$ or $-\frac{1}{2}$, so there are two states, $(4, 2, 2, +\frac{1}{2})$ and $(4, 2, 2, -\frac{1}{2})$.

For $m_\ell = -2$, $m_s = +\frac{1}{2}$ or $-\frac{1}{2}$, so there are two states, $(4, -2, -1, +\frac{1}{2})$ and $(4, -2, -1, -\frac{1}{2})$.

Therefore, there are ten states for $\ell = 2$.

For $\ell = 3$, $m_\ell = 0, \pm1, \pm2, \pm3$.

For $m_\ell = 0$, $m_s = +\frac{1}{2}$ or $-\frac{1}{2}$, so there are two states, $(4, 3, 0, +\frac{1}{2})$ and $(4, 3, 0, -\frac{1}{2})$.

For $m_\ell = +1$, $m_s = +\frac{1}{2}$ or $-\frac{1}{2}$, so there are two states, $(4, 3, 1, +\frac{1}{2})$ and $(4, 3, 1, -\frac{1}{2})$.

For $m_\ell = -1$, $m_s = +\frac{1}{2}$ or $-\frac{1}{2}$, so there are two states, $(4, 3, -1, +\frac{1}{2})$ and $(4, 2, -1, -\frac{1}{2})$.

For $m_\ell = +2$, $m_s = +\frac{1}{2}$ or $-\frac{1}{2}$, so there are two states, $(4, 3, 2, +\frac{1}{2})$ and $(4, 3, 2, -\frac{1}{2})$.

For $m_\ell = -2$, $m_s = +\frac{1}{2}$ or $-\frac{1}{2}$, so there are two states, $(4, 3, -2, +\frac{1}{2})$ and $(4, 3, -2, -\frac{1}{2})$.

For $m_\ell = +3$, $m_s = +\frac{1}{2}$ or $-\frac{1}{2}$, so there are two states, $(4, 3, 3, +\frac{1}{2})$ and $(4, 3, 3, -\frac{1}{2})$.

For $m_\ell = -3$, $m_s = +\frac{1}{2}$ or $-\frac{1}{2}$, so there are two states, $(4, 3, -3, +\frac{1}{2})$ and $(4, 3, -3, -\frac{1}{2})$.

Therefore, there are fourteen states for $\ell = 3$.

Hence, there are a total of $2 + 6 + 10 + 14 = 32$ states.

Generally, there are $2n^2$ states for a given value of n. Here $2n^2 = 2(4)^2 = 32$.

It is common to refer to the total of all the states in a given n as forming a **shell**, and to ℓ levels of that shell as **subshells**. That is, atomic electrons with the same n value are said to be in the same shell. Electrons with the same ℓ value are said to be in the same subshell. The subshells are designated by the letters $s, p, d, f, g, \ldots$ for $\ell = 0, 1, 2, 3, 4, \ldots$, respectively.

The **Pauli exclusion principle** states that no two electrons in a multielectron atom can have the same set of quantum numbers (n, ℓ, m_ℓ, m_s). That is, no two electrons can be in the same quantum state.

Example 28.4   How many possible sets of quantum numbers or electron states are there in the $5f$ subshell? What is the maximum number of electrons that can occupy this subshell?

Solution:

For the $5f$ subshell, $n = 5$ and $\ell = 3$.

For $\ell = 3$, m_ℓ can be $0, \pm1, \pm2, \pm3$, or seven different values, $(2\ell + 1) = (2 \times 3 + 1) = 7$ values.

For each m_ℓ, m_s can be either $+\frac{1}{2}$ or $-\frac{1}{2}$, 2 values.

So there are $2 \times 7 = 14$ sets of quantum numbers or electron states.

According to the Pauli exclusion principle, only one electron can occupy an electron state, so the maximum number of electrons that can occupy the $5f$ subshell is 14.

In a shorthand notation called **electron configuration**, we write the quantum states in increasing order of energy, and designate the numbers of electrons in each level with a subscript. For example, $4p^5$ means that the $4p$ subshell contains five electrons. The electron configuration for the ground state of the element neon (Ne), for example, is $1s^2 2s^2 2p^6$ (both shells $n = 1$, $n = 2$ completely full).

number of electrons in subshell

$4p^5$

$n = 4$ $\ell = 1$

Some energy levels, such as $2s$ and $2p$, have similar energies. We refer to such sets of energy levels that have about the same energy as **electron period**. These electron period are the basis of the **periodic table of elements**. In a periodic table, elements are arranged in horizontal rows, which are called **periods**, in order of increasing atomic mass. Elements of similar chemical properties are arranged in vertical columns called **groups**, or families of elements with similar properties.

4. The Heisenberg Uncertainty Principle (Section 28.4)

The **Heisenberg uncertainty principle** as applied to position (x) and momentum (magnitude p) may be stated as follows: it is impossible to know simultaneously an object's exact position and momentum. This principle overthrows the *deterministic* view of nature in classical physics. The minimum uncertainty of the product of position and momentum is $(\Delta p)(\Delta x) \geq \dfrac{h}{2\pi}$, expressed alternatively in energy and time is $(\Delta E)(\Delta t) \geq \dfrac{h}{2\pi}$.

Note: The value $h/(2\pi)$ is the minimum uncertainty of the product of position and momentum, that is, *at the very best*, the uncertainty of the product is $h/(2\pi)$, that is, it can never be zero.

Example 28.5   A measurement of an electron's speed is 1.0×10^6 m/s and has an uncertainty of $\pm 10\%$. What is the minimum uncertainty in its position?

Solution: Given: $m = 9.11 \times 10^{-31}$ kg, $v = 1.0 \times 10^6$ m/s,   % uncertainty in $v = \pm 10\%$.

Find: Δx.

$v_{max} = v + 0.10v = 1.0 \times 10^6$ m/s $+ (0.10)(1.0 \times 10^6$ m/s$) = 1.1 \times 10^6$ m/s. $v_{min} = 0.90 \times 10^6$ m/s.

So the uncertainty in momentum is

$\Delta p = p_{max} - p_{min} = m\,v_{max} - m\,v_{min} = (9.11 \times 10^{-31}$ kg$)(1.1 \times 10^6$ m/s $- 0.90 \times 10^6$ m/s$) = 1.8 \times 10^{-25}$ kg·m/s.

From the uncertainty principle, $(\Delta p)(\Delta x) \geq \dfrac{h}{2\pi}$,

we have $\Delta x \geq \dfrac{h}{2\pi(\Delta p)} = \dfrac{6.63 \times 10^{-34} \text{ J·s}}{2\pi(1.8 \times 10^{-25} \text{ kg·m/s})} = 5.9 \times 10^{-10}$ m $= 0.59$ nm $\approx$ atomic dimensions.

Example 28.6   The energy of an electron in an atomic state has an uncertainty of about ± 0.0500 eV. What is the life-time (the time the electron remains in that level before making a transition to another level) of that level?

Solution: Given: $\Delta E = 2 \times 0.0500$ eV $= 0.100$ eV $= 1.60 \times 10^{-20}$ J.

Find: Δt.

From $(\Delta E)(\Delta t) \geq \dfrac{h}{2\pi}$, we have $\Delta t \geq \dfrac{h}{2\pi(\Delta E)} = \dfrac{6.63 \times 10^{-34} \text{ J·s}}{2\pi(1.60 \times 10^{-20} \text{ J})} = 4.14 \times 10^{-14}$ s.

---

5.   Particles and Antiparticles (Section 28.5)

A **positron** is a particle that has the same mass as an electron, but possessing a *positive* electronic charge. The oppositely-charged positron is said to be the **antiparticle** of the electron. All subatomic particles have antiparticles. A positron can only be created with the simultaneous creation of an electron in a process called **pair production**. In this process, the energy of a high-energy photon is converted to a positron and an electron and afterwards they have kinetic energies. By mass-energy equivalence, the *threshold energy for pair production* (the minimum energy of the proton) required is $E_{min} = hf = 2m_e c^2 = 1.022$ MeV, that is, $E \geq 1.022$ MeV. A positron and an electron can also be "destroyed" or annihilated in a process called **pair annihilation**. It is the direct conversion of mass into electromagnetic energy (photons), the inverse of pair production so to speak. It is conceivable that antiparticles predominate in some parts of the universe. If so, the atoms of the **antimatter** in this region would consist of negatively charged nuclei (composed of antiprotons and antineutrons), surrounded by positively charged positrons (antielectrons). It would be impossible to distinguish between antimatter and ordinary matter, except, however, if antimatter and ordinary matter should come into contact, that they would annihilate each other with an explosive release of energy.

IV. Mathematical Summary

Momentum of a Photon	$p = \dfrac{E}{c} = \dfrac{hf}{c} = \dfrac{h}{\lambda}$ (28.1)	Defines the momentum of a photon.
de Broglie wavelength of a Moving Particle	$\lambda = \dfrac{h}{p} = \dfrac{h}{mv}$ (28.2)	Defines the deBroglie wavelength of a moving particle (matter wave) in terms of its momentum.
Electron Wavelength When Accelerated through Potential V (nonrelativistic)	$\lambda = \sqrt{\dfrac{1.50}{V}} \times 10^{-9} \text{ m}$ $= \sqrt{\dfrac{1.50}{V}} \text{ nm}$ (28.3)	Calculates the wavelength of an electron when it is accelerated through a potential V (nonrelativistic).
Heisenberg Uncertainty Principle (two forms)	$(\Delta p)(\Delta x) \geq \dfrac{h}{2\pi}$ (28.5) $(\Delta E)(\Delta t) \geq \dfrac{h}{2\pi}$ (28.6)	Expresses the Heisenberg uncertainty principle for momentum-position and energy-time.
Condition for Electron-Positron Pair Production	$E_{min} = hf = 2m_ec^2$ $= 1.022 \text{ MeV}$ (28.7)	Gives the minimum energy (frequency) required to produce an electron-positron pair.

V. Solutions of Selected Exercises and Paired/Trio Exercises

5.   (a) $\lambda = \dfrac{h}{mv} = \dfrac{6.63 \times 10^{-34} \text{ J·s}}{(9.11 \times 10^{-31} \text{ kg})(100 \text{ m/s})} = \boxed{7.28 \times 10^{-6} \text{ m}}$.

     (b) $\lambda = \dfrac{6.63 \times 10^{-34} \text{ J·s}}{(1.67 \times 10^{-27} \text{ kg})(100 \text{ m/s})} = \boxed{3.97 \times 10^{-9} \text{ m}}$.

8.   From Eq. 28.3: $\lambda = \sqrt{\dfrac{1.50}{V}} \text{ nm}$,   so $\lambda \propto \dfrac{1}{\sqrt{V}}$.

     Therefore $\dfrac{\lambda_2}{\lambda_1} = \sqrt{\dfrac{V_1}{V_2}} = \sqrt{\dfrac{250 \text{ kV}}{600 \text{ kV}}} = \boxed{0.645}$.

11.   The initial kinetic energy of the proton is

     $K_o = \frac{1}{2} mv_o^2 = \frac{1}{2}(1.67 \times 10^{-27} \text{ kg})(4.5 \times 10^4 \text{ m/s})^2 = 1.69 \times 10^{-18} \text{ J} = 10.6 \text{ eV}$.

     If the proton accelerates through 10.6 V, it will have a speed of 4.5×10^4 m/s.

     So $V_1 = 10.6$ V and $V_2 = 10.6$ V + 37 V = 47.6 V.

From Eq. 28.3: $\lambda = \sqrt{\dfrac{1.50}{V}}$ nm,   so $\lambda \propto \dfrac{1}{\sqrt{V}}$.

So the percentage difference is

$$\frac{\lambda_2 - \lambda_1}{\lambda_1} = \frac{\lambda_2}{\lambda_1} - 1 = \sqrt{\frac{V_1}{V_2}} - 1 = \sqrt{\frac{10.6\ \text{V}}{37.6\ \text{V}}} - 1 = -0.53 = \boxed{-53\%\ \text{(a decrease)}}.$$

18. (a) The following standing waves can be set up in the well:

$$L = \frac{n\lambda}{2},\ n = 1, 2, 3, \ldots \quad \text{or} \quad \lambda = \frac{2L}{n}.$$

Also the wave function must be zero at $x = 0$ and $x = L$, here we must use sine terms since only $\sin 0 = 0$ at $x = 0$ boundary.

So $\psi_n = A \sin \dfrac{2\pi}{\lambda} x = A \sin \dfrac{n\pi x}{L}$ for $n = 1, 2, 3, \ldots$

(b) $K_n = \dfrac{p^2}{2m} = \dfrac{(h/\lambda)^2}{2m} = \dfrac{h^2}{2m\lambda^2} = \dfrac{h^2}{2m(2L/n)^2} = n^2 \dfrac{h^2}{8mL^2}.$

25. The periodic table arranges elements according to the values of their quantum numbers n and l. Within a group, the outermost electrons of the elements have the same (or very similar) electronic configurations.

26. (a) $2n^2 = 2(2)^2 = \boxed{8}$ and $2(3)^2 = \boxed{18}$.

(b) For $n = 2$:

(2, 1, 1, +1/2);   (2, 1, 1, −1/2);   (2, 1, 0, +1/2);   (2, 1, 0, −1/2);

(2, 1, −1, +1/2);   (2, 1, −1, −1/2);   (2, 0, 0, +1);   (2, 0, 0, −1/2).

For $n = 3$:

(3, 2, 2, +1/2);   (3, 2, 2, −1/2);   (3, 2, 1, +1/2);   (3, 2, 1, −1/2);

(3, 2, 0, +1/2);   (3, 2, 0, −1/2);   (3, 2, −1, +1/2);   (3, 2, −1, −1/2);

(3, 2, −2, +1/2);   (3, 2, −2, −1/2);   (3, 1, 1, +1);   (3, 1, 1, −1/2);

(3, 1, −1, +1/2);   (3, 1, −1, −1/2);   (3, 1, 0, +1/2);   (3, 1, 0, −1/2);

(3, 0, 0, +1/2);   (3, 0, 0, −1/2).

30.

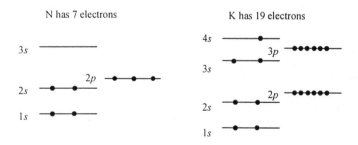

32. (a)

Na has 11 electrons

3s ──●───────────

2s ──●──●── 2p ──●●●●●●──

1s ──●──●──

(b)

Ar has 18 electron

3p ──●●●●●●──

3s ──●──●──

2s ──●──●── 2p ──●●●●●●──

1s ──●──●──

34. $1s^1$ is ⬜hydrogen and $1s^2 2s^2 2p^1$ is ⬜boron.

35. ⬜$1s^3$ since the ground state could contain three electrons without violating the Pauli exclusion principle.

40. From the Heisenberg Uncertainty Principle, $\Delta p \Delta x \geq \dfrac{h}{2\pi}$,

we have $\Delta v \geq \dfrac{\Delta p}{m} \geq \dfrac{h}{2\pi \Delta x m} = \dfrac{6.63 \times 10^{-34} \text{ J·s}}{2\pi[(0.10 - 0.050) \times 10^{-9} \text{ m}](9.11 \times 10^{-31} \text{ kg})} = \boxed{2.3 \times 10^6 \text{ m/s}}$.

44. From the Heisenberg Uncertainty Principle, $\Delta E \Delta t \geq \dfrac{h}{2\pi}$,

we have $\Delta t \geq \dfrac{h}{2\pi \Delta E} = \dfrac{6.63 \times 10^{-34} \text{ J·s}}{2\pi(2 \times 0.0003 \text{ eV})(1.6 \times 10^{-19} \text{ J/eV})} = \boxed{1.1 \times 10^{-12} \text{ s}}$.

49. The rest energy of the electron-positron pair is 2×0.511 MeV $= 1.022$ MeV.

$E = hf = (6.63 \times 10^{-34} \text{ J·s})(2.5 \times 10^{20} \text{ Hz}) = 1.66 \times 10^{-13} \text{ J} = 1.04 \text{ MeV} > 1.022 \text{ MeV}$. So $\boxed{\text{yes}}$.

50. The rest energy of the electron pair is 2×0.511 MeV $= 1.022$ MeV.

From the conservation of energy and momentum, the photons will move in opposite directions with equal

amounts of energy and momentum. So each will carry half the total energy or $\boxed{0.511 \text{ MeV}}$.

55. (a) $p = \dfrac{h}{\lambda} = \dfrac{hf}{c} = \dfrac{E}{c} = \dfrac{(7.5 \times 10^6 \text{ eV})(1.6 \times 10^{-19} \text{ J/eV})}{3.00 \times 10^8 \text{ m/s}} = \boxed{4.0 \times 10^{-21} \text{ kg·m/s}}$.

(b) $\lambda = \dfrac{h}{p} = \dfrac{6.63 \times 10^{-34} \text{ J·s}}{4.0 \times 10^{-21} \text{ kg·m/s}} = \boxed{1.7 \times 10^{-13} \text{ m}}$.

57. The following standing waves can be set up in the well: $2L = \dfrac{n\lambda}{2}$, $n = 1, 2, 3, \ldots$,

or $\lambda = \dfrac{4L}{n}$. Also the wave function must be zero at $x = \pm L$.

So $\psi_n = A \cos \dfrac{2\pi}{\lambda} x = A \cos \dfrac{n\pi x}{2L}$, for $n = 1, 3, 5, \ldots$.

and $\psi_n = A \dfrac{2\pi}{\lambda} = A \sin \dfrac{n\pi x}{2L}$, for $n = 2, 4, 6, \ldots$.

61. $2n^2 = 2(2\ell + 1)$. So $\boxed{\text{yes when } n = 1 \text{ and } \ell = 0}$.

64. They do not have to go off back to back according to momentum conservation, in fact, they have to go off as shown ($\theta < 90°$).

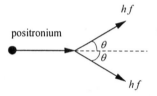

VI. Practice Quiz

1. If the accuracy in measuring the velocity of a particle decreases, the accuracy in measuring its position will
   (a) increase (b) decrease (c) remain the same (d) be uncertain (e) be exact

2. What is the electron configuration for a neutral Li atom?
   (a) $1s^3$ (b) $1s^1 2s^2$ (c) $1s^2 2s^1$ (d) $1s^2 1p^1$ (e) $1s^1 1p^2$

3. An electron is accelerated from rest through a potential difference of 120 V. What is the de Broglie wavelength associated with the electron?
   (a) 0.014 nm (b) 0.12 nm (c) 1.2 mm (d) 1.4 nm (e) 8.6 nm

4. What is the maximum number of electrons that can occupy the d subshell?
   (a) 2 (b) 6 (c) 10 (d) 14 (e) 18

5. The probability of finding an electron at a given location in a hydrogen atom is directly proportional to its
   (a) energy (b) momentum (c) angular momentum
   (d) the wave function (e) the square of the wave function

6. The element krypton has 36 electrons. How many electrons are in the $n = 5$ shell?
   (a) 2 (b) 4 (c) 8 (d) 10 (e) 18

7. If the principal quantum number is 4 which of the following is not an allowed orbital quantum number?

(a) 4   (b) 3   (c) 2   (d) 1   (e) 0

8. A positron is a particle which has

(a) the same mass as an electron   (b) the same mass as a proton   (c) the same mass as a neutron

(d) negative electronic charge   (e) no electronic charge

9. A proton has kinetic energy of 1.0 MeV. If its momentum is measured with an uncertainty of $\pm 1.0\%$, what is the uncertainty in its position?

(a) 2.8×10^{-14} m   (b) 5.6×10^{-14} m   (c) 9.1×10^{-14} m   (d) 2.3×10^{-13} m   (e) 9.1×10^{-13} m

10. Which one of the following particles would tend to have the longest de Broglie wavelength (assuming they travel at the same speed)?

(a) car   (b) basketball   (c) neutron   (d) proton   (e) electron

Answers to Practice Quiz:

1.a 2.c 3.b 4.c 5.e 6.b 7.a 8.a 9.d 10.e

CHAPTER 29

The Nucleus

I. Chapter Objectives

Upon completion of this chapter, you should be able to:

1. distinguish between the Thompson and Rutherford-Bohr models of the atom, specify some of the basic properties of the strong nuclear force, and understand nuclear notation.

2. define the term "*radioactivity*," distinguish among alpha, beta, and gamma decay, and write nuclear decay equations.

3. explain the concepts of activity, decay constant, and half-life of a radioactive sample, and use radioactive decay to find the age of objects.

4. state which proton and neutron number combinations result in stable nuclei, explain the pairing effect and magic numbers in relation to nuclear stability, and calculate nuclear binding energy.

5. gain insight into the operating principles of various nuclear radiation detectors, investigate the medical and biological effects of radiation exposure, and study some of the practical uses and applications of radiation.

II. Key Terms

Upon completion of this chapter, you should be able to define and/or explain the following key terms:

Rutherford-Bohr model	conservation of nucleons	pairing effect
atomic number	conservation of charge	atomic mass unit (u)
nucleon	tunneling	mass defect
strong nuclear force	barrier penetration	total binding energy
proton number	beta (β^-) decay	magic number
mass number	beta (β^+) decay	radiation detector
neutron number	electron capture	roentgen (R)
isotope	gamma decay	rad
nuclide	activity	gray (Gy)
radioactivity	decay constant	rem
alpha particle	half-life	relative biological effectiveness
beta particle	curie (Ci)	(RBE)
gamma ray	becquerel (Bq)	sievert (Sv)
alpha decay	carbon-14 dating	neutron activation analysis

III. Chapter Summary and Discussion

1. Nuclear Structure and the Nuclear Force (Section 29.1)

The **Rutherford-Bohr model** of the atom is essentially a planetary model, with negative electrons orbiting the positively charged nucleus in which almost all the mass of the atom is located. The atomic nucleus is composed of two types of particles -- protons and neutrons -- which are collectively referred to as **nucleons**. Using scattering experiments, the *upper limit* for the nuclear radius is found to be on the order of 10^{-12} m.

The **strong nuclear force**, or simply the nuclear force, is the short range attractive force between nucleons which is responsible for holding the nucleus together against the repulsive electric force of its protons. This force possesses the following properties:

- It is strongly attractive and much larger in relative magnitude than the electrostatic and magnetic forces.
- It is very short-ranged; that is, a nucleon interacts only with its nearest neighbors, over a distance on the order of 10^{-15} m.
- It acts between any two nucleons within a short range, that is, between two protons, a proton and a neutron, or two neutrons.

In a nuclear notation $^A_Z X_N$, Z is the **atomic number** or **proton number** (the number of protons), N is the **neutron number** (the number of neutrons), A is the **mass number** (the number of nucleons), and X is the chemical symbol of the element of the nucleus. The neutron number is often omitted because it can be determined from $N = A - Z$. For example, the $^{13}_6 C$ is for carbon-13, which has 6 protons, $13 - 6 = 7$ neutrons, and 13 nucleons.

Isotopes are nuclei with the same number of protons (atomic number) but a different number of neutrons (neutron numbers or mass numbers), such as $^{12}_6 C$, $^{13}_6 C$, and $^{14}_6 C$. A particular nuclear species or isotope of any element is called a **nuclide**, for example, $^{11}_6 C$, $^{12}_6 C$, $^{13}_6 C$, $^{14}_6 C$, $^{15}_6 C$, and $^{16}_6 C$ are all nuclides of carbon.

Example 29.1 The nuclear notation for cobalt-59 is $^{59}_{27} Co$. If the atom is neutral,

        (a) how many protons does it possess?

        (b) how many neutrons does it possess?

        (c) how many nucleons does it possess?

        (d) how many electrons does it possess?

Solution: $A = 59, \quad Z = 27$.

(a) The number of protons is equal to the atomic number, $Z = 27$ protons.

(b) The number of neutrons is equal to the neutron number, $N = A - Z = 59 - 27 = 32$ neutrons.

(c) The number of nucleons is equal to the mass number, $A = 59$ nucleons.

(d) In a neutral atom, the number of electrons is equal to the number of protons, so there are 27 electrons.

Example 29.2   Which one of the following is an isotope of the hypothetical nuclide $^{39}_{20}X_{19}$?

(a) $^{40}_{20}X_{20}$ (b) $^{39}_{19}Y_{20}$ (c) $^{39}_{21}Z_{18}$.

Solution:

Isotopes are nuclei with the same number of protons and a different number of neutrons, so the answer is (a).

---

2.   Radioactivity (Section 29.2)

The nuclei of some isotopes are not stable and they decay spontaneously with the emission of particles. Such isotopes are said to be *radioactive* or to exhibit **radioactivity**. Five commonly occurring decay modes are:

- **Alpha decay,** in which the nucleus emits an **alpha particle**, a doubly charged $(+2e)$ particle containing two protons and two neutrons (helium nucleus, ^4_2He).

- **Beta decay** (β^-) in which the nucleus emits a **beta particle** or an electron $(^0_{-1}e)$.

- **Gamma decay,** in which an excited nucleus emits a **gamma ray**, or a "particle" or quantum, or a photon, of electromagnetic energy.

- **Beta decay** (β^+) **decay,** in which the nucleus emits a positron $(^0_{+1}e)$.

- **Electron capture,** in which the nucleus *absorbs* one of its atomic orbital electrons.

Alpha decay involves the quantum mechanical process of **tunneling** or **barrier penetration**, that is, there is a finite probability of finding the alpha particle outside the nucleus after penetrates the Coulomb potential energy barrier. In basic beta β^- decay, a neutron decays to a proton and an electron, $^1_0n \rightarrow ^1_1p + ^0_{-1}e$; and in basic beta β^+ decay, a proton decays to a neutron and a positron, $^1_1p \rightarrow ^1_0n + ^0_{+1}e$. (Neutrinos are also involved in beta decays. They will be discussed later and are not shown here for simplicity.)

Two laws apply to all nuclear reactions: the **conservation of nucleons** (the total number of nucleons remains constant) and the **conservation of charge** (the total charge remains constant).

Example 29.3   Write the nuclear equation for the following nuclear decays:

   (a) Alpha decay of $^{214}_{84}$Po.

   (b) Beta (β^-) decay of $^{14}_{6}$C.

   (c) Gamma decay of $^{61}_{28}$Ni* (the * represents an excited nucleus).

Solution:

   For every nuclear reaction, the total number of nucleons and the total charge remain constant.

   (a) If $^{214}_{84}$Po emits an alpha particle ($^4_2$He), the daughter nucleus must have $214 - 4 = 210$ nucleons

   (conservation of nucleons) and $84 - 2 = 82$ protons (conservation of charge). From the periodic table, the

   nucleus that has 82 protons is lead (Pb). Therefore, the nuclear decay equation is $^{214}_{84}$Po $\rightarrow$ $^{210}_{82}$Pb $+ \, ^4_2$He.

   (b) If $^{14}_6$C emits a beta particle or electron ($^0_{-1}$e), the daughter nucleus must have the same number of

   nucleons (14) and $6 + 1 = 7$ protons to conserve nucleons and charge. The element with 7 protons is

   nitrogen (N). Therefore, the nuclear equation is $^{14}_6$C $\rightarrow$ $^{14}_7$N $+ \, ^0_{-1}$e.

   (c) Since a gamma ray (γ) is a photon, the daughter nucleus is still Ni, but at a lower energy level. The

   nuclear equation is $^{61}_{28}$Ni* $\rightarrow$ $^{61}_{28}$Ni $+ \gamma$.

The absorption or degree of penetration of nuclear radiation is an important consideration in applications such as radioisotope treatment of cancer and nuclear shielding around a nuclear reactor or radioactive materials. Alpha particles have large mass and are doubly charged, and generally move slowly. A few centimeters of air or a sheet of paper will usually completely stop them. Beta particles can travel a few meters in air or a few centimeters in aluminum before being stopped. Gamma rays are more penetrating. They can penetrate a centimeter or more of a dense material such as lead.

3.   Decay Rate and Half-Life (Section 29.3)

The **activity** of a radioactive isotope is defined as the number of disintegrations or decays (ΔN) per unit time (Δt), $\dfrac{\Delta N}{\Delta t}$. This activity is proportional to the number of undecayed nuclei present (N), that is, $\dfrac{\Delta N}{\Delta t} = -\lambda N$, where λ is the **decay constant**, which is different for different isotopes. The negative sign indicates the decrease in the number of undecayed nuclei present in a decay process. Solving the activity equation yields the expression for the undecayed nuclei as a function of time, $N = N_0\, e^{-\lambda t}$, where N_0 is the initial number of nuclei at $t = 0$. The **half-life** ($t_{1/2}$) of an isotope is the time required for the number of undecayed nuclei in a sample (hence the radioactive decay rate or activity of the sample) to fall to one-half of its original value. It is related to the decay constant λ

through $t_{1/2} = \dfrac{0.693}{\lambda}$. The SI unit of activity is the **becqerel** (Bq), and 1 Bq = 1 decay/s. Another common unit is the **curie** (Ci), and 1 Ci = 3.70×10^{10} decays/s = 3.70×10^{10} Bq.

Note: When using $N = N_o\, e^{-\lambda t}$, it is not necessary to change time to seconds. You can use any unit (min, h, or y) as long as λ is in the same units (inverse), for example, $(d^{-1} \times d)$ = unitless.

Radioactive isotopes can be used as nuclear clocks. A common radioactive dating method used on materials that were once part of living things is **carbon-14 dating**. This method uses the fact that the carbon-14 content decreases with time once a living organism dies, so measurements of the concentration of carbon-14 in dead matter relative to that in living things can then be used to establish when the organism died.

Example 29.4   A radioactive isotope sample has an half-life of 10 min. What percentage of the sample is left after 55 min?

Solution: Given: $t_{1/2}$ = 10 min, t = 55 min.   Find: $\dfrac{N}{N_o}$.

$\lambda = \dfrac{0.693}{t_{1/2}} = \dfrac{0.693}{10\text{ min}} = 0.0693\text{ min}^{-1}$,   so $\lambda t = (0.0693\text{ min}^{-1})(55\text{ min}) = 3.81$.

From $N = N_o\, e^{-\lambda t}$, we have $\dfrac{N}{N_o} = e^{-\lambda t} = e^{-3.81} = 2.2 \times 10^{-2}$ or 2.2%.

The problem can also be solved with half-life directly. After one half-life, the fraction of nuclei remaining is $\dfrac{N}{N_o} = \dfrac{1}{2} = \dfrac{1}{2^1}$; after two half-lives, the fraction of nuclei remaining is $\dfrac{N}{N_o} = \dfrac{1}{2} \times \dfrac{1}{2} = \dfrac{1}{2^2}$; so after n half-lives, the fraction of nuclei remaining is $\dfrac{N}{N_o} = \dfrac{1}{2^n}$. $n = \dfrac{55\text{ min}}{10\text{ min}} = 5.5$ half-lives,   so $\dfrac{N}{N_o} = \dfrac{1}{2^{5.5}}$ or 2.2%.

Example 29.5   A radioactive sample with a half-life of 10.0 min initially is composed of 1.50×10^{10} nuclei.

   (a) What is the decay constant?

   (b) What is the initial activity in Ci?

   (c) What is the activity in Ci after 2.00 min?

Solution: Given: $N_o = 1.50 \times 10^{10}$ nuclei, $t_{1/2}$ = 10.0 min.

Find: (a) λ (b) $\left|\dfrac{\Delta N}{\Delta t}\right|_o$ (c) $\left|\dfrac{\Delta N}{\Delta t}\right|$.

(a) $\lambda = \dfrac{0.693}{t_{1/2}} = \dfrac{0.693}{10.0\text{ min}} = 0.0693\text{ min}^{-1} = 1.16 \times 10^{-3}\text{ s}^{-1}$.

(b) 1 Ci = 3.70×10^{10} decays/s.

$$\left|\frac{\Delta N}{\Delta t}\right|_{o} = \lambda N_{o} = (1.16 \times 10^{-3} \text{ s}^{-1})(1.50 \times 10^{10}) = 1.74 \times 10^{7} \text{ decays/s} = 4.70 \times 10^{-4} \text{ Ci}.$$

(c) $\lambda t = (0.0693 \text{ min}^{-1})(2.00 \text{ min}) = 0.139$, so $N = N_{o} e^{-\lambda t} = (1.50 \times 10^{10}) e^{-0.139} = 1.31 \times 10^{10}$ nuclei.

Therefore, $\left|\dfrac{\Delta N}{\Delta t}\right| = \lambda N = (1.16 \times 10^{-3} \text{ s}^{-1})(1.31 \times 10^{10}) = 1.52 \times 10^{7}$ decays/s $= 4.11 \times 10^{-4}$ Ci.

4. Nuclear Stability and Binding Energy (Section 29.4)

Experiments show that to lower their total energy in a nucleus, two protons will "pair up", as will two neutrons. This so-called **pairing effect** influences the following criteria for nuclear stability:

(1)   All isotopes with 83 protons or more ($Z \geq 83$) are unstable.

(2)   Most even-even (even number of protons and even number of neutrons) are stable. Many odd-even or even-odd nuclei are stable. Only four odd-odd nuclei are stable, ^{2}H, ^{6}Li, ^{10}Be, and ^{14}N.

(3)   Stable nuclei with less than 40 nucleons ($A < 40$) have approximately the same number of protons and neutrons. Stable nuclei with $A > 40$ have more neutrons than protons.

Example 29.6   Is the sodium isotope $^{22}_{11}$Na likely to be stable?

Solution:

(1)   *Satisfied.* With $Z = 11$, the criterion is satisfied.

(2)   *Not satisfied.* The isotope $^{22}_{11}$Na has an odd-odd nucleus and since it is not one of the four stable odd-odd nuclei, it must be unstable.

Since the masses of nuclei are so small, another unit, the **atomic mass unit (u)** is used to measure nuclear masses. By definition, 1 u = 1.6606×10^{-27} kg, and it is referenced from a neutral atom of ^{12}C, which is taken to have an exact mass of 12.000000 u. According to mass-energy equivalence, 1 u is equivalent to 931.5 MeV, that is I u of mass if converted completely would yield 931.5 MeV of energy. It is found that the sum of masses of the nucleons in a nucleus (bound particles) is less than that when the nucleons are separated as free particles. This mass difference, or **mass defect** (Δm), has an energy equivalence called the total **binding energy** (E_{b}), which is the minimum amount of energy needed to separate a given nucleus apart into its constituent nucleons, $E_{b} = (\Delta m)c^{2}$. The *average binding energy per nucleon, E_{b}/A*, gives an indication of nuclear stability. The greater the E_{b}/A, the more tightly bound are the nucleons in a nucleus, hence the more stable it is.

Example 29.7   What is the total binding energy and average binding energy per nucleon for sodium $^{22}_{11}$Na. The mass of $^{22}_{11}$Na is 21.994435 u.

Solution: Sodium $^{22}_{11}$Na has 11 protons and 11 neutrons.

Given: $^{22}_{11}$Na atomic mass = 21.994435 u, $^{1}_{1}$H mass = 1.007825 u, $^{1}_{0}$n mass = 1.008665 u.

Find: E_b and E_b/A.

The mass of the separated nucleons is $m = 11(1.007825 \text{ u}) + 11(1.008665 \text{ u}) = 22.18139$ u.

So the mass defect is $\Delta m = 22.18139 \text{ u} - 21.994435 \text{ u} = 0.18696$ u.

Therefore the total binding energy is $E_b = (0.18696 \text{ u})(931.5 \text{ MeV/u}) = 174.15$ MeV.

The average binding energy per nucleon is $\dfrac{E_b}{A} = \dfrac{174.15 \text{ MeV}}{22 \text{ nucleons}} = 7.916$ MeV/nucleon.

---

5.   Radiation Detection and Applications (Section 29.5)

A *Geiger counter*, a *scintillation counter*, a *solid state (semiconductor) detector*, a *cloud chamber*, a *bubble chamber*, and a *spark chamber* are all examples of **radiation detectors**.

An important consideration in radiation therapy and radiation safety is the amount, or *dose*, of radiation. Several quantities are used to describe this in terms of *exposure*, *absorbed dose*, and *equivalent dose*. The earliest unit of dosage, the **roentgen** (R) was based on exposure. The roentgen has been largely replaced by the rad (*radiation absorbed dose*), which is an absorbed dose unit. The SI unit for absorbed dose is the **gray** (Gy): 1 Gy = 1 J/kg = 100 rad.

The effective dose is measured in terms of the **rem** unit (*roentgen* or *rad equivalent man*). The different degrees of effectiveness of different particles are characterized by the **relative biological effectiveness (RBE)**. Then, effective dose (in rem) = dose (in rad) × RBE. Another unit of effective dose is the **sievert** (Sv), and effective dose (in Sv) = dose (in Gy) × RBE. Since 1 Gy = 100 rad, it follows that 1 Sv = 100 rem.

Some other applications of radiation in domestic and industrial use are the smoke detector, radioactive tracer (to detect leaks), and in **neutron activation analysis** (to activate compounds for detection purposes).

IV. Mathematical Summary

Decay Rate of a Radioisotope	$\dfrac{\Delta N}{\Delta t} = -\lambda N$ (29.2) $1 \text{ Ci} = 3.70 \times 10^{10}$ decays/s $= 3.70 \times 10^{10}$ Bq	Defines the decay rate of an radioisotope in terms of its decay constant and number of remaining nuclei.
Activity of a Radioisotope	$\text{activity} = \left\|\dfrac{\Delta N}{\Delta t}\right\| = \lambda N$	Defines the activity of a radioisotope.
Number of Undecayed Nuclei	$N = N_0\, e^{-\lambda t}$ (29.3)	Calculates the number of undecayed nuclei as a function of time.
Half-life and Decay Constant	$t_{1/2} = \dfrac{0.693}{\lambda}$ (29.4)	Relates the half-life and the decay constant of a radioisotope.
Total Binding Energy	$E_b = (\Delta m)c^2$ (29.5)	Computes the binding energy of a nucleus from the mass defect between the nucleus and the constituent particles.
Effective Dose	Dose (rem) = does (rad) $\times$ RBE (29.6) Dose (Sv) = dose (Gy) $\times$ RBE (29.7)	Calculates the effective dose by including the relative biological effectiveness (RBE).

V. Solutions of Selected Exercises and Paired/Trio Exercises

6. (a) For ^{24}Mg: $\boxed{12\text{ p}}$, $24 - 12 = \boxed{12\text{ n}}$, and $\boxed{12\text{ e}}$.

   For ^{25}Mg: $\boxed{12\text{ p}}$, $25 - 12 = \boxed{13\text{ n}}$, and $\boxed{12\text{ e}}$.

   (b) For ^{24}Mg: $\boxed{12\text{ p}}$, $24 - 12 = \boxed{12\text{ n}}$, and $\boxed{14\text{ e}}$.

   For ^{25}Mg: $\boxed{12\text{ p}}$, $25 - 12 = \boxed{13\text{ n}}$, and $\boxed{14\text{ e}}$.

   (c) For ^{24}Mg: $\boxed{12\text{ p}}$, $24 - 12 = \boxed{12\text{ n}}$, and $\boxed{11\text{ e}}$.

   For ^{25}Mg: $\boxed{12\text{ p}}$, $25 - 12 = \boxed{13\text{ n}}$, and $\boxed{11\text{ e}}$.

8. The mass number is $8 + 8 = 16$, $8 + 9 = 17$, and $8 + 10 = 18$, respectively.

   They are $\boxed{^{16}_{8}\text{O},\ ^{17}_{8}\text{O},\ ^{18}_{8}\text{O}}$.

14.   (a) $\boxed{^{60}_{27}\text{Co} \rightarrow \,^{60}_{28}\text{Ni} + \,^{0}_{-1}\text{e}}$.   (b) $\boxed{^{226}_{88}\text{Ra} \rightarrow \,^{222}_{86}\text{Rn} + \,^{4}_{2}\text{He}}$.

19. $\alpha\text{–}\beta$: $^{209}_{82}\text{Pb} + \,^{0}_{-1}\text{e} \leftarrow \,^{209}_{81}\text{Tl};$ $^{209}_{81}\text{Tl} + \,^{4}_{2}\text{He} \leftarrow \boxed{^{213}_{83}\text{Bi}}$.

        $\beta\text{–}\alpha$: $^{209}_{82}\text{Pb} + \,^{4}_{2}\text{He} \leftarrow \,^{213}_{84}\text{Po};$ $^{213}_{84}\text{Po} + \,^{0}_{-1}\text{e} \leftarrow \boxed{^{213}_{83}\text{Bi}}$.

20.   (a) $\boxed{^{4}_{2}\text{He}}$.   (b) $\boxed{4(^{1}_{0}\text{n})}$.   (c) $\boxed{\gamma}$.   (d) $\boxed{^{29}_{12}\text{Mg}}$.

22. α–series: $\boxed{^{227}_{89}\text{Ac} \rightarrow \,^{223}_{87}\text{Fr} + \,^{4}_{2}\text{He}};$ $\boxed{^{223}_{87}\text{Fr} \rightarrow \,^{223}_{88}\text{Ra} + \,^{0}_{-1}\text{e}};$

                        $\boxed{^{223}_{88}\text{Ra} \rightarrow \,^{219}_{86}\text{Rn} + \,^{4}_{2}\text{He}};$ $\boxed{^{219}_{86}\text{Rn} \rightarrow \,^{215}_{84}\text{Po} + \,^{4}_{2}\text{He}}$.

     β–series: $\boxed{^{227}_{89}\text{Ac} \rightarrow \,^{227}_{90}\text{Th} + \,^{0}_{-1}\text{e}};$ $\boxed{^{227}_{90}\text{Th} \rightarrow \,^{223}_{88}\text{Ra} + \,^{4}_{2}\text{He}};$

                        $\boxed{^{223}_{88}\text{Ra} \rightarrow \,^{219}_{86}\text{Rn} + \,^{4}_{2}\text{He}};$ $\boxed{^{219}_{86}\text{Rn} \rightarrow \,^{215}_{84}\text{Po} + \,^{4}_{2}\text{He}}$.

29.   (a) $(2.50 \times 10^{6} \text{ decays/s}) \dfrac{1 \text{ Ci}}{3.70 \times 10^{10} \text{ decays/s}} = \boxed{6.76 \times 10^{-5} \text{ Ci}}$.

        (b) $\boxed{2.50 \times 10^{6} \text{ Bq}}$.

32. $\lambda = \dfrac{0.693}{t_{1/2}} = \dfrac{0.693}{18 \times 60 \text{ s}} = 6.42 \times 10^{-4} \text{ s}^{-1};$ $\lambda t = (5.92 \times 10^{-4} \text{ s}^{-1})(3600 \text{ s}) = 2.31$.

     So $\left| \dfrac{\Delta N}{\Delta t} \right| = \lambda N = \lambda N_0 \, e^{-\lambda t} = \dfrac{\Delta N_0}{\Delta t} \, e^{-\lambda t} = (10 \text{ mCi}) \, e^{-2.31} = \boxed{1 \text{ mCi}}$.

34. $\boxed{\text{Tc greater by 70 times}}$.

37.   After n half-lives, $\lambda t = \dfrac{0.693}{t_{1/2}} \, nt_{1/2} = 0.693n$.

     So $N = N_0 \, e^{-\lambda t} = N_0 \, e^{-0.693n} = \dfrac{N_0}{(e^{0.693})^n} = \dfrac{N_0}{2^n} = \left(\dfrac{1}{2} \right)^n N_0$.

40.   Since $\left| \dfrac{\Delta N}{\Delta t} \right| = \lambda N = \lambda N_0 \, e^{-\lambda t} = \dfrac{\Delta N_0}{\Delta t} \, e^{-\lambda t},$ $\dfrac{\Delta N/\Delta t}{\Delta N_0 / \Delta t} = 0.20 = e^{-\lambda t}$.

     So $t = -\dfrac{\ln 0.20}{\lambda} = -\dfrac{t_{1/2} \ln 0.20}{0.693} = -\dfrac{(5.3 \text{ y}) \ln 0.20}{0.693} = \boxed{12 \text{ y}}$.

42.	(a) $\boxed{1.05 \times 10^{-2} \text{ s}^{-1}}$.	(b) $\boxed{219 \text{ s}}$.

45.	$\lambda t = \dfrac{0.693}{t_{1/2}} \, t = \dfrac{0.693}{1600 \text{ y}} \, (2100 \text{ y} - 1898 \text{ y}) = 0.0875.$

From $N = N_0 \, e^{-\lambda t}$, we have $\dfrac{N}{N_0} = e^{-\lambda t} = e^{-0.0875} = 0.916.$

So the amount of radium that would remain is $(0.916)(10 \text{ mg}) = \boxed{9.2 \text{ mg}}$.

53.	(a) $\boxed{{}^{17}_{8}\text{O}}$ because of an unpaired neutron beyond a magic number.

(b) $\boxed{{}^{42}_{20}\text{Ca}}$ because of a magic number difference making ${}^{40}_{20}\text{Ca}$ more stable (both are paired).

(c) $\boxed{{}^{10}_{5}\text{B}}$ because of lack of pairing.

57.	$E_b = \Delta mc^2 = (m_n + m_p - m_D)c^2$,

so	$m_D = m_p + m_n - E_b/c^2 = 1.007276 \text{ u} + 1.008665 \text{ u} + \dfrac{2.224 \text{ MeV}}{931.5 \text{ MeV/u}} = \boxed{2.013553 \text{ u}}.$

60.	$E_b = \Delta mc^2 = (8m_H + 8m_n - m_O)c^2 = [8(1.007825 \text{ u}) + 8(1.008665 \text{ u}) - 15.994915 \text{ u}](931.5 \text{ MeV/u})$

$= 127.6 \text{ MeV}.$

$\dfrac{E_b}{A} = \dfrac{127.6 \text{ MeV}}{16 \text{ nucleon}} = \boxed{7.98 \text{ MeV/nucleon}}$

66.	For Al: $E_b = (13m_H + 14m_n - m_{Al})c^2 = [13(1.007825 \text{ u}) + 14(1.008665 \text{ u}) - 26.981541 \text{ u}](931.5 \text{ MeV/u})$

$= 225.0 \text{ MeV}.$ So $\dfrac{E_b}{A} = \dfrac{225.0 \text{ MeV}}{27 \text{ nucleon}} = 8.33 \text{ MeV/nucleon}.$

For Na: $E_b = (11m_H + 12m_n - m_{Na})c^2 = [11(1.007825 \text{ u}) + 12(1.008665 \text{ u}) - 22.989770 \text{ u}](931.5 \text{ MeV/u})$

$= 186.6 \text{ MeV}.$ So $\dfrac{E_b}{A} = \dfrac{186.6 \text{ MeV}}{23 \text{ nucleon}} = 8.11 \text{ MeV/nucleon}.$

So the nucleons are more tightly bound in $\boxed{\text{Al}}$ on average.

74.	(a) $E = m \times \text{Dose} = (0.20 \text{ kg})(1.25 \text{ rad}) = \boxed{0.25 \text{ J}}$.

(b) Dose (in rem) = Dose (in rad) × RBE = (1.25 rad)(4) = 5.0 rem, which exceeds the maximum

permissible radiation dosage with background radiation included. So the answer is $\boxed{\text{yes}}$.

79.   (a) $E_b = (7.075 \text{ MeV/nucleon})(4 \text{ nucleon}) = \boxed{28.3 \text{ MeV}}$.

(b) $E_b = (2m_p + 2m_n - m_{He})c^2$,

so $m_{He} = 2m_p + 2m_n - E_b/c^2 = 2(1.007276 \text{ u}) + 2(1.008665 \text{ u}) - \dfrac{28.3 \text{ MeV}}{931.5 \text{ MeV/u}} = \boxed{4.001501 \text{ u}}$.

83.   (a) $\boxed{{}^{0}_{-1}e}$.   (b) $\boxed{{}^{222}_{86}Rn}$.   (c) $\boxed{{}^{237}_{94}Pu}$.   (d) $\boxed{\gamma}$.   (e) $\boxed{{}^{0}_{+1}e}$.

87.   (1) All nuclei with $Z \geq 83$ are unstable.

(2) All odd–odd nuclei are unstable except for ${}^2H$, ${}^6Li$, ${}^{10}B$, and ${}^{14}N$.

(3) Stable nuclei with $A < 40$ have about equal numbers of protons and neutrons.

(4) Stable nuclei with $A > 40$ have an excess number of neutrons which becomes more obvious as A increases.

So $\boxed{\text{(c) and (d)}}$ are stable.

91.   (a) $\dfrac{\text{nucleon}}{\text{volume}} = \dfrac{A}{4\pi R^3/3} = \dfrac{3A}{4\pi R^3} = \dfrac{3A}{4\pi R_o^3 A} = \dfrac{3}{4\pi R_o^3} = \dfrac{3}{4\pi(1.2 \times 10^{-15} \text{ m})^3}$

$\approx 1.4 \times 10^{44} \text{ nucleons/m}^3$.

(b) 1 nucleon has a mass of 1.66×10^{-27} kg:

So $\rho = (1.4 \times 10^{44} \text{ nucleon/m}^3) \dfrac{1.66 \times 10^{-27} \text{ kg}}{1 \text{ nucleon}} = \boxed{2.3 \times 10^{17} \text{ kg/m}^3}$.

VI.   Practice Quiz

1.   How many nucleons are there in ${}^{15}_{8}O$?

(a) 7   (b) 8   (c) 15   (d) 22   (e) 23

2.   Which particle has the least mass?

(a) alpha   (b) beta (β^-)   (c) beta (β^+)   (d) nucleon   (e) gamma

3.   A radioactive sample has a half-life of 4.0 min. What fraction of the sample is left after 20 min?

(a) 1/32 . (b) 1/25   (c) 1/16   (d) 1/5   (e) zero

4.   An atom has 98 protons and 249 nucleons. If it undergoes beta (β^-) decay, what are the numbers of protons and nucleons, respectively, in the daughter nucleus?

(a) 96, 245   (b) 98, 250   (c) 96, 245   (d) 99, 249   (e) 100, 249

5. The half-life of radioactive iodine-137 is 8.0 days. How many iodine nuclei are necessary to produce an activity of 1.0 μCi?

(a) 2.9×10^9 (b) 4.6×10^9 (c) 3.7×10^{10} (d) 7.6×10^{12} (e) 8.1×10^{13}

6. What is the average binding energy per nucleon for $^{197}_{79}$Au? The mass of $^{197}_{79}$Au is 196.96656 u.

(a) 6.8 MeV   (b) 7.3 MeV   (c) 7.7 MeV   (d) 7.9 MeV   (e) 8.3 MeV

7. The nuclear equation, $^{227}_{89}$Ac $\rightarrow$ $^{223}_{87}$Fr + $^{4}_{2}$He, is for what type of decay?

(a) alpha.   (b) beta (β^-).   (c) gamma.   (d) electron capture.   (e) beta (β^+).

8. An X-ray technician takes an average of ten X-rays per day and receives 2.5×10^{-3} rad per X-ray. What is the total dose in rem the technician receives in 250 working days?

(a) 2.50 rem   (b) 5.00 rem   (c) 6.25 rem   (d) 7.75 rem   (e) 9.00 rem

9. Which one of the following nuclei is likely to be unstable?

(a) $^4_2$He   (b) $^{27}_{13}$Al   (c) $^{13}_{6}$C   (d) $^{24}_{11}$Na   (e) $^2_1$H

10. The binding energy of a nucleus is directly related to

(a) radioactivity.   (b) mass defect.   (c) too many neutrons.

(d) conservation of nucleons.   (e) alpha decay.

Answers to Practice Quiz:

1. c 2. e 3. a 4. d 5. c 6. d 7. a 8. c 9. d 10. b

CHAPTER 30

Nuclear Reactions and Elementary Particles

I.   Chapter Objectives

Upon completion of this chapter, you should be able to:

1.   use charge and nucleon conservation to write nuclear reaction equations, and understand and use the concepts of Q value and threshold energy to analyze nuclear reactions.

2.   understand the process of nuclear fission, the nature and the cause of a nuclear chain reaction, and the basic principles involved in the operation of nuclear reactors.

3.   explain the fundamental difference between fusion and fission, calculate energy releases in fusion reactions, and understand how fusion might eventually provide a source of electric energy.

4.   explain why the neutrino is necessary to account for observed beta decay data, specify some of the physical properties of neutrinos, and write complete beta decay equations.

5.   understand the quantum mechanical description of forces, and classify the various forces according to their strengths, properties, ranges, and virtual particles.

6.   classify the various elementary particles into families and understand the different properties of the various families of elementary particles.

7.   become familiar with the quark model and quark properties, and understand how the quark model accounts for the properties of baryons and mesons.

8.   become familiar with current attempts to unify the four fundamental forces and understand why elementary particle interactions might hold the key to the very early evolution of the universe.

II.   Key Terms

Upon completion of this chapter, you should be able to define and/or explain the following key terms:

nuclear reactions	fuel rods	magnetic confinement
Q-value	core	inertia confinement
endoergic	coolant	neutrino
exoergic	control rods	fundamental forces
threshold energy	moderator	virtual particle
cross section	breeder reactor	exchange particle
fission reaction	LOCA	photon
chain reaction	meltdown	meson
critical mass	fusion reaction	muon

nuclear reactor	plasma	pion
weak nuclear force	tauon	gluons
W particle	baryons	color force
graviton	quarks	unification theories
elementary particles	quark model	electroweak force
leptons	quark confinement	grand unified theory
hadrons	color charge	superforce

III. Chapter Summary and Discussion

1. Nuclear Reactions (Section 30.1)

In **nuclear reactions**, a particular nuclide is converted into a different one, which in general is a nuclide of a completely different element. Some reactions can be produced by energetic particles from radioactive sources or from **particle accelerators**, in which particles are accelerated to high speeds. The general form of a nuclear reaction equation is $A + a \rightarrow B + b$, which is written as $A(a, b)B$, where upper case letters represent the nuclei and the lowercase letters represent the particles, A and a are the reactants and B and b are the products.

The **Q value** of a reaction or decay represents the energy released ($Q > 0$) or absorbed ($Q < 0$) in the process; it appears as a change in the total mass of the system, $Q = (m_A + m_a - m_B - m_b)c^2 = (\Delta m)c^2$. When the Q value of a reaction is positive ($Q > 0$), the mass of the reactants is greater than the mass of the products, energy is released in the form of kinetic energy, and the reaction is said to be **exoergic**. When the Q value of a reaction is negative ($Q < 0$), the mass of the products is greater than the mass of the reactants, energy is absorbed, and the reaction is said to be **endoergic**. All naturally occurring radioactive decay processes are exoergic.

In an endoergic reaction, the minimum energy required for the reaction to happen is called the **threshold energy** and is given by $K_{min} = \left(1 + \dfrac{m_a}{m_A}\right)|Q|$. A measure of the probability that a particular reaction will occur is called the **cross section** of the reaction.

Note: The mass-energy equivalence of 1 u equivalent to 931.5 MeV is often used in calculating Q values and threshold energies.

Example 30.1   Is the following reaction exoergic or endoergic?

$$^{7}\text{Li} \quad + \quad ^{1}\text{p} \quad \rightarrow \quad ^{4}\text{He} \quad + \quad ^{4}\text{He}$$

(7.016005 u) (1.007825 u)   (4.002603 u) (4.002603 u)

Solution: Given: $m_A = 7.016005$ u, $m_a = 1.007825$ u, $m_B = 4.002603$ u, $m_b = 4.002603$ u.

Find: If exoergic or endoergic.

$$Q = (\Delta m)c^2 = (m_A + m_a - m_B - m_b)c^2$$

$$= (7.016005 \text{ u} + 1.007825 \text{ u} - 4.002603 \text{ u} - 4.002603 \text{ u})(931.5 \text{ MeV/u}) = +17.35 \text{ MeV}.$$

Since the Q value is positive, the reaction is exoergic.

Example 30.2 Find the threshold energy for the following reaction.

$$^{13}\text{C} \quad + \quad ^{1}\text{p} \quad \rightarrow \quad ^{13}\text{N} \quad + \quad ^{1}\text{n}$$

$$(13.003355 \text{ u}) \ (1.007825 \text{ u}) \quad (13.005739 \text{ u}) \ (1.008665 \text{ u})$$

Solution: Given: $m_A = 13.003355$ u, $m_a = 1.007825$ u, $m_B = 13.005739$ u, $m_b = 1.008665$ u.

Find: K_{min}.

$$Q = (\Delta m)c^2 = (m_A + m_a - m_B - m_b)c^2$$

$$= (13.003355 \text{ u} + 1.007825 \text{ u} - 13.005739 \text{ u} - 1.008665 \text{ u})(931.5 \text{ MeV/u}) = -3.003 \text{ MeV}.$$

So the threshold energy is $K_{min} = \left(1 + \dfrac{m_a}{m_A}\right)|Q| = \left(1 + \dfrac{1.007825 \text{ u}}{13.003355 \text{ u}}\right)(3.003 \text{ MeV}) = 3.236 \text{ MeV}.$

The threshold energy is more than 3.003 MeV because products must have kinetic energies to conserve linear momentum.

---

2. Nuclear Fission (Section 30.2)

In a nuclear **fission reaction**, a heavy nucleus divides into two lighter nuclei whose total mass is less than that of the heavy nuclei, with the emission of two or more neutrons. The energy liberated in a fission reaction is about 1 MeV per nucleon in fission products. Since there are billions of nuclei in even a tiny sample of fissionable material, the amount of energy released in nuclear fission can be enormous. A sustained release of nuclear energy can be accomplished by a **chain reaction**, in which neutrons from one fission reaction initiate more fission reactions, and the process multiplies, the number of neutrons doubling or tripling with each generation. When this occurs uniformly with time, we have very fast exponential growth of released energy (or a bomb). To have a sustained chain reaction, there must be a minimum, or **critical mass** of the fissionable material. Basically, enough mass is needed so the neutrons from the previous reaction do not escape without causing more fission reactions.

Example 30.3 Estimate the energy released in the reaction, $^{235}_{92}\text{U} + ^{1}_{0}\text{n} \rightarrow ^{140}_{54}\text{Xe} + ^{94}_{38}\text{Sr} + 2(^{1}_{0}\text{n})$.

Solution:

There are $140 + 94 = 234$ nucleons in the fission products, so the energy release is approximately $(1 \text{ MeV/nucleon})(234 \text{ nucleons}) \approx 234 \text{ MeV}$. This is the energy released per initial ^{235}U nucleus decay.

Example 30.4   Find the energy released in the following neutron-initiated fission reaction.

$$^1\text{n} \quad + \quad ^{235}\text{U} \quad \rightarrow \quad ^{141}\text{Ba} \quad + \quad ^{92}\text{Kr} \quad + \quad 3(^1\text{n})$$
$$(1.008665 \text{ u}) \ (235.043925 \text{ u}) \qquad (140.91420 \text{ u}) \ (91.9252 \text{ u}) \ (1.008665 \text{ u})$$

Solution:

The mass difference between the reactants and the products is

$\Delta m = 1.008665 \text{ u} + 235.043925 \text{ u} - 140.91420 \text{ u} - 91.9252 \text{ u} - 3(1.008665 \text{ u}) = 0.1872 \text{ u}$.

So the energy released is $(0.1872 \text{ u})(931.5 \text{ MeV/u}) = 174.4 \text{ MeV}$.

Currently, the only type of practical **nuclear reactor** is based on the fission chain reaction. There are four key elements to a reactor: fuel rods, coolant, control rods, and a moderator. **Fuel rods** are tubes packed with pellets of enriched uranium oxide (or other fissionable material) located in the reactor core. A **coolant** is usually ordinary water (H_2O) or heavy water (D_2O) flowing around the fuel rods to remove the energy released from the fission chain reaction. The chain reaction and energy output of a reactor are controlled by means of boron or cadmium **control rods**, which can be inserted into or withdrawn from the reactor core. (Boon and cadmium are very effective in absorbing neutrons.) The water flowing around the fuel rods acts not only as a coolant, but also as a **moderator**, which slows down the speed of the neutrons from fission before they initiate more fission reactions. Other materials, such as graphite (carbon), may be used as moderators.

A **breeder reactor** produces more fissionable materials than it consumes. U-235 is consumed as in a regular reactor and fissionable Pu-239 is produced from neutron reactions with the non-fissionable isotope od uranium, U-238.

Should the coolant be lost or stop flowing through the core of a reactor, the reactor will overheat. This could result in a **LOCA** (loss-of-coolant accident) and the **meltdown** of the core. The huge amount of energy released can cause explosions and the release of radioactive materials to the environment. Therefore, a safety shutdown must be started when a loss of coolant is detected and emergency cooling water must be provided to keep the radioactive fragments from overheating.

3. Nuclear Fusion (Section 30.3)

In a nuclear **fusion reaction**, light nuclei fuse together to form a heavier nucleus with less total mass than the original nuclei, with the release of energy. At the centers of some stars, four hydrogen nuclei (^{1}H) fuse to form helium (^{4}He). The energy released per fusion is about 24.7 MeV. Due to the light mass of H and He, fusion releases more energy than fission on a per unit mass basis.

Fusion requires high temperatures that ionize the fusion material into a gas of positive ions and negative electrons called **plasma**. Due to the high temperature of plasma, plasma confinement is a major problem. The problem of plasma confinement is being approached in several different ways. Among these are magnetic confinement and inertial confinement. In **magnetic confinement**, magnetic fields are used to hold the plasma in a confined space, a so-called magnetic bottle. In **inertial confinement**, pulsed laser, electron, or ion beams would be used to implode hydrogen fuel pellets, producing compression and high temperatures.

Example 30.5 Find the energy released in the following fusion reaction.

$$^2\text{H} \quad + \quad ^2\text{H} \quad \rightarrow \quad ^3\text{He} \quad + \quad ^1\text{n}$$
$$\text{(2.014102 u)} \quad \text{(2.014102 u)} \quad\quad \text{(3.016029 u)} \quad \text{(1.008665 u)}$$

Solution:

The mass difference between the reactants and the products is

$\Delta m = 2.014102 \text{ u} + 2.014102 \text{ u} - 3.016029 \text{ u} - 1.008665 \text{ u} = 3.510 \times 10^{-3}$ u.

So the energy released is $(3.510 \times 10^{-3} \text{ u})(931.5 \text{ MeV/u}) = 3.270$ MeV.

---

4. Beta Decay and the Neutrino (Section 30.4)

In a beta decay, there is an apparent violation of conservation of energy, linear momentum, and angular momentum (spin). To account for this apparent problem, an additional particle called the **neutrino** (v_e) was proposed. Neutrinos have zero rest mass (recent experiments show possible small mass) and interact with matter by the weak nuclear force (to be discussed later). Neutrinos (v_e) are produced in β^+ decays and antineutrinos ($\bar{v}_e$) are produced in β^- decays. The complete basic reactions are

β^- decay: $\quad\quad$ $n \rightarrow p + e^- + \bar{v}_e$ $\quad\quad\quad\quad\quad$ β^+ decay $\quad$: $\quad$ $p \rightarrow n + e^+ + v_e$.

5.   Fundamental Forces and Exchange Particles (Section 30.5)

There are only four known **fundamental forces**: the *gravitational force*, the *electromagnetic force*, the *strong nuclear force*, and the *weak nuclear force*.

In quantum mechanics, forces visualized as being transmitted by the exchange of particles. These particles are created within the time allowed by the uncertainty principle and are called **virtual particles**. The fundamental forces are considered to be carried by these virtual **exchange particles**. The greater the mass of an exchange particle, the more energy required to create it, and the shorter the range of the particle. Therefore, the range of the force produced by this particle is inversely related to its mass.

The exchange particle for the electromagnetic force is a (virtual) **photon**. Since a photon has zero mass, its range is infinite. The strong nuclear force is associated with the **meson**. The relationship between the mass of the meson and the range of the strong nuclear force is given by $R = c \Delta t = \dfrac{h}{2\pi m_m c}$, where m_m is the mass of the exchange particle. The mass of π mesons (**pions**) is found to be from 247 to 264 times that of an electron, so its range is about 10^{-15} m, which indicates a very short-ranged force. The **W particle** carries the **weak nuclear force**. The W particle has a mass of about 100 times that of a proton, which explains the extremely short range ($\approx \times 10^{-17}$ m) and the weakness of the weak force. The weak force is the only force that acts on neutrinos, which explains why they are so difficult to detect. The **Graviton** is the massless exchange particle for the gravitational force and interacts very weakly with matter, making it very difficult to detect. There is still no firm evidence of the existence of this massless particle. A summary of the relative strength, range, and exchange particles of the four fundamental forces are given in the following table:

Force	Relative Strength	Action Distance	Exchange Particle
Strong Nuclear	1	Short range ($\approx 10^{-15}$ m)	Pion (π meson)
Electromagnetic	10^{-3}	Inverse square (infinite)	Photon
Weak Nuclear	10^{-8}	Extremely short range ($\approx 10^{-17}$ m)	W particle
Gravitational	10^{-45}	Inverse square (infinite)	Graviton

Example 30.6   The π^0 meson has a mass of 264 times that of an electron. What is the range of the force mediated by this particle?

Solution: Given: m_m (meson) $= 264 m_e = 264(9.11 \times 10^{-31}$ kg$) = 2.41 \times 10^{-28}$ kg.

Find: R.

$$R = \frac{h}{2\pi m_m c} = \frac{6.63 \times 10^{-34} \text{ J·s}}{2\pi (2.41 \times 10^{-28} \text{ kg})(3.00 \times 10^8 \text{ m/s})} = 1.46 \times 10^{-15} \text{ m}.$$

---

6. Elementary Particles (Section 30.6)

Elementary particles are fundamental particles, or the building blocks, of atoms. **Leptons** are particles that interact by the weak nuclear force and there are six of them, electrons, **muons**, **tauons**, and three types of neutrinos. Each lepton has its antiparticle and so there are twelve different leptons in all.

Hadrons are particles that interact by the strong nuclear force, for example, protons, neutrons, and **pions**. The hadrons are subdivided into **baryons** and **mesons**. Baryons include the nucleons-proton and neutron and have half integer spin values ($\frac{1}{2}$ or $\frac{3}{2}$). Mesons, which include the pion, have integer spin values (0 or 1).

7. The Quark Model (Section 30.7)

and

8. Force Unification Theories and the Early Universe (Section 30.8)

Quarks are elementary particles that make up hadrons. Quarks combine only in two ways, either in three or in quark-antiquark pairs. Three quark combinations are called baryons, and quark-antiquark combinations are called mesons.

There are six *flavors* or types of quarks: up (u), down (d), strange (s), charm (c), top (t), and bottom (b). Quarks have fractional electronic charges of either $-e/3$ or $+2e/3$. Quarks interact by the strong nuclear force, but they are also subject to the weak force. A weak force acting on a quark changes its flavor, and gives rise to the decay of hadrons.

The exchange particle for quarks is the **gluon**. To give the strong force a field representation, each quark is said to possess an analog of electric charge which is the source of the "gluon field." Instead of charge, their property is called **color charge** (no relationship with ordinary color). The force between quarks of different color is sometimes called **color force**. Each quark can come in one of the three possible colors: red, blue, and green. There are corresponding anticolors and antiquarks. When a quark emits or absorbs a gluon, it changes its color.

The electromagnetic force and the weak force are two parts of a single **electroweak force**. Attempts to unify the various forces are called **unification theories**. A theory that would merge the strong nuclear force and the electroweak force is called the **grand unified theory** (GUT). Perhaps all forces are part of a single **superforce**, which is currently a primary theoretical challenge in elementary particle physics.

IV. Mathematical Summary

Q Value (for reaction $A + a \rightarrow B + b$)	$Q = (m_A + m_a - m_B - m_b)c^2$ $= (\Delta m)c^2 \qquad (30.3)$	Defines the Q value of a nuclear reaction in terms of the mass defect.
Threshold Energy	$K_{min} = \left(1 + \dfrac{m_a}{m_A}\right)\lvert Q \rvert \qquad (30.4)$	Calculates the minimum energy required (threshold energy) for an endoergic reaction.
Range of Exchange Particle	$R = c\,\Delta t = \dfrac{h}{2\pi m_m c} \qquad (30.5)$	Defines the range of an exchange particle.

V. Solutions of Selected Exercises and Paired/Trio Exercises

3. $K_{min} = \left(1 + \dfrac{m_a}{M_A}\right)\lvert Q \rvert.$

   As the target mass M_A becomes very large, $K_{min} \rightarrow \lvert Q \rvert$.
   If $Q = 0$, the incident particle would not need any kinetic
   energy to start the reaction.

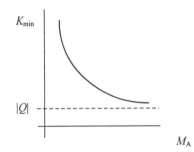

4. (a) $\boxed{{}^{41}_{19}\text{K}}$.   (b) $\boxed{{}^{135}_{52}\text{Te}}$.   (c) $\boxed{4({}^{1}_{0}\text{n})}$.

   (d) $\boxed{{}^{14}_{7}\text{N}}$.   (e) $\boxed{\text{n}}$.

8. $Q = (m_C + m_p - m_{He} - m_B)c^2 = (13.003355\ \text{u} + 1.007825\ \text{u} - 4.002603 - 10.012938\ \text{u})(931.5\ \text{MeV/u})$

   $= -\ \boxed{4.06\ \text{MeV}}$.

   It requires 4.06 MeV.

12. $Q = (m_O + m_n - m_C - m_{He})c^2 = (15.994915\ \text{u} + 1.008665\ \text{u} - 13.003355\ \text{u} - 4.002603\ \text{u})(931.5\ \text{MeV/u})$

   $= -2.215\ \text{MeV}.$

   $K_{min} = \left(1 + \dfrac{m_a}{M_A}\right)\lvert Q \rvert = \left(1 + \dfrac{1.008665}{15.994915}\right)(2.215\ \text{MeV}) = \boxed{2.35\ \text{MeV}}.$

14. $\boxed{3.24\ \text{MeV}}$.

19. $Q = (m_{\text{H-3}} + m_p - m_{\text{H-2}} - m_d)c^2 = (3.016049 \text{ u} + 1.007825 \text{ u} - 2.014102 \text{ u} - 2.014102 \text{ u})(931.5 \text{ MeV/u})$

        $= -4.033 \text{ MeV}$.

        $K_{\min} = \left(1 + \dfrac{m_a}{M_A}\right)|Q| = \left(1 + \dfrac{1.007825}{3.016049}\right)(4.033 \text{ MeV}) = \boxed{5.38 \text{ MeV}}$.

21.   For the first reaction: $(K_1)_{\min} = \left(1 + \dfrac{m_a}{M_A}\right)|Q| = \left(1 + \dfrac{1}{15}\right)|Q_1|$,

        For the second reaction: $(K_2)_{\min} = \left(1 + \dfrac{1}{20}\right)|Q_2|$.

        So $\dfrac{(K_1)_{\min}}{(K_2)_{\min}} = \dfrac{16/15}{21/20}\dfrac{|Q_1|}{|Q_2|} = \dfrac{16/15}{21/20} 3 = 3.05$.

        Therefore $\boxed{\text{the first reaction by 3.05 times}}$.

23.   (a) $\pi R^2 = \pi R_o^2 (A)^{2/3} = \pi(1.2 \times 10^{-15} \text{ m})^2 (12)^{2/3} = 2.37 \times 10^{-29} \text{ m}^2 = (2.37 \times 10^{-29} \text{ m}^2)\dfrac{1 \text{ b}}{10^{-28} \text{ m}^2} = \boxed{0.24 \text{ b}}$.

        (b) $\pi R^2 = \pi R_o^2 (A)^{2/3} = \pi(1.2 \times 10^{-15} \text{ m})^2 (56)^{2/3} = 6.62 \times 10^{-29} \text{ m}^2 = \boxed{0.66 \text{ b}}$.

        (c) $\pi R^2 = \pi R_o^2 (A)^{2/3} = \pi(1.2 \times 10^{-15} \text{ m})^2 (208)^{2/3} = 1.59 \times 10^{-28} \text{ m}^2 = \boxed{1.6 \text{ b}}$.

        (d) $\pi R^2 = \pi R_o^2 (A)^{2/3} = \pi(1.2 \times 10^{-15} \text{ m})^2 (238)^{2/3} = 1.74 \times 10^{-28} \text{ m}^2 = \boxed{1.7 \text{ b}}$.

30.   (a) 231 nucleons are involved in the fission process.

        So the energy released is (1 MeV/nucleon)(231 nucleon) = $\boxed{231 \text{ MeV}}$.

        (b) 238 nucleons are involved in the fission process.

        So the energy released is (1 MeV/nucleon)(238 nucleon) = $\boxed{238 \text{ MeV}}$.

37.   Neutrino has zero mass and so $E = pc = \dfrac{hc}{\lambda}$.

        Therefore $\lambda = \dfrac{hc}{E} = \dfrac{(6.63 \times 10^{-34} \text{ J·})(3.00 \times 10^8 \text{ m/s})}{(2.65 \times 10^6 \text{ eV})(1.6 \times 10^{-19} \text{ J/eV})} = \boxed{4.69 \times 10^{-13} \text{ m}}$.

40.   In a β^- decay: $_Z^A p \rightarrow {}_{Z+1}^A d + {}_{-1}^0 e$.

        So $Q = (m_P - m_D - m_e)c^2 = \{(m_P + Zm_e) - [m_D + (Z+1)m_e]\}c^2 = (M_P - M_D)c^2$.

42. $\boxed{0.71 \text{ MeV}}$.

49. From $R = \dfrac{h}{2\pi mc}$, we have $m = \dfrac{h}{2\pi Rc} = \dfrac{6.63 \times 10^{-34} \text{ J·s}}{2\pi(10^{-15} \text{ m})(3.00 \times 10^{8} \text{ m/s})} = \boxed{3.5 \times 10^{-28} \text{ kg}}$.

52. From the Heisenberg Uncertainty Principle, $(\Delta E)(\Delta t) \geq \dfrac{h}{2\pi}$,

   we have $\Delta t \geq \dfrac{h}{2\pi \Delta E} = \dfrac{6.63 \times 10^{-34} \text{ J·s}}{2\pi(140 \times 10^{6} \text{ eV})(1.6 \times 10^{-19} \text{ J/eV})} = \boxed{4.71 \times 10^{-24} \text{ s}}$.

57. All hadrons contain quarks and/or antiquarks. Quarks are not believed to exist freely outside the nucleus.

60. Charge: $\frac{2}{3}e + \frac{2}{3}e - \frac{1}{3}e = +e$.

66. (a) $Q = (m_{\text{H-2}} + m_{\text{H-3}} - m_{\text{He}} - m_n)c^2 = (2.014102 \text{ u} + 3.016049 \text{ u} - 4.002603 \text{ u} - 1.008665 \text{ u})(931.5 \text{ MeV/u})$

   $= \boxed{17.6 \text{ MeV}}$.

   (b) $Q = (2m_{\text{H-2}} - m_{\text{H-3}} - m_p)c^2 = [2(2.014102 \text{ u}) - 3.016049 \text{ u} - 1.007825 \text{ u}](931.5 \text{ MeV/u}) = \boxed{4.03 \text{ MeV}}$.

67. $^{14}_{6}\text{C} \rightarrow {}^{14}_{7}\text{N} + {}^{0}_{-1}\text{e}$.

   $Q = (m_{\text{C}} - m_{\text{N}})c^2 = (14.003242 \text{ u} - 14.003074 \text{ u})(931.5 \text{ MeV/u}) = \boxed{0.156 \text{ MeV}}$.

   The mass of the electron in the products is not included in the calculation because it is already included in the mass of the N atom (See Exercise 30.40).

VI.   Practice Quiz

1. What is the Q-value of the following reaction?

   $\qquad$ ^{14}N $\quad + \quad$ ^{4}He $\quad \rightarrow \quad$ ^{1}p $\quad + \quad$ ^{17}O

   (14.003074 u) (4.002603 u)   (1.007825 u) (16.999131 u)

   (a) 0   (b) -1.279×10^{-3} MeV   (c) 1.279×10^{3} MeV   (d) -1.191 MeV   (e) 1.191 MeV

2. The meson particles are made up of

   (a) three quark combinations. $\qquad\qquad$ (b) three antiquark combinations.

   (c) quark-antiquark combinations. $\qquad$ (d) two quark combinations.

   (e) two antiquark combinations.

3. The fuel for nuclear fission is

(a) hydrogen.   (b) helium.   (c) uranium.   (d) neutron.   (e) any radioactive material.

4. What is the energy released (positive) or absorbed (negative) in the following reaction.

$$^{3}H \quad + \quad ^{3}H \quad \rightarrow \quad ^{4}He \quad + \quad 2(^{1}n)$$

(3.016049 u) (3.016049 u)   (4.002603 u) (1.008665 u)

(a) 0   (b) −0.0122 MeV   (c) 0.0122 MeV   (d) −11.3 MeV   (e) 11.3 MeV

5. Which one of the following is an example of a lepton?

(a) electron   (b) proton   (c) neutron   (d) pion   (e) gluon

6. Calculate the range of a hypothetical force if the electron were a virtual exchange particle.

(a) 3.9×10^{-13} m   (b) 1.2×10^{-12} m   (c) 2.4×10^{-12} m   (d) 7.3×10^{-4} m   (e) 1.5×10^{-3} m

7. The charge of some quarks and antiquarks is

(a) 0.   (b) $e/3$.   (c) $e/4$.   (d) $e/2$.   (e) e.

8. In a nuclear fission, the mass of the reactants compared with the mass of the products is

(a) zero.   (b) greater.   (c) smaller.   (d) the same.   (c) infinite.

9. A chain reaction can occur

(a) in any uranium core.   (b) when critical mass is reached.   (c) in the center of the Sun.

(d) when the control rods are fully inserted.   (e) when the coolant is too hot.

10. What is the threshold energy for the following reaction?

$$^{14}N \quad + \quad ^{4}He \quad \rightarrow \quad ^{1}p \quad + \quad ^{17}O$$

(14.003074 u) (4.002603 u) (1.007825 u) (16.999133 u)

(a) 0   (b) −1.193 MeV   (c) 1.193 MeV   (d) −1.534 MeV   (e) 1.534 MeV

Answers to Practice Quiz:

1. d 2. c 3. c 4. e 5. a 6. a 7. b 8. b 9. b 10. e